DNA Markers

DNA Markers
Protocols, Applications, and Overviews

Edited by
Gustavo Caetano-Anollés
University of Tennessee
Ornamental Horticulture and Landscape Design
Knoxville, Tennessee

Peter M. Gresshoff
University of Tennessee
Plant Molecular Genetics
Knoxville, Tennessee

 **WILEY-VCH**

New York • Chichester • Weinheim • Brisbane • Singapore • Toronto

Address All Inquiries to the Publisher
Wiley-Liss, Inc., 605 Third Avenue, New York, NY 10158-0012

While the authors, editors, and publisher believe that drug selection and dosage and the specification and usage of equipment and devices, as set forth in this book, are in accord with current recommendations and practice at the time of publication, they accept no legal responsibility for any errors or omissions, and make no warranty, express or implied, with respect to material contained herein. In view of ongoing research, equipment modifications, changes in governmental regulations and the constant flow of information relating to drug therapy, drug reactions and the use of equipment and device, the reader is urged to review and evaluate the information provided in the package insert or instructions for each drug, piece of equipment, or device for, among other things, any changes in the instructions or indication of dosage or usage and for added warnings and precautions.

Library of Congress Cataloging-in-Publication Data

DNA markers : protocols, applications, and overviews / edited by
Gustavo Caetano-Anolles, Peter M. Gresshoff.
    p. cm.
Includes bibliographical references and index.
ISBN 0-471-16067-9 (pbk. :alk. paper)
    1. DNA fingerprinting. 2. Nucleic acid hybridization.
I. Caetano-Anolles, Gustavo, 1955- II. Gresshoff, Peter M., 1948-

QP624.D147   1998
572.8'6--dc21   97-22138
                                                  CIP

The text of this book was printed on acid-free paper.

10 9 8 7 6 5 4 3 2 1

Contents

Applications

Preface

It has been only over four decades since discovery of DNA[1] and a decade since minisatellites opened the door to genetic testing by discriminating human individuals[2]. However, the past few years have witnessed an unprecedented revolution in molecular genetics and biotechnology. This revolution has been fueled in part by the discovery of recombinant DNA[3] and the polymerase chain reaction[4], which in turn brought innumerable techniques to the bench of the biologist, and by the general public awareness of the power that molecular biotechnology *sensu lato* can have on our everyday life. Being in the "genome era" of genetics[5], we are now facing the challenge of proactive control of our future, by genetic diagnostics, gene therapy and proper monitoring and control of biodiversity. Molecular biotechnology will no doubt provide important tools for a successful "man in harmony with its planet", capable of evaluating and managing resources and change in a responsible manner. DNA markers are important tools in such endeavor because they can be made to measure constitution, diversity and evolution of genetic material, serving as suitable indicators of wealth, health and future of life.

DNA markers: Protocols, Applications and Overviews is an attempt to introduce the reader into established and novel nucleic acid marker systems, providing experimental protocols that can be used by both the novice or the experienced. A second aim is to provide overviews of the application of nucleic acid markers to specific problems. It is impossible to cover all possible subjects here, so we have settled with only few but enticing areas of research where current trends can be envisioned.

We have produced this volume as a bench-top guide to those willing to become practitioners or willing to explore alternate DNA marker methodologies. The format of the book is rather informal, providing a description of techniques that are exceptionally diverse in nature. Each chapter generally opens with an overall description of the topic and then presents detailed protocols on selected methods of nucleic acid analysis. Methods depict only few of techniques available in this rapidly evolving field.

For the ease of the day-to-day practitioners, scientific disciplines often develop specialized terms and nomenclature. This is particularly evident in the area of DNA markers, especially when their origin is based on nucleic acid amplification. This volume is not exempt of the plethora of terms and jargon normally used in nucleic acid analysis. However, an effort has been made to make sense of the various and sometimes confusing methodologies available (*see* Chapter 1 and Glossary).

In closing, we would like to thank all authors for their timely manuscripts and their patience during the editing process. In particular, we would like to thank Angela Karp and Keith Edwards for accepting the challenge of providing an overview to the compendium of methods and applications. We would also like to thank Gloria Caetano-Anollés for helping to compile Appendix material and her support, Robert N. Trigiano for proofreading and Janice Crockett for secretarial assistance.

We hope the reader will find this volume helpful and informative.

Gustavo Caetano-Anollés
Peter M. Gresshoff

University of Tennessee at Knoxville, October 1996

[1] Watson JD and Crick FHC (1953) Molecular structure of nucleic acids: a structure for deoxyribose nucleic acid. *Nature* 171:737-738.
[2] Jeffreys AJ, Wilson V and Thein SL (1985) Individual-specific "fingerprints" of human DNA. *Nature* 317:76-79.
[3] Cohen S (1993) Bacterial plasmids: their extraordinary contribution to molecular genetics. *Gene* 135: 67-76.
[4] Mullis KB, Ferré F and Gibbs RA (1994) *The polymerase chain reaction.* Birkhäuser, Berlin.
[5] Watson JD (1993) Looking forward. *Gene* 135:309-315.

Contributors

Frederick M. Ausubel
  Department of Molecular Biology,
  Massachusetts General Hospital,
  Boston, MA 02114, USA

Kenneth L. Beattie
  Oak Ridge National Laboratory, 4500S,
  MS-6123, PO Box 2008, Oak Ridge, TN
  37631-6123, USA

Michael T. Boyce-Jacino
  Molecular Tool Inc., Alpha Center,
  Hopkins Bayview Research Campus,
  5210 Eastern Avenue, Baltimore, MD
  21224, USA

Gustavo Caetano-Anollés
  Department of Ornamental Horticulture
  and Landscape Design, The University
  of Tennessee, PO Box 1071, Knoxville,
  TN 37901, USA

Perry B. Cregan
  Soybean and Alfalfa Research
  laboratory, Bldg. 011, HH-19, USDA-
  ARS, BARC-West, Belstville, MD
  20705-2350, USA

Ranajit Chakraborty
  Human Genetics Center, School of
  Public Health, The University of Texas
  Houston Health Science Center, PO Box
  20334, Houston, TX 77225, USA

Frans J. de Bruijn
  MSU-DOE Plant Research Laboratory,
  A306 Plant Biology Bldg., Michigan
  State University, East Lansing, MI
  48824, USA

Axel Diez
  J. W. Goethe-Universität, Dermatologie,
  D-60590 Frankfurt, Germany

Eliana Drenkard
  Department of Molecular Biology,
  Massachusetts General Hospital,
  Boston, MA 02114, USA

Ismail M. Dweikat
  Department of Agronomy, 1150 Lilly
  Hall of Life Sciences, Pudue University,
  West Lafayette, IN 47907-1150, USA

Keith J. Edwards
  IACR-Long Ashton Research Station,
  Department of Agricultural Sciences,
  University of Bristol, Long Ashton,
  Bristol BS18 9AF, United Kingdom

Andrea Ender
  J. W. Goethe-Universität, Zoologisches
  Institut, Abt. Ökologie und Evolution,
  Siesmayerstr. 70, D-60054 Frankfurt,
  Germany

Mark G. Erlander
R. W. Johnson Pharmaceutical Research
Institute, 3535 General Atomics Ct.,
Suite 100, San Diego, CA 92121, USA

José E. Galindo
R. W. Johnson Pharmaceutical Research
Institute, 3535 General Atomics Ct.,
Suite 100, San Diego, CA 92121, USA

Jane Glazebrook
Center for Agricultural Biotechnology,
University of Maryland Biotechnology
Institute, College Park, MD 20742, USA

David Grant
Department of Agronomy, G403
Agronomy Hall, Iowa State University,
Ames, IA 50011-1010, USA

Peter M. Gresshoff
Plant Molecular Genetics, Institute of
Agriculture, The University of
Tennessee, PO Box 1071, Knoxville,
TN 37901, USA

Hongqing Guo
R. W. Johnson Pharmaceutical Research
Institute, 3535 General Atomics Ct.,
Suite 100, San Diego, CA 92121, USA

J. Perry Gustafson
USDA-ARS, Plant Genetics Research
Unit, Curtis Hall, University of
Missouri, Columbia, MO 65211, USA

Heike Hadrys
J. W. Goethe-Universität, Zoologisches
Institut, Abt. Ökologie und Evolution,
Siesmayerstr. 70, D-60054 Frankfurt,
Germany

Steven R. Head
Molecular Tool Inc., Alpha Center,
Hopkins Bayview Research Campus,
5210 Eastern Avenue, Baltimore, MD
21224, USA

Hans Holzmann
J. W. Goethe-Universität, Dermatologie,
D-60590 Frankfurt, Germany

Arne Huvar
R. W. Johnson Pharmaceutical Research
Institute, 3535 General Atomics Ct.,
Suite 100, San Diego, CA 92121, USA

Günter Kahl
J. W. Goethe-Universität, Biozentrum,
Pflanzliche Molekularbiologie, Marie-
Curiestr. 9 / N200, N-60439 Frankfurt,
Germany

Angela Karp
IACR-Long Ashton Research Station,
Department of Agricultural Sciences,
University of Bristol, Long Ashton,
Bristol BS18 9AF, United Kingdom

Martin Kuiper
Keygene n.v., Agro Business Park 90,
PO Box 216, 6700 AE Wageningen,
The Netherlands

Frank Kullmann
Sidney Kimmel Cancer Center, 11099
North Torrey Pines Road, Suite 290, La
Jolla, CA 92037, USA

Nori Kurata
Laboratory of Plant Genetics, National
Institute of Genetics, Yata 1111,
Mishima, Shizuoka 411, Japan

Sally A. Mackenzie
Department of Agronomy, 1150 Lilly
Hall of Life Sciences, Pudue University,
West Lafayette, IN 47907-1150, USA

Françoise Mathieu-Daude
Sidney Kimmel Cancer Center, 11099
North Torrey Pines Road, Suite 290, La
Jolla, CA 92037, USA

Michael McClelland
Sidney Kimmel Cancer Center, 11099
North Torrey Pines Road, Suite 290, La
Jolla, CA 92037, USA

Tina C. McIntosh
Molecular Tool Inc., Alpha Center,
Hopkins Bayview Research Campus,
5210 Eastern Avenue, Baltimore, MD
21224, USA

Graham Moore
Cereals Research Department, John
Innes Centre, Colney Lane, Norwich
NR4 7UH, United Kingdom

David M. O'Malley
Forest Biotechnology Group, North
Carolina State University, Raleigh, NC
26795, USA

Ghislaine M.-C. Poirier
R. W. Johnson Pharmaceutical Research
Institute, 3535 General Atomics Ct.,
Suite 100, San Diego, CA 92121, USA

Daphne Preuss
Department of Molecular Genetics and
Cell Biology, University of Chicago,
Chicago, IL 606373, USA

Charles V. Quigley
Soybean and Alfalfa Research
laboratory, Bldg. 011, HH-19, USDA-
ARS, BARC-West, Belstville, MD
20705-2350, USA

Jan L. W. Rademaker
MSU-DOE Plant Research Laboratory,
A306 Plant Biology Bldg., Michigan
State University, East Lansing, MI
48824, USA

J. Antoni Rafalski
DuPont Agricultural Products,
Biotechnology Research, PO Box
80402, Wilmington, DE 19880, USA

Jennifer E. Reynolds
Molecular Tool Inc., Alpha Center,
Hopkins Bayview Research Campus,
5210 Eastern Avenue, Baltimore, MD
21224, USA

Takuji Sasaki
Rice Genome Research Program,
STAFF Institute, 446-1, Ippaizuka,
Kamiyokoba, Tsukuba, Ibaraki 305,
Japan

Bernd Schierwater
J. W. Goethe-Universität, Zoologisches
Institut, Abt. Ökologie und Evolution,
Siesmayerstr. 70, D-60054 Frankfurt,
Germany

Werner Schroth
J. W. Goethe-Universität, Zoologisches
Institut, Abt. Ökologie und Evolution,
Siesmayerstr. 70, D-60054 Frankfurt,
Germany

Pablo A. Scolnik
E.I. DuPont de Nemours and Co. Inc.,
Central Research and Development,
Wilmington, DE 19880-0402, USA

Randy Shoemaker
Department of Agronomy, G403
Agronomy Hall, Iowa State University,
Ames, IA 50011-1010, USA

Stephen Smith
Pioneer Hi-Bred International Inc., 7300
NW 62nd Avenue, PO Box 1004,
Johnston, IA 50131-1004, USA

Bruno Streit
J. W. Goethe-Universität, Zoologisches
Institut, Abt. Ökologie und Evolution,
Siesmayerstr. 70, D-60054 Frankfurt,
Germany

Jennifer Tench
R. W. Johnson Pharmaceutical Research
Institute, 3535 General Atomics Ct.,
Suite 100, San Diego, CA 92121, USA

Julie M. Vogel
DuPont Agricultural Products,
Experimental Station, Wilmington, DE
19880-0402, USA

Thomas Vogt
Sidney Kimmel Cancer Center, 11099
North Torrey Pines Road, Suite 290, La
Jolla, CA 92037, USA

Pieter Vos
Keygene n.v., Agro Business Park 90,
PO Box 216, 6700 AE Wageningen,
The Netherlands

Leslie P. Vrolijk
Molecular Tool Inc., Alpha Center,
Hopkins Bayview Research Campus,
5210 Eastern Avenue, Baltimore, MD
21224, USA

Pamela C. Wagaman
    R. W. Johnson Pharmaceutical Research
    Institute, 3535 General Atomics Ct.,
    Suite 100, San Diego, CA 92121, USA

Jackson S. Wan
    R. W. Johnson Pharmaceutical Research
    Institute, 3535 General Atomics Ct.,
    Suite 100, San Diego, CA 92121, USA

Kurt Weising
    J. W. Goethe-Universität, Biozentrum,
    Pflanzliche Molekularbiologie, Marie-
    Curiestr. 9 / N200, N-60439 Frankfurt,
    Germany

John Welsh
    Sidney Kimmel Cancer Center, 11099
    North Torrey Pines Road, Suite 290, La
    Jolla, CA 92037, USA

Ross Whetten
    Forest   Biotechnology   Group,
    Department of Forestry, North Carolina
    State University, Raleigh, NC 26795,
    USA

Jessica Zhu
    R. W. Johnson Pharmaceutical Research
    Institute, 3535 General Atomics Ct.,
    Suite 100, San Diego, CA 92121, USA

DNA markers: a global overview

Angela Karp and Keith J. Edwards
IACR-Long Ashton Research Station, Department of Agricultural Sciences, Bristol, BS18 9AF, UK

> *"The most wonderful mystery of life may well be the means by which it created so much diversity from so little physical matter"*
> B. O. Wilson (1992)

Introduction

Qualitative and quantitative information on diversity is an essential aspect of many fields in biology, both fundamental and applied: Ecology, Evolutionary Biology, Taxonomy, Agronomy, Breeding and Conservation are examples. Many different molecular techniques can be used to provide markers of diversity. They constitute a tool box from which "implements" may be selected to address problems in any of the above fields. This volume presents a compilation of chapters which collectively detail protocols and applications of DNA markers from across the range currently available. Authors have highlighted the specific attributes of the methodology they describe and have discussed limitations, potential problems and applications of the procedure. Newcomers to the field may feel overwhelmed by the number of different DNA marker systems available. In fact, questions concerning which DNA markers to apply to a given problem are pertinent ont only to those who seek to take on board these technologies for the first time but also to those who aim to keep abreast of new developments in this rapidly evolving area. In this introductory overview chapter we provide a framework in which different DNA markers can be compared, both from the viewpoint of the technologies themselves and their applications.

A quick tour through the forest of acronyms

The wide array of different molecular techniques that can be used to detect polymorphism at the DNA level are well (although not exclusively) covered by the different contributions in this volume. A reader glancing down the contents might wonder

1

at the need for so many diferent marker systems and question what the differences are between them. Here, we provide a simple guide to the different methodologies and their various acronyms.

Most molecular markers fall into one of three basic categories of techniques that use either hybridization or are based on the **polymerase chain reaction** (PCR): (1) hybridization-based (non-PCR) techniques, (2) arbitrarily-primed PCR and other PCR-based multi-locus profiling techniques, and (3) sequence targeted and single locus PCR. Some techniques are also derivatives or combinations of other methodologies.

Category 1: Hybridization-based (non-PCR) techniques

Category I includes **restriction fragment length polymorphism** (RFLP) analysiswhere probes are hybridized to filters containing DNA which has been digested with restriction enzymes. The resultant fragments are separated by gel electrophoresis and transferred onto filters by Southern blotting [*see* Grant and Shoemaker (Chapter 2)]. Hybridization can also be carried out with probes for minisatellite or microsatellite sequences to give **variable number of tandem repeats** (VNTR) and **oligonucleotide fingerprinting** [*see* Weising and Kahl (Chapter 3) and Chakraborty (Chapter 20)].

Category 2: Arbitrarily-primed PCR and other PCR-based multi-locus profiling techniques

The development of the PCR removed the necessity for probe hybridization steps. In Category 2 we can group together all PCR-based techniques which use 'arbitrary' or semi-arbitrary primers for amplification of DNA products, including those derived from RNA. A common feature of these techniques is the lack of requirement for sequence information from the genome under investigation. The range of different approaches in this category differ in the length and sequence of the primers used, the stringency of the PCR conditions and the method of fragment separation and detection.

In one group within this category, single arbitrarily chosen primers are used in PCR conditions in which the primer will initiate synthesis even when the match with the template is imperfect. DNA techniques of this kind have been collectively termed **multiple arbitrary amplicon profiling** (MAAP) [*see* Caetano-Anollés (Chapter 7)] and **arbitrarily amplified DNA** (AAD) [*see* Schierwater et al. (Chapter 21)]. In these cases, each amplified product will be derived from a region of the genome that contains two short segments which share sequence similarity to the single primer and which are on opposite strands and sufficiently close together for the amplification to work. They include **random amplified polymorphic DNA** (RAPD) analysis in which the amplification products are separated on agarose gels in the presence of ethidium bromide and visualised under ultraviolet light [*see* Rafalski (Chapter 5)], **arbitrarily primed PCR** (AP-PCR) [*see* Vogt et al. (Chapter 4)] and **DNA amplification fingerprinting** (DAF) [*see* Caetano-Anollés (Chapter 7)] in which the products are separated on polyacrylamide gels.

In a second subgroup, the primers used are semi-arbitrary in that they are based upon restriction enzyme sites or sequences that are interspersed in the genome, such as repetitive elements, transposable elements and microsatellites. The use of primers based on restriction sites is the basis of the technique of **selective restriction fragment amplification** (SRFA) or **amplified fragment length polymorphism** (AFLP), as it is now most commonly known. In AFLP, DNA is restricted with two restriction enzymes, adaptors are ligated, and then PCR is carried out with generic primers which comprise a common part corresponding to the adaptors and restriction site and a unique part corresponding to selective bases [see Vos and Kuiper (Chapter 8)].

Because of the high mutability of **simple sequence repeats** (SSR), there are many versions in which 'microsatellites' are used as primers in **microsatellite (repeat)-primed PCR** (MP-PCR). The following are described by Vogel and Scolnik (Chapter 9): (1) un-anchored **single SSR primer amplification reaction** (SPAR) which is very similar to RAPD but the primers are SSR-based—however, polymorphisms are apparently not SSR-based; (2) **inter-SSR amplification** (ISA or inter-ISSR PCR) in which two SSR primers anchored at the 5' or 3' end are used—this techniques is more reproducible than SPAR or RAPD but variation is between SSR rather than at the SSR; (3) **randomly amplified microsatellite polymorphism** (RAMP) which is based on the random distribution of nucleotide sequences immediately flanking an SSR and is performed between a 5' anchored mono-, di-, or a tri-repeat and a arbitrary decamer primer—this approach does reflect variation in SSR; and (4) **selective amplification of microsatellite polymorphic loci** (SAMPL), an SSR-based modification of AFLP in which amplification is performed with one labeled SSR primer (anchored through the use of compound repeats) and one unlabeled adaptor primer.

Primers may also be based on repeat sequences such as Alu or SINE or bacterial repetitive elements in interspersed repeat-based PCR strategies, e.g. rep-PCR [see Rademaker and de Bruijn (Chapter 10)]. The success of these approaches depends upon the repeats being well represented and dispersed in the genome and also on their being in sufficient places where adjacent copies of the repeat occurs in inverse orientation at distances in which PCR amplification is still possible.

All of the above marker systems scan the genome in slightly different ways, depending upon the primers and PCR conditions used. Limitations in the efficiency with which polymorphisms can be detected and in the reproducibility of the amplification products also vary in magnitude between the different techniques. These are described in detail by the authors in the corresponding chapters. It is, primarily, the recognition of these limitations that has led to the development of many modifications and combinations of the basic procedures. Additional concerns have been the desire to identify and clone SSR loci for use as single-locus markers, as in RAMP and SAMPL, above.

A number of modifications in primer design and PCR conditions for DAF reactions have been developed and are described by Caetano-Anollés (Chapter 7). These include the use of mini-hairpin primers in DAF reactions (mhpDAF) and **template endonuclease cleavage MAAP** (tecMAAP) which involves the digestion of template DNA with restriction enzymes prior to amplification with one or more arbitrary primers. In **arbitrary signatures from amplification profiles** (ASAP), DAF products are scanned again using mini-hairpin or primers complementary to interspersed repeat sequences. In the related techniques of **randomly amplified microsatellite polymorphisms** (RAMPO) [*see* Weising and Kahl (Chapter 3)], also termed **randomly amplified microsatellites** (RAMS) [*see* Schierwater et al. (Chap. 21)], SSR-complementary oligonucleotides are hybridized to PCR products generated by arbitrary primers (in this case RAPD). Increased polymorphism detection can also be obtained by combining arbitrary or semi-arbitrary primed PCR with **denaturing gradient gel electrophoresis** (DGGE) [*see* Dweikat and Mackenzie (Chapter 6)].

Arbitrarily primed approaches can also be applied to the amplification of DNA derived from RNA. The **RNA arbitrarily primed PCR** (RAP-PCR) method is, as the name indicates, AP-PCR on RNA and is probably best suited to situations where changes in expression of several different genes is expected [*see* Vogt et al. (Chapter 4)]. In **differential display** (DD), carefully selected 10-mers are used and rare differentially expressed RNA detected [Galindo et al. (Chapter 15)].

Category 3: Sequence targeted and single locus PCR

"DNA markers are a short-cut to sequence information" and increased efficiency in screening is exchanged for incomplete sequence information [*see* O'Malley and Whetten (Chapter 16)]. A limitation of arbitrarily amplified DNA is the lack of allelic information, both in terms of dominance (although some of the procedures described above do give co-dominant markers) and in terms of the assignment of alleles to loci. These problems are overcome with PCR directed to specific single locus targets, for which a necessary prerequisite is knowledge of the sequence of the target or flanking target regions. In plants and animals there are three sources of potential sequences for a PCR-targeted approach: the chloroplast (cpDNA), mitochondrial (mtDNA) and nuclear (nDNA) genomes. Nucleotide sequences contain both phylogenetic and frequency information and are thus extremely important markers for ecological and evolutionary studies. Sequencing is, however, labor and resource intensive and often insufficient polymorphisms are detected at the cultivar, breed or race level. DNA sequencing to identify genetic variation for making markers is rather inefficient—only 0.5 to 2 nucleotide differences are expected per kb depending upon species and gene (genome) [*see* O'Malley and Whetten (Chapter 16)]. Where polymorphic sequences have been identified, however, a number of simple PCR-based assays can be used to provide useful easy-to-screen markers. In **cleaved amplified polymorphic sequence** (CAPS) analysis, sometimes referred to as PCR-RFLP, PCR amplified DNA fragments are digested with a restriction enzyme to reveal restriction site polymorphisms. Starting with mapped and sequenced *Arabidopsis thaliana* DNA fragments, a set of 18 primer pairs have been designed to provide CAPS markers that can be identified using agarose gel electrophoresis [*see* Drenkard et al. (Chapter 12)]. Similarly, 7 pairs of primers were designed for CAPS bacterial typing in *Staphylocossus epidermidis* from 5 previously sequenced genes. CAPS markers were also developed in loblolly pine for the lignin biosynthetic enzyme 4-coumarate CoA ligase (4CL) and used for mapping [*see* O'Malley and Whetten (Chap 16)].

Where markers are required for highly sensitive detection of specific variants at a single locus a number of techniques can be used, such as *allele specific oligonucleotide* (ASO) hybridization, *amplification refractory mutation system* (ARMS) and *oligonucleotide ligation assay* (OLA). Such diallelic DNA markers are reviewed, along with othe similar methodologies, by Reynolds et al. (Chapter 13). A solid phase microtitre assay which utilized ELISA-like detection, termed *genetic bit analysis* (GBA) is described by the authors.

If SSR loci are cloned and sequenced, primers to the flanking regions can be designed to produce a *sequence-tagged microsatellite site* (STMS), or SSR marker as they are often simply called. SSRs are highly attractive markers because each primer pair (typically) identifies a single locus which, because of the high mutability of SSR loci may have many alleles. However, the initial development of STMS can be difficult and time-consuming. Once isolated, several different methods can be used for the detection of SSR allele length variation [*see* Cregan and Quigley (Chapter 11)]. Minisatellites are generally very difficult to clone by virtue of their size but if they can be isolated with sufficient flanking sequence for primer design, they provide single locus markers similar to STMS [*see* Charkraborty (Chap. 20)].

A brief look at DNA marker applications

Just as the wide range of DNA marker technologies (and their acronyms) might well appear bewildering, the broad sweep of applications for which these markers can be deployed is all embracing. Rather than dissect the different applications and examine them in turn, many of which are elegantly described in different chapters in this volume, in this section we will consider DNA marker applications from the simple basis of the two major ways in which they are used: (1) as genetic markers for mapping and tagging traits of interest, and (2) as indicators of genetic diversity.

DNA markers for mapping and tagging traits of interest

The fundamental attration of DNA-based assays is the immense number of 'characters' they reveal, far exceeding those obtainable using morphological or biochemical methods. Character differences revealed as DNA polymorphisms can be used for investigating the organization of genomes and for the construction of dense genetic maps [*see* O'Malley and Whetten (Chapter 16), Moore and Gustafson (Chapter 17), Kurata and Sasaki (Chapter 18)]. These in turn provide detailed blue-prints for strategies of gene isolation by map-based cloning, marker-assisted selection and introgression and the dissection of complex traits.

Twenty years ago all genetic maps consisted of only phenotypic (morphological) markers and/or isoenzymes and, since such markers were always in short supply, these maps consisted of no more than a few markers per chromosome. Since then, with the advance of DNA marker technologies and the availability of almost unlimited numbers of DNA markers, most genetic maps now consist of several hundreds of markers [*see* O'Malley and Whetten (Chapter 16), Kurata and Sasaki (Chapter 18)]. It should always be remembered, however, that there is little point in constructing a genetic map unless it is of some practical value for either breeding studies or map-based cloning projects. The technologies described in this book enable genetic maps to be considered not just as an end-point in themselves, but as one step in a longer process (*see* Chapters 16-18). Such a progression is to be encouraged, as should the generation of genetic maps which can be used in different crosses by several teams. This last observation is born out by an examination of which genetic maps have proved to be the most useful over time and which continue to be enriched with new markers. The most obvious example can be found in maize where through the efforts of Ben Burr and co-workers two inbred mapping populations have been generated and widely distributed to the maize community. The information generated from using these populations is continually collated and made freely available to the research community via the world wide web (WWW), a development that has resulted in over 3,000 co-dominant markers (RFLP and SSR) having been mapped so far.

The generation of high density maps in plant species such as *Arabidopsis*, rice and tomato means that these maps now can be used for map-based cloning purposes [*see* Kurata and Sasaki (Chap. 17)]. Briefly, there are two methods that can be used to identify genes via a map-based approach. In the first approach termed chromosome walking, molecular markers which have been shown to flank the trait of interest are used to screen a *y*east *a*rtificial *c*hromosome (YAC) or *b*acterial *a*rtificial *c*hromosome (BAC) libraries. End fragments are isolated from the positive clones and these are used in turn to identify the next clone along. The whole process is then repeated until the gene of interest is identified. In the second approach termed chromosome landing (first suggested by Tanksley and co-workers) very large mapping populations are used in combination with a very large number of molecular markers. Markers which map to the region of interest are then used to screen large insert YAC or BAC libraries. Using this approach, individual clones will hybridize to several markers simultaneously without the need for chromosome walking. This second approach is technically simpler and less prone to problems such as chimeric clones, however, it does require a marker density at least ten times that required for chormosome walking. It is only recently, through the advent of techniques such as MAAP and AFLP, that such marker densities have been achieved. This together with the generation of large insert YAC and BAC clones should mean that chromosome landing will become more important as a tool for the map-based cloning of agronomically important genes.

Marker asisted selection is based upon the principle that if a gene(s) conferring a trait of interest is linked to an easily identifiable molecular marker, it may be more efficient to select in a breeding program for the marker than for the trait itself. To carry out introgression assisted by molecular markers, a sufficient number of polymorphic markers must be identified for the whole genome to be sufficiently well assayed. Application of markers to introgression programs can result in a reduction in the number of breeding cycles by improving selection efficiency, particularly at the early stage.

Most traits of agronomic and economic importance are classified as multigenic or quantitative. The use of molecular markers to identify *q*uantitative *t*rait *l*oci (QTL) has the potential to enhance the efficiency of complex trait selection in plant breeding.

Sophisticated high resolution methods are now available for QTL mapping in inbred organisms and innovative and statistically sophisticated approaches have been developed for outbreeders [see O'Malley and Whetten (Chapter 16)].

O'Malley and Whetten point out the distinction between QTL mapping as an extension of the high resolution genetic analysis already possible in crops with high density maps and the broader concept of the use of DNA markers as a novel means of complex trait dissection in undomesticated outbred long-lived plants such as trees. They review the four ways in which complex trait dissection may be carried out: (1) segregation analysis in inbred lines, (2) linkage analysis in outbred lines, (3) association studies in populations, and (4) allele sharing methods, with particular reference to their application in forest trees which present unique problems for genetical analysis because of outbreeding and limited domestication. Genetic analysis of complex traits and populations using new molecular marker technologies provides a powerful approach to understanding the organization and distribution of genetic resources in managed and natural populations.

DNA markers as indicators of diversity

Application of molecular techniques to diversity questions must take into account whether the data derived from the specific methodology will provide suitable information for answering the question being addressed. Diversity questions can be addressed at the species, population and within population (individual) level [see Schierwater et al. (Chapter 21)].

At the species level, the identification of taxonomic units and the determination of the uniqueness of species is essential information for conservation, systematic, ecological and evolutionary studies [see Rademaker and de Bruijn (Chapter 10), Schierwater et al. (Chapter 21)]. Molecular markers can provide information that can help define the distinctiveness of species and their ranking according to the number of close relatives and their phylogenetic position. They also have much to offer to the resolution of problems concerning hybridization and polyploidy.

At the population level, diversity questions may concern artificial or cultivated populations, including accessions, collections, germplasm and breeding lines. Extensive studies have shown that DNA markers provide highly efficient and informative ways of characterizing diversity at this level. DNA markers can help resolve how many different genetic classes are present and the genetic similarities among them, how much diversity is present in those classes and their evolutionary relationships with wild relatives (see Chapters 3-11 and 19). Much of ex situ conservation, germplasm and breeding line characterization is concerned with questions of this kind. Knowledge of genetic relationships among genotypes is useful in plant breeding programs because it permits the organization of germplasm, including elite lines, and provides for more efficient parental selection [see Smith (Chapter 19)]. Furthermore, exotic germplasm is an important source of genes with highly qualitative effects on traits such as biotic and abiotic stress resistance. Transfer of such genes in breeding programs can be enhanced through marker-assisted selection with backcrossing. The increase in availability of markers for marker-assisted approaches is thus likely to spin off into an increase in the use of exotic germplasm in breeding, thereby widening the gene pool from which the breeders are selecting. Improved characterization of germplasm collections will in turn facilitate structured access, thus making diversity within the collections more accessible.

Diversity questions relating to natural populations are fundamental to *in situ* conservation, forestry, ecology, population biology and behavioural studies [*see* Schierwater et al. (Chapter 21)]. DNA markers can provide information that can help in determining how populations of given species are distributed, how genetically distinct different populations are, how much genetic variation is present in and among populations, and in establishing whether there is gene flow or migration between them. For effective conservation and sustainable utilization practices, population management principles have to be established and, here, information on genetic diversity is needed to define appropriate geographical scales for monitoring and mangement, to establish gene flow mechanisms, to identify the origin of individuals (e.g. to determine the role of migration) and to monitor the effects of management practices. DNA markers can make substantial contributions to all investigations of this kind.

Within population questions (individual questions) are extremely important for the management of small numbers of individuals in *ex situ* collections, for establishment of identities in cultivar, breed or clonal identification and for paternity testing and forensics. DNA markers can provide information on individual identity, on who breeds with whom, on degrees of relatedness and on how genetic variation is distributed within populations [*see* Rademaker and de Bruijn (Chapter 10), Smith (Chapter 19), Chakraborty (Chapter 20)].

A question of which markers for which application?

Given the myriad of different DNA marker technologies and the wide diversity of applications they can be used for, an obvious problem that has to be faced is how to choose the most appropriate DNA marker for a specific investigation. Even when programs involving DNA markers are already being carried out, investigators will be faced with this problem whenever a new attractive DNA marker technique is developed. There are a large number of factors involved in this question, some of which concern the technology itself, others of which relate more to aspects of the problem under investigation and yet others to circumstances of the investigator. It is therefore not realistic to try to provide a set of discrete answers that would fit all possible scenarios. Instead, what we will do here is to examine the main elements that should be brought into the argument whenever this question is tackled.

Technological and resource considerations

It is not the aim of this overview to compare all the DNA marker systems described in the separate contributions in terms of their technological features and resource requirements. Authors have, themselves, compared different methodologies within their overview sections in each chapter. Here, we have simply chosen some representative methodologies of the major classes and compared them on the basis of: (1) development costs, (2) running costs (i.e. cost per assay), (3) number of samples that could be run in a research laboratory, (4) level of skill required, (5) whether the procedure is automatable or not, (6) whether radioactive labeling is a necessity; and (7) reliability. These comparisons are shown in Table 1.

TABLE 1. Comparison of the main features of different DNA marker techniques. All the techniques covered by this volume are not shown but, rather, features of representative methods have been summarized.

Feature	RFLP	MAAP (e.g. RAPD)	AFLP	STMS	Sequencing
Development costs ($/sample)	medium (100)	low (none)	low (none)	high (500)	high (500)
Running costs ($/sample)	high (2)	low (1)	medium (1.5)	low (1)	high (2)
Samples/day (in research lab)	20	50	50	50	20
Level of skill required	low	low	medium	low-medium	high
Automation	difficult	yes	yes	yes	yes
Radioactivity necessary	yes-no	no	yes-no	yes-no	yes-no
Reliability	high	low-medium	high	high	high
Dominant or codominant	codominant	dominant	dominant (codominant)	codominant	codominant
Polymorphism	medium	medium	medium	high	medium-low (sequence dependent)

Data and marker informativeness

In Table 1, we have added rows which indicate whether markers are dominant or co-dominant and their general level of polymorphism. These are both important considerations when selecting markers for specific applications. However, informativeness is more than just a question of dominance and polymorphism, it also concerns what the data can be used for, which is itself related to the ways in which it can be analyzed.

Shared bands are scored as presence/absence and converted into measurements of similarity, or dissimilarity, depending on the measure used (e.g. simple matching, Jaccard, Dice, etc.). Such measures of 'genetic distance' provide an important method of expressing divergence (or similarity) between sequences, individuals or taxa. All possible comparisons between the entities screened are then used to construct a matrix of pairwise distances and analyzed using one of several different clustering algorithms such as UPGMA (unweighed pair-group method using arithmeric averages) and Neighbour-Joining, or by using principal coordinate analysis (PCA). The results are presented as phenogram or principal coordinate plot, respectively, providing graphic representations of the similarity between groups of entities, or operational taxonomic units (OTUs).

Nucleotide and restriction site data can be analyzed using such phenetic approaches (based on measure of overall distance/similarity) but they are also appropriate for cladistic analyses in which two samples are placed together in the same cluster on the basis not of high genetic similarity but because they share one or more given markers. The resultant dendrograms (called cladograms) are reconstructions of phylogeny, and parsimony or maximum likelihood approaches can be used to select the best tree. The OTUs in such studies are usually selected representatives of higher taxa, or species, but cladistic methods may be used whenever recombination between OTUs is negligible, for example, for well isolated con-specific populations and organellar genes. Arbitrarily amplified DNA data (e.g. RAPD) is not usually thought suitable for cladistic analyses because of problems of band identity and homology.

Data on gene and allele frequencies are more relevant to the investigation of the diversity that resides in natural populations. Here, data are computed using population genetic statistics, the most common of which are F-statistics.

When is this relevant to the question of choosing markers for specific applications? With respect to the use of markers for mapping and tagging traits of interest, the relevance of data informativeness relates mostly to levels of polymorphisms and how easy it is to derive allelic information, both in terms of identification of heterozygotes and of assigning alleles to loci. This is obviously harder to achieve with arbitrary profiling techniques compared with RFLP or single locus markers, such as STMS. For the construction of high density maps, the most highly polymorphic markers, that are easy to obtain in sufficient numbers, are obviously desirable. Assays that are quick and which involve informative (i.e. codominant polymorphic) markers will be preferred and reproducibility and robustness are important.

Despite their labor intensive nature, RFLPs are still viewed as among the most promising of markers for mapping, largely because of their polymorphism, allelic information and robustness [see Grant and Shoemaker (Chapter 2), O' Malley and Whetten (Chapter 16), Kurata and Sasaki (Chapter 19)]. This is not to say, however, that RFLPs would be the marker of choice in a newly initiated mapping project in a species in which few probes were already available. MAAP markers (and their variations) and AFLP, have made significant contributions to map construction, even in those crops with extensive RFLP maps, and in specific mapping applications such as in bulk segregant analysis (BSA).

Such marker systems would be an attractive choice for those starting from scratch, because of the relative speed with which maps can be constructed (*see* Chapters 4-9 and 16-18). In particular, AFLP enable the construction of high density maps with only modest efforts [*see* Vos et al. (Chapter 8)]. Microsatellite markers (STMS) are co-dominant, polymorphic, widespread in the genome and and also detectable by simple PCR-based assays that are automatable. Therefore, they promise a further breakthrough for mapping and tagging applications [*see* Cregan and Quigley (Chapter 11)]. STMS are likely to make major contributions to backcross, introgression and QTL mapping programs. Here, however, development costs are very high and it is only recently that reliable SSR isolation procedures have been demonstrated. Markers based on known sequence differences, such as CAPS markers, will also become important wherever more DNA sequence data becomes available. All these and other related issues, are discussed in various chapters of this volume with respect to the different marker technologies.

With respect to the use of DNA markers for diversity assessment, there is also a trade-off between multi-locus and single locus markers. Techniques that generate multi-locus profiles (e.g. VNTR, MAAP, AFLP, etc.) provide information on numerous loci, but the information content on a single locus is low. The data derived from arbitrarily amplified DNA, AFLP and VNTR multi-locus fingerprinting approaches have their strengths in distinguishing individuals. Major applications of these approaches are thus in establishing identities, in determining parentage, in fingerprinting and distinguishing genotypes and varieties below the species level such as in cultivars and clones (*see* Chapters 3-10).

RFLPs are co-dominant and thus, although they too are useful markers for fingerprinting, they can be used for population studies and diversity classification, provided that sufficient polymorphisms can be detected at the level and genome being studied. If they are recorded as restriction site data, RFLPs are useful for phylogenetic studies, as too are CAPS (*see* Chapters 2 and 12). Disadvantages of RFLPs are that only relatively low input can be achieved, labor and material costs are relatively high, quantities of high quality DNA are required and automation is difficult [*see* Grant and Shoemaker (Chapter 2)].

Sequence tagged approaches, such as PCR-sequencing and STMS, provide markers that are more limited in number in the genome (with respect to practical use) but that are more informative for the locus concerned. DNA sequences are the only molecular markers that contain a comprehensible record of their own history. In addition to revealing the groupings of individuals into different classes, appropriate analysis based on sequences (or restriction site and CAPS) can thus provide hypothesis on the *evolutionary* relationships between the different categories that are classed together. This is important, for example in germplasm characterization, when precise knowledge on evolutionary relationships with wild species is required and in population studies for questions concerning the history of populations.

STMS markers are highly informative for diversity studies both in terms of identity and assignment of alleles and co-dominance. Coupled with their high polymorphic rate, ease of assay, and multiplexing potential, these features make STMS highly attractive markers for genotyping, idenfication and population studies (*see* Chapters 9-11). STMS are not without disadvantages, however. The mutational mechanisms by which alleles arise makes adoption of the two major models used in population genetics difficult (the infinite alleles and the stepwise mutation models), although distance measures have recently been derived for microsatellite alleles. The accuracy with which true homology can be inferred for different genotypes becomes less as genetic similarities become wider, due to the increasing possibility that different forward and back mutation events may result in alleles of the same size (homoplasy). A further problem with STMS is the occurrence of null alelles as a result of mutation in the primer site, which can result in classification of heterozygotes as homozygotes.

In summary, all DNA markers have both advantages and disadvantages but for diversity studies it appears that whilst multi-locus profiling techniques have their strength in fingerprinting, identification and characterization, co-dominant markers such as RFLPs, CAPS, STMS and sequence haplotypes (when available) are preferable for genotyping and population studies and questions concerning evolutionary relationships, whether at the species or population level, are most accurately answered with sequences (or restriction site data).

Concluding remarks on marker choices

Given the complexity of features that need to be considered, marker choices for different applications, whether for mapping or diversity questions, could be debated for hours by scientists in the field. A consensus may well be reached, but individuals would probably disagree with fervor. Readers will find both consensus and apparent contradictions even within this volume, but this is the nature of the field.

Ultimately, the crux of the issue is: *What is the question being addressed and how important is the accuracy of the answer?* The former question will dictate the range of choices available. 'What is the acuracy' will narrow the range down. Where accuracy is of paramount importance, e.g. the data must stand up in a court of law or will be used in a decision-making process, choice will be confined to marker systems that provide the highest quality (statistically validated) information for the question being addressed. If, however, there is leniency in the accuracy required, e.g. only estimates are needed or the marker is used in an initial screen and in the next stage of the investigation may be replaced by another marker type, then the choice becomes greater. All the factors discussed in this brief overview, and in the chapters to come, should be taken into consideration, but the final weight given will depend upon the answer to the overriding question highlighted above.

A final remark—whatever next?

Has the DNA marker field been exhausted? We doubt it. Current developments suggest that the explosion in marker development will continue at least for the forseeable future. The next significant step will undoubtedly be the development of non-gel based screening systems. This is already in process, to some extent, with technologies such as ASOs and GBA but, int the near future, the microchip-based hybridization systems currently being developed should become common place [*see* Beattie (Chapter 14)]. They will no doubt make a huge impact. Wide-scale application of existing marker technology is mainly limited to companies and crops that can afford the considerable capital/consumable expenditure necessary. Whilst it may be acceptable to charge several hundred dollars for a genetic screen of a a single human, such costs are not likely to be accepted for most plant species, especially when the large numbers that need to be screened in a normal breeding/diversity program are considered. The problem becomes even greater in broader biodiversity studies, often in plant species with no obvious immediate economic attributes. The development of low cost screening technologies will mean that DNA marker technologies will become common place within both large and small companies or research laboratories and on expensive/commodity based crops as well as on endangered species or these plants that are simply of scientific interest.

Acknowledgments

IARC receives grant-aided support from the Biotechnology and Biological Sciences Research Council of the United Kingdom.

Molecular hybridization

David Grant and Randy Shoemaker
USDA-ARS-FCR and Department of Agronomy, Iowa State University, Ames, IA 50011-1010, USA

Introduction

DNA hybridization as a tool for detecting specific DNA sequences after electrophoretic separation was first described by Southern (1975). Since then this technique has become an indispensable tool for molecular biologists. Because of its usefulness, literally hundreds of variations and modifications of the original procedure have been developed, as well as many different hybridization-based techniques. In this chapter we will consider only one of these variations, restriction fragment length polymorphism (RFLP) analysis.

Restriction enzymes are site-specific DNases that cleave a DNA molecule whenever the recognition sequence, which is usually a 4-6 base palindrome, is present. Because of the enzyme's sequence specificity, digestion of a particular DNA results in a reproducible array of fragments. RFLP, or length differences in homologous fragments between different DNA, are caused by changes in the primary sequence of the DNA. These length differences can be the result of:
- a point mutation resulting in the loss or gain of a restriction enzyme cut site
- an insertion or deletion of DNA between two restriction enzyme cut sites
- a deletion which overlaps a restriction enzyme site
- a DNA rearrangement where one end of the rearranged segment resides between two restriction enzyme sites

When such differences occur, they can be detected by DNA hybridization and used as molecular markers in fingerprinting or genetic studies.

The procedural outline for RFLP analysis is shown in Figure 1 and is discussed in more detail in the remainder of this chapter. The specific procedures we describe are not intended to preclude consideration of any of the many variants developed by others. They are, however, a set of protocols which are known to work in concert and which can be used as a starting point in setting up RFLP analyses.

Isolate DNA to be Analyzed

Choose a DNA isolation procedure that works for the species being analyzed. The main requirements are that the DNA be of high average molecular weight and that it be pure enough to be cleaved with restriction enzymes.

Digest DNA with Restriction Enzyme

Restriction enzymes are site-specific nucleases that reproducibly cleave DNA. Over 100 are currently available with many different sequence specificities.

Perform Agarose Gel Electrophoresis

Agarose is a long chain polysaccharide that forms a mesh of molecules when it is melted and allowed to solidify. During electrophoresis this mesh acts as a molecular sieve for the restriction fragments, which are separated by size with the smallest fragments moving fastest through the gel.

Transfer DNA from Gel to Membrane

Transfer of the DNA from the gel to a membrane in a form suitable for hybridization analysis is known as Southern transfer or blotting. This step consists of
- denaturation of the DNA to single strands by treatment with alkali
- transfer of the DNA to a membrane by capillary action and
- permanently fixing the DNA to the membrane by UV treatment and/or drying.

During this step water and DNA move together from the gel to the stack of dry paper towels (see Figure 3 below). However, since the DNA can't pass through the membrane, at the end of the transfer process the membrane contains all of the DNA originally in the gel in the exact spatial orientation that it had in the gel.

Hybridize with Labeled Probe

DNA hybridization takes advantage of the fact that DNA is normally a double stranded molecule with the two strands having complementary base sequences and held together by relatively weak hydrogen bonds between the base pairs. Since the DNA on the membrane is single stranded, the parts of the DNA molecules that are   involved in interstrand base pairing are exposed. When the membrane is incubated with probe under conditions that promote the formation of hydrogen bonds between complementary base pairs, the probe will hybridize wherever it finds a DNA sequence on the membrane that is complementary to it.

Wash Membrane to Remove Unhybridized Probe

After hybridization there will be probes stuck nonspecifically to the membrane as well as weakly hybridized to DNA fragments where short segments of homology occurred by chance. Both of these events cause unwanted background and they can be removed by high temperature washes in a low salt solution. Conditions are chosen so that true probe:target hybrids are stable but the nonspecifically bound probes are not.

Autoradiography
At this point, the membrane has tiny amounts of radioactivity bound to it. This can be detected by placing a piece of photographic film in contact with the membrane, a process called autoradiography. The film will be exposed at the regions that are close to radioactivity. This exposed sheet of film, with its variously placed bands, is the raw data of an RFLP experiment.

Reuse of Membrane
Membranes can usually be reused up to ten times. Before reuse any previously hybridized probe should be removed by denaturing the probe:target DNA hybrid. After prehybridization, the membrane is then ready for another round of hybridization.

FIGURE 1. Flow diagram of RFLP analysis.

Materials

Equipment

RFLP analysis requires few specific apparati. The choice of manufacturer and model is relatively unimportant and will usually be made based on local availability, cost or personal preference.
1. Horizontal agarose gel electrophoresis apparatus; the usual sizes are 20×25 cm and 10×16 cm for minigels.
2. Electrophoresis power supply providing a minimum output of 200 V at 250 mA.
3. Oven for hybridization at 60-65°C large enough to hold a bottle rotator.
4. Drying oven at 80°C.
5. Bottle rotator (optional; see below).
6. Shaking incubator for post-hybridization membrane washes at 60-65°C.
7. Shaker for treating gels before transfer of DNA to membrane.
8. Fluorometer for DNA quantitation.

Solutions

1. 20× SSC: 3 M NaCl and 0.3 M sodium citrate; adjust to pH 7.0 with NaOH, autoclave and store at room temperature. Use distilled water (dH_2O) to prepare all solutions.
2. 20× SSPE (can optionally substitute SSPE for SSC in all steps): 3.6 M NaCl, 0.2 M Na_2HPO_4 and 0.02 M EDTA; adjust to pH 7.4, autoclave and store at room temperature.
3. 50× TAE: 1 M Tris base, 57.1 mL/L glacial acetic acid, 50 mM EDTA.

4. Gel loading mix: Formulation A: 250 mg Ficoll 400, 100 μL of 1 M Tris-HCl pH 8 and 5 μL 10% bromophenol blue in 1 mL dH$_2$O; use 3 μL per 25 μL digest. Formulation B: 500 μL glycerol, 500 μL dH$_2$O and 5 μL 10% bromophenol blue; use 5 μL per 25 μL digest.

5. 50× Denhardt's solution (Denhardt 1966): 5 g Ficoll, 5 g polyvinylpyrolidone (PVP) and 5 g bovine serum albumin (BSA) in 500 mL of dH$_2$0; filter sterilize twice using a 0.22 μm-pore membrane and store at 4°C.

6. NaP$_i$ solution (1 M): 0.5 M each monobasic and dibasic sodium phosphate. This will result in about pH 6.7. Filter sterilize using a 0.22 μm-pore membrane and store at room temperature.

7. Hybridization solution: 50 mL dH$_2$0, 30 mL 20× SSC, 5 mL 20% SDS and 4 mL 1 M NaP$_i$; mix until the solution turns cloudy white, then heat about 30 s in microwave until solution becomes clear. Add 6.5 mL 50× Denhardt's solution and 1 mL of 10 mg/mL herring sperm (sheared 6 times through a 18 gauge needle, heated to 100°C for 5 min and quick chilled on ice just before use).

8. OLB (Feinberg and Vogelstein 1983, 1984): Prepare the following stock solutions. Solution O: 1.25 M Tris-HCl pH 8, 0.125 M MgCl$_2$; Solution A: 1 mL Solution O and 5 μL each dATP, dGTP and dTTP (each 0.1 M in 3 mM Tris-HCl pH 7, 0.2 mM EDTA); Solution B: 2 M HEPES pH 6.6 (titrated with NaOH); Solution C: 22.5 U d(N)$_6$ suspended in 3 mM Tris-HCl pH 7, 0.2 mM EDTA. To prepare OLB, mix solutions A, B and C in the ratio 100:250:150. Store at -20°C.

Experimental protocol

Genomic DNA to be used for Southern analysis can be prepared by any one of a number of methods. We normally use the CTAB method (Saghai-Maroof et al. 1984) with modifications described by Keim and Shoemaker (1988). In most cases the exact isolation procedure is not critical and the choice of method will usually be made based on the requirements for the specific species being studied.

Restriction enzyme digestion

A) Genomic DNA

One measure of DNA suitability for Southern analysis is the average molecular weight of the isolated DNA. This is most easily determined by electrophoresis of 1-2 μg undigested DNA on an agarose gel. Ideally, the sample will present a majority of the DNA migrating as a relatively tight band near the top of the lane with little "smear" below it. Figure 2A shows examples of good and marginally acceptable genomic DNA.

Autoradiogram signal strength is determined to a large extent by the amount of DNA loaded in the lane. About 5-10 μg of DNA per lane is usually sufficient for Southern analysis of organisms that have haploid genomes similar in size to that of humans ($\sim 3 \cdot 10^9$ base pairs). If the genome to be analyzed is substantially larger than this, the amount of DNA per lane will have to be increased in proportion to the difference in genome size.

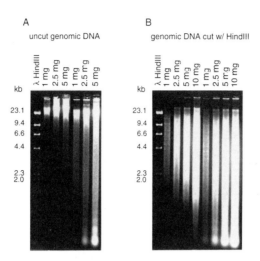

FIGURE 2. A. Undigested soybean genomic DNA. B. *Hind*III digested soybean genomic DNA. Both panels show electrophoresis of DNA on an 0.8% agarose gel. In both panels lanes 2-4 show high quality DNA samples while lanes 5-7 show DNA samples with marginal quality.

DNA concentration is best determined by fluorometry using the dye Hoechst 33258 (Labarca and Paigen 1980). Many DNA preparations contain impurities which do not interfere with restriction enzyme digestion, but cause erroneously high A_{260} readings. In cases where it is important that several lanes be loaded equally, it is helpful to run a check gel to verify the relative DNA concentrations of all samples. Figure 2B shows soybean genomic DNA digested with *Hind*III. Note that it is difficult to use digested DNA samples to determine relative loadings when the samples are of marginal quality. It will simplify subsequent steps if working DNA stocks are prepared at the same concentration.

B) Plasmid DNA
We isolate plasmid DNA using the procedure described by Xiang et al. (1994) but any standard minipreparation method should be suitable. When testing samples on agarose gels use 0.1-0.5 μg per lane.

C) Digestion
Use 1-2 units (U) of restriction enzyme per μg of DNA. This amount is usually sufficient for complete digestion but in some cases may have to be increased. Reaction volumes should be kept as low as practical. We have found that 25 μL works well. If the DNA is too dilute for this volume, it is acceptable to digest in a larger volume and concentrate the sample afterwards (by ethanol precipitation, SpeedVac, etc.). It is also possible to use a comb with thicker teeth or thicker gels to accommodate larger volumes but these can result in wider bands and a subsequent loss in band resolution. Digestion is normally done for 4 h at a temperature for optimal activity of the

restriction enzyme. Many restriction enzymes lose their activity within a few hours, so extending the digestion for longer times, while not harmful, will not usually help.

Add the reagents in the following order: water, buffer, DNA and enzyme. When many digests are to be done at one time it is easier to prepare a master mix of water, buffer and enzyme. This mix can be aliquoted in each tube and the DNA then added. Digests can be prepared at room temperature but the restriction enzyme or master mix should be stored on ice before use.

Sometimes it is necessary to digest a DNA sample with more than one enzyme. This can be accomplished by sequential digestions with ethanol precipitation between them in order to change the salt concentrations, or by using one of the 'universal' cocktails offered by many of the commercial enzyme suppliers. In cases where the two enzymes do not have common reaction conditions, we have found that it usually works to simply use a 1:1 mix of the two specific reaction cocktails along with the two enzymes.

After the digestion has been completed a gel loading buffer is added. This mixture serves to: (i) stop the enzymatic reactions and disassociate the enzyme from the DNA; (ii) add a tracking dye for monitoring loading and the progress of the electrophoresis; and (iii) increase the density of the sample so it will stay in the well. After the gel loading buffer has been added, samples can be stored at 4°C for at least a week.

NOTE: Enzymes are very susceptible to inactivation at the air:water interface. Mix solutions by gently flicking the tube once or twice. If needed, a brief 1-2 s centrifugation step will collect all the liquid at the bottom of the tube.

Agarose gel electrophoresis

Agarose gel electrophoresis is usually conducted in a submerged horizontal tray. Choose a tray whose length will allow the required band resolution. As a general guideline, at least 7.5 cm of gel down-current of the comb is needed for electrophoresis of genomic DNA. A longer gel will allow greater separation between closely spaced bands.

Genomic DNA is usually electrophoresed in 0.8-1% agarose. Several running buffers are commonly used for DNA electrophoresis but we have found that the best results are obtained using TAE.

Our usual gel tray is 20×25 cm. It takes at least 250 mL of melted agarose solution to make a gel deep enough to allow the comb to be easily removed. Up to 300 mL can be used without any adverse effect on autoradiogram band width.

In many cases it is important to know the molecular weight of the DNA bands. A useful DNA marker for genomic DNA is *Hind*III-digested phage λ DNA. Use 1 μg/lane if the marker bands are to be visualized with ethidium bromide (EtBr) or 300 ng/lane if they will be detected by autoradiography (see below).

EtBr can be added to the gel at 0.5 μg/mL to make the DNA visible during electrophoresis. It is not necessary to add EtBr to the running buffer. It is also possible to omit the EtBr from the gel and do a post-electrophoresis staining before photography.

For genomic DNA, electrophoresis is carried out at 30V for 18-20 h. It is important to avoid the use of high voltages as they will cause poor resolution of the fragment bands. Under these conditions, the bromophenol blue will migrate about 8 cm and the leading edge of the DNA about 10-12 cm.

A permanent record of the gel can be made by using Polaroid Type 667 film or equivalent. The gel should be illuminated using a 302 nm transilluminator and photographed through both a UV haze filter (Kodak Wratten no. 2B) and an orange filter (Kodak Wratten no. 23A) to attenuate the background due to non-bound ethidium bromide.

NOTE: EtBr is a potent mutagen and probably a carcinogen. Always wear gloves when handling stained gels and dispose of solutions and stained gels appropriately.

Blotting

The following steps transfer the DNA from the agarose gel to a suitable membrane, where it is ready for hybridization with a probe.
1. Cut off gel above the wells and below the bottom of DNA smear.
2. Conduct the following steps in a glass or plastic tray on a shaker at room temperature. The size of the tray is chosen to easily hold and remove the gel. A 8 × 12 in tray requires 500 mL of each solution.
   (i) Depurinate DNA in 0.25 M HCl for 7 min. Bromophenol blue turns yellow in this step.
   (ii) Rinse with dH$_2$O for 15-30 s.
   (iii) Denature DNA in 1 M NaCl, 0.5 M NaOH, or 0.3 M NaCl, 0.3 M NaOH for 10-15 min. Bromophenol blue turns blue. This step can be repeated with fresh solution.
   (iv) Rinse with dH$_2$O for 15-30 s.
   (v) Raise salt concentration with 1.5 M NaCl for 5 min.
3. Southern transfer: We have had good success using a modification of the downward capillary transfer method of Zhou et al. (1994). Figure 3 shows a cartoon of an assembled blotting stack. For clarity the two pieces of 1/4 in thick hardboard we use for the weight on top of the stack have been omitted from the figure. The stack is assembled on a large piece of plastic wrap, which is used to completely enclose the stack during transfer. Plastic wrapping of the stack is done to keep the gel and membrane from drying out. The 2-3 in stack of dry paper towels act as a sink for the liquid being transferred by capillary action from the gel. We normally use plain brown bifold towels but any kind of absorbent paper is suitable.
   Both nitrocellulose and Nylon membranes have been used for DNA transfers. We normally use ZetaProbe GT charge-modified Nylon membrane (BioRad) because of its durability and ease of reuse. The protocols in this chapter have been optimized for this type of membrane.

Both the filter paper and membrane are sized to be ¼-½ in larger than the gel (*see* Figure 3) and are wetted in dH₂O before assembling the stack. The gel is placed on the membrane right side up and centered so that there is a border of membrane all around the gel. It is important that all bubbles be removed from between every layer of the blotting stack as DNA transfer will not take place through a bubble. A 10 mL pipette (or similar round smooth item) can be gently rolled over each layer as it is assembled to remove any bubbles below it.

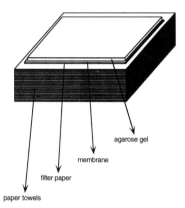

agarose gel

membrane

filter paper

paper towels

FIGURE 3. Diagram of stack for Southern transfer.

8. Transfer overnight at room temperature. Transfer in 1 h can be accomplished by placing a sponge soaked in 10× SSC on top of the gel instead of the weight (Zhou et al. 1994).
9. Disassemble blotting stack and wash the membrane. Up to 12 membranes can be washed using 200 mL in a 8×12 in tray. Wash in 5× SSC for 5-10 min or 0.1 M Tris-HCl pH 7, 1 M NaCl for 15 min, with shaking at room temperature.
10. Remove membrane and quickly blot dry using clean paper towels.
11. Fix the DNA to the membrane. Place the damp membrane between paper towels and bake it at 80°C for at least 1 h. Overnight baking does not improve fixation but will not harm the DNA or membrane. If desired, before baking crosslink the DNA to the membrane by exposing both sides to the 302 nm transilluminator for 15 s at a distance of about 20 cm.

Probe labeling

A) PCR amplification of plasmid insert.
   We normally use a PCR-amplified plasmid insert to prepare a probe for labeling, although PCR amplification of specific regions of genomic DNA can also be used.
   1. Add the following reagents:
      • 0.5 µL DNA (miniprep diluted 1:100 in dH₂0).
      • 2.5 µL *Taq* polymerase magnesium-free buffer (provided by enzyme supplier).
      • 2 µL 25 mM MgCl₂.

- 4 μL dNTPs (stock is 1.25 mM each in dH$_2$O).
- 2.5 μL each P$_1$ and P$_2$ primer (stocks are 20 μM in dH$_2$O). Primer sequences for vector pBS+ are CGCAATTAACCCTCACTAAAGGG (P$_1$) and GTAATACGACTCACTATAGGGCG (P$_2$). If a different vector has been used to clone the probe it will be necessary to choose suitable primers for that vector.
- 1 U *Taq* polymerase.
- Complete to 25 μL of dH$_2$O.

2. Amplify DNA using the following thermal cycle profile: 93°C for 2 min; 40 cycles of 60°C for 2 min, 72°C for 3 min, and 93°C for 45 s; 60°C for 2 min; and 72°C for 7.5 min.

3. A 1% agarose minigel can be used to check the success of the amplification and to estimate the yield of amplified product. A volume of 2 μL of the completed PCR reaction should contain about 50 ng of the desired product.

B) Labeling of probe

Several methods have been described for labeling a probe prior to hybridization (nick translation, end labeling, random priming) as well as several nonradioactive probe detection alternatives. We have had the best success using the OLB protocol developed by Feinberg and Vogelstein (1983) with ^{32}P as the label.

1. Add the following reagents:
   - 50-100 ng of PCR reaction product.
   - Complete to 32.5 μL of dH$_2$O; mix well, incubate at 100°C for 3-5 min, and quick chill on ice.
   - 10 μL OLB.
   - 2 μL BSA (stock is 10 mg/mL in dH$_2$O).
   - 5 μL ^{32}P-dCTP (3000 Ci/mmole, 10 μCi/μl).
   - 3 U Klenow fragment of DNA polymerase I.

2. Incubate at room temperature for 3-4 h.

We normally label undigested phage λ DNA by the OLB procedure and add 5-10 pg to probe labeling reactions after the denaturation step.

Hybridization

A) Prehybridization

Prehybridize new membranes for 3 h to overnight at 65°C. Previously used membranes (see below) can be prehybridized for as little as 30 min.

We normally carry out the prehybridization and hybridization steps in plastic bottles (500 mL; 7 cm diameter × 14 cm height) on a rotator in a convection oven, although several other arrangements (Seal-a-Meal bags, closed trays, etc.) are also acceptable. If roller bottles are not available a shaker should be used to keep the solution and probe in constant movement over the membrane to ensure an even signal.

Multiple membranes can be hybridized in a single bottle. Use 15 m L of solution for 1-2 membranes and 25-30 mL for 3-6 membranes. The exact volumes are not critical but it is important that the membranes not be allowed to dry out. In general it is good to keep the solution volume to a minimum in order to have a high probe concentration and to reduce the amount of radioactive waste that must be discarded.

B) Hybridization
1. Denature the probe by heating to 98-100°C for 5 min.
2. Quick chill on ice.
3. Add probe to the prehybridization solution already in the bottle. Mix well before allowing the concentrated probe to come into contact with the membrane.
4. Incubate at 65°C in an oven for 16-20 h. If the probe concentration or amount of DNA on the membrane is low this hybridization step can be extended up to 48 h.

Post-hybridization washes

It is necessary to do several post-hybridization washes to remove all the nonspecifically bound probe. These steps are very important for controlling both general background and signal intensity since the final quality of the autoradiogram is largely determined by their success. All solutions should be preheated to the appropriate temperature before use.

A) Low stringency wash
Incubate the membrane in a glass or plastic tray and shake gently at 60-65°C for 15-20 min. Higher temperature may be needed if the hybridization step was done for an extended time. Multiple membranes, including those hybridized with different probes, can be washed together. Use about 1 L of 2× SSC, 0.4% SDS for every 4 membranes.

B) High stringency wash
Wash as above in 0.6× SSC, 0.12% SDS at 60-63°C for 15-20 min. Test the efficacy of non-hybridized probe removal with a Geiger counter. Experience has shown that readings of <2000 cpm are needed for good quality autoradiograms. If at the end of the first high stringency wash the membrane background is still too high, one or more of the following steps will usually reduce it to acceptable levels:
(i)   Repeat the high stringency wash with fresh solution.
(ii) Increase the incubation temperature to 65°C.
(iii) Lower the salt concentration of the wash solution to 0.1× SSC, 0.1% SDS.

Once the washes are completed, the membrane is quickly blotted dry with paper towels and wrapped in plastic wrap or enclosed in Seal-A-Meal bags. It is important that the membrane not be allowed to dry out as this will permanently fix the probe to the membrane and prevent its reuse (see below). The high stringency wash can be repeated after the first autoradiographic exposure if the background is too high.

Autoradiography

For maximum sensitivity, autoradiography should be performed at -70°C (Laskey and Mills 1977). Exposures typically range from a few hours to 7 d. Longer exposures are helpful in detecting weak signals but the signal to background noise ratio degrades with increasing exposure times.

Intensifying screens are used for autoradiography of ^{32}P. These screens capture some of the beta particles that pass through the film and convert their energy to a photon which can then enhance exposure of the film. These screens are matched to specific X-ray film. Good results can be obtained with either Kodak X-OMAT AR film and a DuPont Cronex Lightning Plus-T intensifying screen or DuPont Reflection NEF-496 film and a DuPont NEN Reflection intensifying screen.

Almost any light-tight arrangement can be used for autoradiography. However, it is more convenient to use one of the commercially available cardboard or metal cassettes which are available in sizes that accommodate most X-ray film dimensions. The membrane is placed in the cassette sandwiched between a sheet of film and an intensifying screen. Since the plastic film used to wrap the membrane often will stick to X-ray film, a sheet of white typing paper can be placed between the film and the wrapped membrane.

Figure 4 shows examples of soybean DNA hybridized with a ^{32}P -labeled RFLP probe.

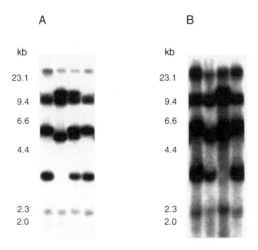

FIGURE 4. Examples of autoradiograms. 8 µg of soybean genomic DNA was digested with *Hind*III and hybridized with the RFLP probe pA664. A. DNA of quality similar to that used in lanes 2-4 of figure 2A. B. DNA of quality similar to that used in lanes 5-7 of figure 2A.

Reusage of membranes

It is often convenient to hybridize a membrane with several different probes. In these cases the first probe must be removed before reuse. This is most easily accomplished by removing the hybridized probe using high pH.
1. Strip probe with 100 mM NaOH, 0.4% SDS solution for 5 min.
2. Wash with 2 quick rinses of dH$_2$O.
3. Restore high salt in 5× SSC solution for 20 min.

Use 50-100 mL of solution per membrane. All steps are done with shaking in a glass or plastic tray at room temperature.

Comments

Analysis of DNA variation by RFLPs has several strengths and weaknesses when compared to other commonly used molecular marker methods (Table 1). In general, RFLP analysis is most often used for genetic analyses where the number of samples is moderately low (<300-400). If many samples must be analyzed or information for many markers is needed, it may be appropriate to use one of the amplification-based techniques that have been recently developed and which are described in other chapters in this book.

TABLE 1. Strengths and weaknesses of RFLPs as molecular markers.

Strengths	Weaknesses
• All alleles are seen simultaneously • The DNA polymorphisms act as codominant genetic markers • No prior information on DNA sequence or allelic makeup is needed • The techniques are simple and robust and thus easily reproducible between experiments and laboratories	• Has relatively low throughput • Labor and materials costs are higher than those of many other techniques • Requires relatively large amounts of high quality DNA • Is gel based and thus not easily automatable

Acknowledgments

The authors thank Jennifer Lee and Cindy Clark for help in compiling these procedures. The autoradiograms in Figure 4 were kindly provided by Jennifer Lee.

References

Denhardt DT (1966) *Biochem Biophys Res Commun* 23:641-646.
Feinberg AP and Vogelstein B (1983) *Anal Biochem* 132:6-13.
Feinberg AP and Vogelstein B (1984) *Anal Biochem* 137:266-267.
Keim P, Olson TC and Shoemaker RC (1988) *Soybean Genet Newsletter* 15:150-152.
Labarca C and Paigen K (1980) *Anal Biochem* 102:344-352.
Laskey RA and Mills AD (1977) *FEBS Letters* 82:314-316.
Saghai-Maroof MA, Biyashev RM, Yang GP, Zhang Q and Allard RW (1984) *Proc Natl Acad Sci USA* 81:8014-8018.
Southern EM (1975) *J Mol Biol* 98:503-527.
Xiang C, Wang H, Shiel P and Guerra DJ (1994) *Biotechniques* 17:30-32.
Zhou MY, Xue D, Gomez-Sanchez EP and Gomez-Sanchez CE (1994) *Biotechniques* 16:58-59.

Hybridization-based microsatellite fingerprinting of plants and fungi*

Kurt Weising and Günter Kahl
Plant Molecular Biology, Department of Biology, Biozentrum, University of Frankfurt, D-60439 Frankfurt, Germany

Introduction

The technique of "classical" DNA fingerprinting is a derivative of RFLP analysis, but differs from the latter by the type of hybridization probe applied to reveal genetic polymorphisms. To obtain a typical multilocus DNA fingerprint, probes are used which create complex banding patterns by recognizing multiple genomic loci simultaneously. Each of these loci is characterized by more or less regular arrays of tandemly repeated DNA motifs that occur in different numbers at different loci.

Two categories of such multilocus probes are mainly used. The first category comprises cloned DNA fragments or synthetic oligonucleotides which are complementary to so-called *minisatellites*, tandem repeats of a basic motif about 10 to 60 bp long. Genetic fingerprinting using minisatellites was first achieved about 12 years ago (Jeffreys et al. 1985a,b). Since then, minisatellites have been cloned from many organisms including plants (e.g. rice: Winberg et al. 1993; *Arabidopsis thaliana*: Tourmente et al. 1994; tomato: Broun and Tanksley 1993, 1996, Phillips et al. 1994), and synthetic minisatellite probes have been designed (Vergnaud 1989, Rogstad 1993). The second category of probes is exemplified by short, synthetic oligonucleotides which are complementary to *microsatellites*, i.e. tandem repeats of about one to five base pairs (synonyms: simple sequence repeats, SSRs; short tandem repeats, STRs). The use of oligonucleotide probes is compatible with an in-gel hybridization approach that is faster and more efficient than conventional Southern blotting (Ali et al. 1986, Schäfer et al. 1988, Epplen 1992). RFLP

---

*Dedicated to Prof. Dr. h.c. Herbert Oelschläger (Frankfurt, Germany) on occasion of his 75th birthday.

analysis with microsatellite-complementary oligonucleotides is also referred to as "oligonucleotide fingerprinting" or "hybridization-based microsatellite fingerprinting".

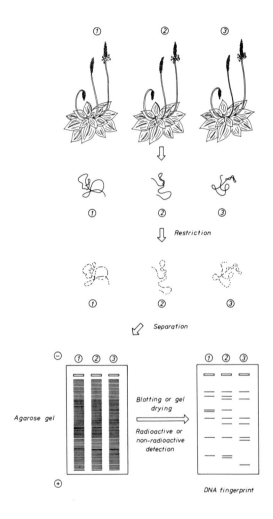

FIGURE 1. Methodology of hybridization-based microsatellite fingerprinting. Genomic DNAs of the organisms to be analyzed (exemplified here by three individuals of *Plantago media* L.) are isolated, purified, and subsequently digested with an appropriate restriction endonuclease. The restriction fragments are separated by agarose gel electrophoresis, the DNA denatured, and the gel is either dried or blotted onto a hybridization membrane. Hybridization of the blot or dried gel with a mini- or microsatellite-complementary, (non)radioactively labeled oligonucleotide probe results in polymorphic banding patterns that may be visualized by autoradiography or chemiluminescence.

In recent years, both minisatellite and microsatellite probes have been applied to RFLP fingerprinting of numerous animal, plant and fungal species (reviewed by Rosewich and McDonald 1994, Weising et al. l995a). High levels of polymorphism between related genotypes were often observed, and the technique found its way to diverse areas of genome analysis, including paternity testing, genotype identification and population genetics. The present chapter provides a detailed protocol for hybridization-based microsatellite fingerprinting of plants and fungi. The following steps are involved (*see* Figure l).

1. Isolation of genomic DNA suitable for RFLP analysis.
2. Complete digestion of genomic DNA with an appropriate restriction enzyme.
3. Electrophoretic separation of the restriction fragments in agarose gels.
4. Denaturation and immobilization of the separated DNA fragments within the gel (or, alternatively, blotting onto a membrane).
5. Hybridization of the dried gel (or the membrane) to (non)radioactively labeled, microsatellite-complementary oligonucleotide probes.
6. Detection of hybridizing fragments (i.e. fingerprints) by autoradiography or by chemiluminescence, and documentation by photography.
7. Evaluation of banding patterns.

A. Digestion of Genomic DNA
   with Restriction Endonucleases

B. Gel Electrophoresis and Hybridization with
   Labeled Microsatellite-complementary Probes

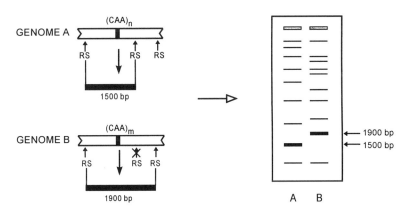

FIGURE 2. RFLP fingerprinting with microsatellite-complementary probes: molecular basis of polymorphism. RFLPs between genome A and B may originate from several kinds of mutations: (1) variable number of tandem repeats (VNTR) polymorphism within the target sequence of the probe (i.e. [CAA]$_m$ vs. [CAA]$_n$), (2) VNTR polymorphism of a second mini- or microsatellite adjacent to the probe target sequence, (3) restriction site (RS) mutation, (4) a combination of some or all of these events. In the example given in the figure, the observed RFLP (1900 vs. 1500 bp) is caused by a restriction site mutation.

The probes used in oligonucleotide fingerprinting most probably recognize microsatellite-like target sequences. However, the molecular basis of the observed polymorphisms is not yet perfectly clear. While sequenced microsatellites were generally found to be in a size range of about 20-100 bp, polymorphic RFLP fragments detected by microsatellite probes are often much larger (up to more than 10 kb), and their size may vary by several kilobases. Thus, variable numbers of tandemly repeated microsatellite units are not likely to be the only underlying cause for the observed polymorphisms. Since microsatellites are often embedded in other types of repeats (e.g. Lohe and Brutlag 1986, Zischler et al. 1992, Phillips et al. 1994, Broun and Tanksley 1993, 1996), minisatellites or interspersed repeats in the vicinity of the probe target sequences may play an important role in the generation of polymorphisms. A further source of variability is provided by restriction site variation (Figure 2).

Microsatellite-complementary oligonucleotides are not only useful for RFLP fingerprinting, but also produce informative banding patterns when hybridized to PCR products generated by arbitrary primers (Richardson et al. 1995, Cifarelli et al. 1995, Ender et al. 1996). A protocol is given below for microsatellite hybridization to such PCR products, following a strategy which we call random amplified microsatellite polymorphism (RAMPO) analysis.

Experimental protocols

DNA isolation

Numerous problems may be encountered during DNA isolation from plant and fungal species. These include DNA degradation and co-isolation of viscous polysaccharides, polyphenols and other secondary plant compounds which cause damage to DNA and/or inhibit restriction enzymes and DNA polymerases. DNA preparations often tend to be brownish due to the oxidation of polyphenols to quinone compounds. Quinones in turn are powerful oxidizing agents that damage proteins and DNA. As a consequence, yields of high molecular weight DNA from plants are often poor, and there is no single protocol which is optimally suited for each plant species. Accordingly, hundreds of different DNA isolation methods have been described that nevertheless rely on few basic principles (for review see Milligan 1992, Weising et al. 1995a).

Different kinds of experiments demand different levels of DNA purity. Thus, relatively pure high molecular weight DNA is required for RFLP fingerprinting, while PCR of specific target amplicons may also work on rather crude preparations. In our hands, a modified CTAB procedure based on the protocol of Doyle and Doyle (1987) is the method of choice for obtaining good quality total DNA from many plant species, and also from fungi (Weising et al. 1991a,b, 1995a). CTAB is a cationic detergent which solubilizes membranes and forms a complex with DNA. After cell disruption and incubation with hot CTAB isolation buffer, proteins are extracted using chloroform/isoamylalcohol, and the CTAB-DNA complex is precipitated with isopropanol. The DNA pellet is washed, dried and redissolved in TE buffer. Depending on the species, additional purification steps may be necessary to remove RNA, polysaccharides, polyphenols, and other contaminating substances. Here we include

RNase treatment and ammonium acetate precipitation which removes RNA and some polysaccharides. Recommendations on how to deal with other contaminants are given below.

A) Materials
1. Liquid nitrogen.
2. Isolation buffer: 2% [w/v] cetyltrimethylammonium bromide (CTAB), 1.4 M NaCl, 20 mM EDTA, 100 mM Tris-HCl pH 8.0, 0.2% β-mercaptoethanol (added just before use).
3. Chloroform-isoamyl alcohol (24:1).
4. RNase A solution: 10 mg/mL RNase A in 10 mM Tris-HCl, 15 mM NaCl (pH 7.5); boiled for 15 min, cooled to room temperature and stored at -20°C.
5. 100 % isopropanol.
6. Washing solution: 76% ethanol, 10 mM ammonium acetate.
7. TE buffer:10 mM Tris-HCl, 1 mM EDTA (pH 8.0).
8. 7.5 M ammonium acetate.

B) Experimental protocol
1. Grind up to 3 g of fresh, or 0.5 g of lyophilized or dried plant material to a fine powder, using liquid nitrogen, mortar and pestle. Liquid nitrogen is not essential for lyophilized tissue, but facilitates the grinding procedure considerably. Fresh tissue should not thaw before being added to the isolation buffer, since cellular enzymes may rapidly degrade the DNA.
2. Transfer the powder into 15 mL of pre-warmed (60°C) isolation buffer in a capped polypropylene tube. Suspend clumps using a spatula.
3. Incubate for 30-60 min at 60°C in a water bath. Mix every 10 min. The optimal incubation temperature and time may vary between different tissues and/or species.
4. Add 1 volume of chloroform-isoamyl alcohol, cap the tube and extract for 10 min on a rotary shaker, or by hand. Mixing should be done gently but thoroughly enough to ensure emulsification of the phases.
5. Centrifuge for 10 min (5,000 g, room temperature). Re-extract the aqueous phase once with fresh chloroform-isoamyl alcohol and centrifuge again.
6. Transfer the final aqueous phase to a glass centrifuge tube using a large-bore pipette. Add heat-treated RNase A to a final concentration of 100 μg/mL. Mix and incubate at room temperature for 30 min.
7. Add 0.6 volumes of ice cold isopropanol, and mix gently but thoroughly by inverting the tube several times. If the DNA/CTAB complex precipitates as a fibrous network, lift it from the solution using, for example, a bent Pasteur pipette as a hook, transfer it to the washing solution (step 8), and dry (step 9). If this is not the case, collect the amorphous precipitate by centrifugation (5000 g, 10 min at 4°C).
8. Add 20 mL of washing solution, gently agitate the pellet for a few minutes, and collect by centrifugation (5000 g, 10 min at 4°C).
9. Invert the tubes and drain on a paper towel for about 1 h. Pellets should neither contain residual ethanol nor should they be too dry. In both cases redissolving may be difficult.

10. Add an appropriate volume of TE buffer and allow the pellets to dissolve overnight (4°C) without agitation.
11. Add 0.5 volumes of 7.5 M ammonium acetate solution, mix, and chill on ice for 15 min.
12. Centrifuge for 30 min (10,000 g, 4°C). Transfer supernatant to a new tube, add 2 volumes of 96% ethanol, mix by inversion, and store for 1 h at -20°C.
13. Centrifuge for 10 min (5,000 g, 4°C), wash pellet in 70% ethanol, and centrifuge again. Drain final pellet and dissolve in an appropriate volume of TE buffer.

C) Comments
1. The way how plant and fungal tissue is collected and stored prior to DNA isolation may considerably influence the results (for review see Sytsma et al. 1993, Blackwell and Chapman 1993, Weising et al. 1995a). Whenever possible, freshly harvested, young leaf or hyphal material should be used. If secondary compounds present in leaves are a problem, flower petals (if available) may be a useful alternative (Lin and Ritland 1995). For short periods of time, fresh tissue may be stored on ice. Storage for longer periods of time requires either freezing, lyophilization, or rapid drying by means of, for example, silicagel or anhydrous calcium sulfate (Liston and Rieseberg 1990).
2. The protocol given above can be scaled down to mg amounts of tissue. If less than 300 mg of fresh tissue are used, homogenization (using an exactly fitting conical pestle) as well as all incubation, precipitation and centrifugation steps are best performed within microfuge tubes.
3. The detrimental influence of polyphenols and their oxidation products may be counteracted by different strategies: (i) inclusion of polyphenol adsorbants such as bovine serum albumin (BSA) and soluble polyvinylpyrrolidone (e.g. PVP-40) in the isolation buffer (Couch and Fritz 1990, Bult et al. 1992); (ii) inclusion of phenoloxidase inhibitors such as diethyldithiocarbamic acid (DIECA) in the isolation buffer (Paterson et al. 1993); and (iii) inhibition of polyphenol oxidation by including antioxidants such as sodium bisulfite, cysteine, ascorbic acid or high concentrations of β-mercaptoethanol in the isolation buffer (Kanazawa and Tsutsumi 1992, Fulton et al.1995).
4. Polysaccharides make DNA preparations highly viscous and inhibit the activity of restriction enzymes (Do and Adams 1991). Strategies to remove polysaccharides include: (i) purifying the crude DNA preparation by passage through ion exchange columns (Weising et al. 1991a, Do and Adams 1991); (ii) precipitation of DNA with polyethylene glycol (e.g. PEG 8000), leaving polysaccharides in the supernatant (Rowland and Nguyen 1993, Li et al. 1994); (iii) exploiting the differential ethanol precipitation of DNA and polysaccharides from aqueous solutions containing different salt concentrations (Fang et al. 1992, Michaels et al. 1994); and (iv) treatment of DNA preparations with pectic enzymes (Rether et al. 1993). Since plant and fungal polysaccharides are diverse, these techniques may work in some species but not in others, and a suitable strategy has to be determined by trial and error.
5. More general strategies to purify crude DNA preparations from undesired contaminants include ultracentrifugation in cesium chloride density gradients

(Murray and Thompson 1980, Sambrook et al. 1989) and preparative electrophoresis in low-melting point agarose (Bult et al. 1992). Though cesium chloride centrifugation is expensive and time-consuming, it is the method of choice for high quality DNA.

6. In our hands, the CTAB method described above works well for many plant species as well as for mycelia of filamentous fungi. See Weising et al. (1995a) for a comprehensive survey of DNA isolation protocols for fungi and plants.

DNA restriction

The quality of an RFLP fingerprint is highly dependent on the accurate and complete restriction of the DNA samples. Several parameters influence the choice of an appropriate restriction enzyme for oligonucleotide fingerprinting experiments:

(i)   *Length of recognition sequence:* Restriction enzymes with a 4-base recognition motif are generally preferable for DNA fingerprinting of human, plant and animal genomes, while 6-base cutters are more suitable for the less complex genomes of fungi (Meyer et al. 1991). Though the statistical probability of a particular 4-base motif to occur in DNA is 1:256, the observed fragment distribution of eukaryotic genomic DNA restricted by 4-base cutters and visualized by mini- or microsatellite probes usually ranks between 200 bp and more than 20 kb (see below, Fig. 5). The generation of fragments that are larger than expected from the cutting frequencies of the enzyme could be the consequence of a low sequence heterogeneity in the probe target regions.

(ii) *Probe sequence:* It is obvious that the selected enzyme should not recognize sequence elements that are part of the hybridization probe.

(iii) *Sensitivity to cytosine methylation:* Many restriction enzymes will not cut if a methylated cytosine is present in its target sequence. Since DNA fingerprinting aims at the examination of DNA 'sequence variability' rather than DNA 'modification', the enzyme of choice should be insensitive to cytosine methylation.

(iv) *Cost:* If all other parameters are met, one may choose the cheapest enzyme available. We found *Taq*I, *Rsa*I, and the 6-cutter *Dra*I (recognition sequence: 5'-AAATTT-3') to be a good choice for fingerprinting plant genomes.

A) Materials
   1. Restriction enzymes.
   2. 10× restriction buffer (usually supplied by the enzyme manufacturer).

B) Experimental protocol
   1. Set up the restriction mix by combining into a reaction tube distilled water, 10× restriction buffer to a final 1× concentration, genomic DNA (5-10 μg), and enzyme (5-8 units/μg DNA). Preparing a "master mix" of water, buffer and enzyme saves pipette tips and time. Distribute this mix into the reaction tubes, then add the DNA.
   2. Mix carefully, centrifuge for a few seconds to collect the ingredients at the bottom of the tube, and incubate for 3 h to overnight at the recommended incubation temperature (37°C for most enzymes).

3. At the end of the incubation period, the sample can either be used directly for electrophoresis (see below), stored at -20°C, or ethanol-precipitated.

C) Comments
1. Digestion should be performed under the conditions recommended by the enzyme manufacturer. About 5-10 μg of DNA are usually sufficient for one fingerprint analysis, depending on the probe-species combination and genome size. Fingerprinting very large genomes may require >10 μg DNA per lane.
2. A minimum of 5 units of enzyme per μg of genomic DNA should be used to ensure complete digestion. Control reactions including λ DNA in the sample may help to evaluate completeness of digestion. Inclusion of 100 μg/mL BSA or 4 mM spermidine in the restriction buffer, as well as diluting the DNA sample prior to restriction may overcome inhibitory effects of, for example, plant secondary compounds present in the DNA preparation. After digestion, diluted samples may be concentrated by ethanol precipitation and redissolving in an appropriate volume of electrophoresis buffer.

Separation and immobilization of restriction fragments

Restriction fragments are separated by electrophoresis on horizontal agarose gels, stained with ethidium bromide and photographed. The DNA is then denatured, and gels are either dried, or blotted onto a nylon membrane. In our hands, dried gels are preferable for radioactive (^{32}P) fingerprinting and detection procedures, while Southern blots yield better results with nonradioactive detection.

A) Materials
1. Horizontal agarose gel electrophoresis set.
2. Power supply.
3. Transilluminator (or other UV light source).
4. Vacuum gel dryer.
5. Electrophoresis buffer: 45 mM Tris-borate, 1 mM EDTA, pH 8.0 (0.5× TBE), or any other buffer system recommended by Sambrook et al. (1989).
6. Loading buffer: 0.25% bromophenol blue, 0.25% xylene cyanol and 30% glycerol in electrophoresis buffer or water.
7. Agarose: 0.7-1.2% electrophoresis grade agarose (e.g. Seakem LE) dissolved in electrophoresis buffer.
8. Staining solution: 1 μg/mL ethidium bromide in electrophoresis buffer or water.
9. Denaturation buffer: 0.5 M NaOH, 0.15 M NaCl.
10. Neutralization buffer: 0.5 M Tris-HCl, 0.15 M NaCl, pH 7.

B) Experimental protocol
1. Suspend agarose at the desired concentration in an appropriate amount of electrophoresis buffer and dissolve it completely by boiling in a microwave oven for 2-4 × 2 min.
2. Allow to cool to 60°C, pour the agarose into a gel mold, insert a slot-forming comb, and remove air bubbles with a pipette. After the agarose has solidified, remove the

comb, and insert the gel mold into an electrophoresis apparatus filled with buffer, so that the gel is submerged.

3. Add 0.2 volumes of loading buffer to the DNA samples, mix, centrifuge for a few seconds, and slowly load the samples into the submerged slots using a micropipette with disposable tips.

4. Electrophorese at 1-2 V/cm for 24-48 h, until the bromophenol blue approaches the lower end of the gel. To minimize diffusion, gels should preferably be run in the cold (4°C).

5. After electrophoresis, stain the gel with ethidium bromide for 15-60 min. Rinse briefly in water, and photograph on a transilluminator (302 nm) using an orange filter. Placing a fluorescent ruler alongside the gel facilitates alignment of molecular weight markers with fingerprint bands in autoradiograms.

6. Transfer the gel to a tray filled with several volumes of denaturation buffer. Incubate for 30-45 min at room temperature under slight agitation.

7. Decant denaturation buffer, rinse the gel twice with distilled water, and incubate in several volumes of neutralization buffer for 1 h with slight shaking.

8. Transfer the gel onto two sheets of filter paper (e.g. Whatman 3MM) soaked with neutralization buffer, and cover the gel with plastic wrap. Place the gel onto a commercial gel dryer, with the plastic wrap facing upward. Insert a cooling trap between the gel dryer and the vacuum pump.

9. Apply vacuum (without heat) for 1-3 h, until the gel is evenly flat. Then turn on the heater to 60°C, and dry for 1 h. Turn off vacuum and check whether the filter paper is completely dry. If not, continue for another 30 min. The duration of the process depends on gel thickness and the power of the vacuum pump.

10. Store dried gels under desiccated conditions at room temperature.

C) Comments

1. *Safety precautions:* Ethidium bromide is a powerful mutagen and carcinogen. Use gloves, and dispose ethidium bromide waste appropriately (Sambrook et al. 1989). UV light is also highly mutagenic and destructive. Eyes and skin should be protected properly when using a transilluminator.

2. Applying too much DNA per slot may result in overloading effects such as "smiling".

3. Aliquots of the digested samples should be checked on test gels for completeness of restriction and even loading of lanes.

4. Several lanes per gel should be loaded with molecular weight markers. Marker positions are visualized by staining and documented by photography. Small amounts of marker (<50 pg/lane) may also be included in the sample lanes (Taggart and Ferguson 1994). Rehybridization to a marker-complementary probe then results in reliable in-lane markers. This strategy circumvents the frequently encountered problem of between-lane variation in gels (*see* "Evaluation of fingerprint patterns"). For nonradioactive detection, labeled markers (e.g. with digoxigenin) are commercially available.

5. Since large restriction fragments are often more informative than small ones, relatively low gel concentrations are usually applied (0.7-1.2%). In general, gels

should be run long and slow, in order to minimize band smiling and optimize resolution of fingerprints.

6. Gel drying has several advantages over conventional Southern blotting: it is faster, prehybridization steps can be omitted, a "100% transfer efficiency" is obtained, and signal strength is about 5 times higher than using blot hybridization (Miyada and Wallace 1987). However, Southern blotting as described by Sambrook et al. (1989) is preferable for nonradioactive hybridization and detection procedures. Among the membranes tested by us, Hybond N performed best.
   NOTE: The NaCl concentrations of denaturing and neutralizing solutions should be raised to 1.5 M and 3 M, respectively, if the gel is meant for blotting.

7. It is also possible to dry the gel, immediately after staining. Denaturation and neutralization of DNA are then performed within the dried gel. In this case, the incubation times for both denaturation and neutralization step can be reduced to 15 min each.

8. Fully dried gels may be stored at room temperature for years. For a few weeks, partially rehydrated gels may also be kept in high salt solutions ($6\times$ SSC). Prolongued storage under aqueous conditions, however, results in successive rehydration and destruction of the gel.

Generation and labeling of probes

Microsatellite-complementary oligonucleotide probes may be synthesized with an automated DNA-synthesizer, or purchased commercially. For radioactive analyses, oligonucleotides are end-labeled with polynucleotide kinase using γ-^{32}P-ATP as a phosphate donor (Ali et al. 1986, Schäfer et al. 1988, Epplen 1992), and purified by ion exchange chromatography on Whatman DE-52. For nonradioactive analysis, end-labeled oligonucleotides may either be purchased commercially (digoxigenated oligonucleotides are available from most manufacturers), or prepared with the help of a commercial kit (e.g. Boehringer Mannheim).

A) Materials
1. Disposable plastic columns (e.g. Econo columns™, BioRad).
2. Oligonucleotide probe: 10 pmoles dissolved in 3.5 μL distilled water.
3. $10\times$ kinase buffer: 670 mM Tris-HCl pH 8.0, 100 mM MgCl$_2$, 100 mM dithiothreitol. This buffer is usually supplied by the manufacturer of the polynucleotide kinase.
4. γ-^{32}P-ATP: Stabilized aqueous solution (5000 Ci/mmol, 10 mCi/mL)
5. T4 polynucleotide kinase.
6. 0.5 M EDTA, pH 8.
7. TE buffer: 10 mM Tris-HCl, 1 mM EDTA, pH 8.
8. $10\times$ TE buffer: 100 mM Tris-HCl, 10 mM EDTA, pH 8.
9. 0.2 M NaCl in TE buffer.
10. 0.5 M NaCl in TE buffer.
11. Whatman DE-52 cellulose: To prepare DE-52 equilibrated in TE buffer, suspend dry DE-52 in $10\times$ TE buffer and allow it to swell overnight. Let it settle and change buffer. Repeat this procedure several times until the pH of the suspension

approaches pH 8.5. Then equilibrate the DE-52 material several times with 1× TE buffer and store it in this buffer at 4°C until use.

B) Experimental protocol

1. Dissolve 10 pmoles of oligonucleotide in 3.5 μL of distilled water in a reaction tube. Add 1 μL of 10× kinase buffer, followed by 5 μL (50 μCi) of γ-^{32}P-ATP. Close the tube, mix cautiously, and spin for a few seconds in a microfuge.
2. Add 2-4 units of polynucleotide kinase directly to the mixture, and incubate for 30 min on ice (some suppliers also recommend incubation at 37°C).
3. In the meantime, prepare DE-52 columns. Fill a disposable plastic column with 0.2-0.4 mL of DE-52 equilibrated in TE buffer. Wash with several volumes of TE buffer, close the outlet (e.g. with a plastic stopper) and store until use.
4. Stop labeling reaction by adding 1 μL of 0.5 M EDTA, and add 90 μL of TE buffer to facilitate subsequent handling.
5. Remove the stopper from the column, put a small flask below the outlet, and apply the labeled oligonucleotide solution to the column. Wash once with 4 mL of TE buffer, and once with 4 mL of 0.2 M NaCl in TE buffer to elute unincorporated γ-^{32}P-ATP.
6. Replace the flask by a 50 mL Falcon™ tube and discard the flow-through washing solutions (radioactive waste). Elute the oligonucleotide from the column with 2× 0.5 mL of 0.5 M NaCl in TE buffer. Discard the column (radioactive waste). Store the labeled oligonucleotide at -20°C until use.

C) Comments

1. *Safety precautions:* Radioactive labeling should be carried out in a separate laboratory, and behind appropriate shielding (e.g. 1 cm of plexiglass). Gloves should be worn, and a hand monitor must be routinely used to check for radioactive contamination. Radioactive waste will have to be collected and disposed in a legal and environmentally safe way.
2. Tandemly repeated, microsatellite-complementary probes may also be obtained by end-to-end ligation of suitable oligos, followed by cloning and/or PCR (e.g. Vergnaud 1989, Ledwith et al. 1990). In a method described by Rogstad (1993, 1994), template oligonucleotides and their complement consisting of at least two tandem copies of the desired repeat were mixed at equimolar ratios and subjected to PCR. A subset of oligo pairs annealed unevenly, allowing the *Taq* polymerase to fill the overhangs. After about 15 cycles, long stretches of tandemly repeated PCR products were formed. A desired size range (e.g. 300-500 bp) was isolated by gel electrophoresis and electroelution. These so-called "PCR-STR" probes were labeled by asymmetric PCR in the presence of a ^{32}P-labeled nucleotide (Rogstad 1993), but can as well be labeled by nick translation or random priming (Sambrook et al. 1989).
3. Minisatellite probes can be labeled by conventional nick translation or random priming (Sambrook et al. 1989). Higher sensitivities are obtained if single-stranded probes are generated. This may be achieved by either *in vitro* transcription from cloned inserts (Carter et al. 1989) or by linear amplification of cloned minisatellites

using one primer only, and including a labeled deoxynucleotide in the PCR (Rassmann and Tautz 1992).

Gel hybridization and autoradiography

In this step, the immobilized DNA is hybridized with a labeled probe, unbound probe is washed off, and hybridizing restriction fragments are detected by autoradiography or chemiluminescence. Below we present a radioactive and a digoxigenin-based variant of the general protocol. The quality of the resulting fingerprint patterns is influenced by the stringency of hybridization, which depends on various parameters including the GC content and concentration of the probe, temperature of hybridization and washing steps, and buffer composition (e.g. salt concentration and the inclusion of formamide). In the present protocol, annealing temperatures (T_m) are calculated according to the so-called "Wallace rule" (Miyada and Wallace 1987, Ali et al. 1986, Schäfer et al. 1988, Epplen 1992), i.e. with 2°C for each AT pair and 4°C for each GC pair, respectively (assuming 1 M NaCl and an oligonucleotide length of about 16 bp). Hybridization is carried out at T_m -5°C. According to Miyada and Wallace (1987), these conditions result in 100% stringency, i.e. no mismatches are allowed. Though this might not hold for all oligonucleotides containing microsatellite motifs (Grady et al. 1992), we found hybridization results perfectly reproducible if the conditions were kept constant between experiments.

A) Materials
1. Roller bottle hybridization oven and suitable flasks.
2. Hybridization buffer: 5× SSPE, 5× Denhardt's solution, 0.1% SDS, 10 μg/mL fragmented and denatured *E. coli* DNA. Prepare from stock solutions and sterilize by filtration.
3. 20× SSPE stock solution: 3 M NaCl, 0.2 M sodium phosphate pH 7.4, 0.02 M EDTA.
4. 100× Denhardt's stock solution: 2% polyvinylpyrrolidone (PVP-40), 2% BSA and 2% Ficoll. Sterilize by filtration and store in aliquots at -20°C.
5. 20% SDS stock solution.
6. *E.coli* DNA: 2.5 mg/mL in 10 mM Tris-HCl and 1 mM EDTA (pH 8.0); store in aliquots at -20°C and denature by heating (100°C, 5 min) prior to addition to the hybridization buffer.
7. Probe: ^{32}P-labeled oligonucleotide; add to an appropriate amount of hybridization buffer at a concentration of 0.5 pmol/mL.
8. 6× SSC (washing solution): 0.9 M NaCl, 0.09 M sodium citrate, pH 7.0.

B) Experimental protocol
1. Remove plastic wrap cautiously, and soak the dried gel in a tray filled with distilled water. After a few minutes, the filter paper will detach. Remaining pieces of paper may cause background and should be cautiously wiped off the gel using gloves.
2. Transfer the gel into a new tray filled with 6× SSC and incubate for 5 min.

3. Wind the gel onto a disposable pipette, transfer it into a hybridization flask filled with 6× SSC and unroll it to the inner wall of the flask. Discard the 6× SSC, and add 10 mL of hybridization buffer including the labeled probe.
4. Hybridize for 3 h to overnight at T_m -5°C.
5. After hybridization, carefully decant the probe into a 50 mL Falcon tube. The probe may be reused several times. Store at -20°C.
6. Fill the hybridization flask up to one half with 6× SSC, close it, and wash off most of the unbound probe by gentle shaking. Decant the washing solution (radioactive waste). Use gloves to remove the gel from the flask, transfer it to a tray filled with 6× SSC, and wash the gel in this tray at room temperature (3× 30 min).
7. Transfer the gel to another tray containing 6× SSC prewarmed to hybridization temperature. Wash for 1-2 min (stringent "hot wash"; Miyada and Wallace 1987).
8. Transfer the gel onto a sheet of plastic wrap, and drain excess liquid with filter paper.
9. Cover the gel with plastic wrap and insert it into an X-ray cassette with or without intensifying screen (see comments). In the darkroom, place a sheet of X-ray film between gel and intensifying screen. If screens are used, store the cassette at -80°C.
10. After an exposure of several hours to several days (depending on signal strength), remove the cassette from the freezer, let it warm up to room temperature, and develop the film as recommended by the supplier. Handle X-ray films carefully since they are sensitive to scratching, especially when wet. Document dry autoradiograms by photography.

C) Comments
1. Because microsatellites are ubiquitous components of all eukaryotic genomes, inclusion of "non-specific" eukaryotic DNA in the buffer may lead to an increase rather than a decrease of background. Therefore, such DNA should either be omitted from the hybridization buffer (Westneat et al. 1988), or prokaryotic DNA (e.g. from *E.coli)* which typically lacks tandem-repetitive sequences should be used for blocking (Ali et al. 1986, Schäfer et al. 1988).
2. Prehybridization is not necessary for in gel-hybridization with oligonucleotides, as dried gels have little tendency to bind single-stranded DNA probes unspecifically.
3. Hybridization may either be performed in sealed plastic bags in a shaking water bath, or in glass cylinders in a roller bottle oven. Though the equipment needed for oven hybridization is more expensive, an oven is preferable for two reasons: (i) radioactivity is effectively shielded by the glass walls of the hybridization flask; and (ii) post-hybridization washing steps can be performed within the flask, thus reducing contamination risks.
4. To avoid background problems, gels should neither run dry during hybridization nor washing.
5. Autoradiography is best performed in X-ray cassettes. Signal strength and appropriate exposure time (several hours to several days) may be estimated using a hand monitor. If signals are weak, intensifying screens may be included in the cassettes. At low temperatures, these screens emit photons upon receiving radioactive β-particles, thereby increasing signal strength several fold (Sambrook et al. 1989). However, band sharpness is reduced if intensifying screens are used. If

strong signals are measured, it is therefore conceivable to perform the exposure at room temperature without screens.

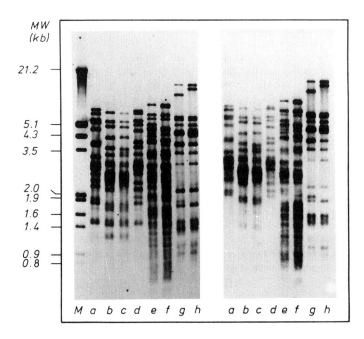

FIGURE 3. Nonradioactive (left) and radioactive (right) detection of DNA fingerprints: Comparison of results. Genomic DNA of different plant species was restricted with *TaqI*, and two identical sets of samples were simultaneously electrophoresed on agarose gels (5 μg per lane). One gel was blotted onto a Nylon membrane (Hybond N) and hybridized to a digoxigenin-labeled [GATA]$_4$ probe (left panel). The second gel was dried and hybridized to a ^{32}P-labeled [GATA]$_4$ probe (right panel). Nonradioactive detection was done with an antidigoxigenin antibody conjugated to alkaline phosphatase that catalyzed the decomposition of the chemiluminogenic substrate, AMPPD. Radioactive detection was done by standard autoradiography without intensifying screens. a-d: four different cultivars of chickpea *(Cicer arietinum* L.); e-f: two individuals of a wild tomato, *Lycopersicon peruvianum;* g-h two varieties of cultivated tomato, L. *esculentum.* M: digoxigenated size markers (Boehringer Mannheim; molecular weight is given in kilobases, kb)

Blot hybridization and chemiluminescent signal detection

Nonradioactive hybridization and detection methods possess several inherent advantages: (i) health hazards and waste disposal problems are minimized; (ii) the longer stability of probes implicates a more convenient planning of experiments; (iii) no special laboratory facilities and containments are required; and (iv) signal detection is considerably faster, at least in case of chemiluminesence-based methods (Tizard et al. 1990, Bierwerth et al. 1992).

The protocol described below is based on the procedures of Zischler et al. (1989a) and Bierwerth et al. (1992). Digoxigenated oligonucleotide probes bound to their target sequences are recognized by a specific antibody which is in turn conjugated to an alkaline phosphatase molecule. The enzyme dephosphorylates a chemiluminogenic substrate, AMPPD [also called PPD; 3-(2'-spiroadamantane)-4-methoxy-4-(3'-phosphoryloxy) phenyl-1,2-dioxetane, disodium salt]. In the course of the highly exothermic two-step reaction initiated by the enzymatic cleavage, AMPPD decomposes with concurrent emission of visible light (477 nm) which can be detected by exposure to X-ray films (Bronstein and McGrath 1989, Tizard et al. 1990, Höltke et al. 1995).

Pros and cons of radioactive versus chemiluminescent fingerprint protocols are summarized in Table 1. A comparison of ^{32}P- and PPD-detected oligonucleotide fingerprints using the same membrane is shown in Figure 3.

TABLE 1. Typical features of radioactive and digoxigenin-based hybridization procedures.

Feature	DIG-hybridization and chemiluminescence	^{32}P-hybridization
Isotope laboratory required	No	Yes
Health risk and waste problem	Low	Medium
Steps involved in signal detection	Several blocking, washing and incubation steps	Few; autoradiography
Signal development	Within one hour	From hours to days
Sensitivity	Medium to high	High
Background problems	Medium	Usually low
Stability of probes	Up to one year	2-4 weeks
In-gel hybridization possible	No	Yes
Reuse of membranes or dried gels	5-10 times	5-10 times

A) Materials
Same as for gel hybridization and autoradiography with the following additions:
1. Blocking solution: 0.5 % blocking reagent (Boehringer Mannheim) in 6× SSC. Dissolve by heating; prepare in advance and allow to cool to room temperature.
2. Solution A: 0.9 M NaCl, 0.1 M Tris-HCl pH 7.5.
3. Solution B: 0.9 M NaCl, 0.1 M Tris-HCl pH 9.5, 0.05 M MgCl$_2$ (sterilized by filtration, or made up from sterile stock solutions; insoluble precipitates will form upon autoclaving).
4. Antibody solution: Polyclonal sheep antidigoxigenin antibody conjugated to alkaline phosphatase (Boehringer Mannheim); stock solution, diluted 1:5000 in solution A (final concentration 0.15 U/mL).
5. PPD solution: PPD stock solution (Boehringer Mannheim), diluted 1:200 in solution B (final concentration 50 μg/mL). Diluted solutions can be reused several times.

B) Experimental protocol
1. Incubate the membrane in blocking solution at room temperature for 1 h.

2. Transfer the membrane to a hybridization flask as described above (DNA side facing inward), add 10 mL of hybridization buffer excluding the probe, and pre-hybridize overnight at T_m -5°C.

3. Decant pre-hybridization buffer and add 10 mL of hybridization buffer including 5 pmol/mL of probe.

4-7. Hybridize and wash as described above for radioactive probes.

8. After the final wash, incubate membrane for 1 h at room temperature in blocking solution. This step prevents unspecific binding of the antibody to the membrane.

9. Rinse membrane shortly in solution A, and spread it on a glass plate (DNA side facing upward). Pipette the antibody solution onto the membrane surface (ca. 5 mL/100 cm^2) and incubate for 15 min at room temperature. Make sure that the antibody solution covers the membrane completely.

10. Discard antibody solution. Remove excess antibody by washing in solution A in a tray (3× 15 min).

11. Incubate membrane in solution B (2× 15 min), then spread it on a glass plate (DNA side facing upward) and pipette the PPD solution onto the membrane surface (ca. 5 mL/100 cm^2). Incubate for 5 min under gentle agitation.

12. Decant the PPD solution, shortly drain the membrane on filter paper to remove excess substrate (membrane should not become dry!), seal it in a transparent plastic bag (e.g. Saran wrap), incubate at 37°C for 15 min and transfer to an X-ray cassette.

13. In a darkroom, expose membrane to X-ray film (preferably Kodak XAR or equivalent) for 2 min to 1 h at room temperature. Assess optimal exposure time by test exposures. Signal strength increases during the first 24 h and slowly levels off afterwards (Bierwerth et al. 1992).

C) Comments

1. Because the intermediate of the enzyme-catalyzed dephosphorylation of AMPPD, AMP⁻D, is moderately stable, chemiluminescent glowing can last for more than a week (Tizard et al. 1990, Bierwerth et al. 1992). However, strong bands weaken after some days, and weak bands appear relatively stronger than before. This effect is due to substrate limitations as well as to the catalytic rather than stoichiometric nature of signal generation. It can be reverted by reincubating the membrane with the substrate.

2. In general, nonradioactive DNA fingerprinting variants produce more unspecific background noise than those based on [32]P. Since background is often due to residual antibodies, membranes should be washed carefully after the antibody incubation step. Background may also be reduced by increasing washing times, including 0.1% SDS in the first washing step, diluting the PPD solution, and centrifuging the antibody solution in a microfuge prior to use (Neuhaus-Url and Neuhaus 1993). Careful handling of membranes is also critical. Membranes should never be touched without gloves, and microbial contamination of buffers or membranes must be avoided (microbes may contain acidic phosphatase activities).

3. After signal detection and documentation, gels and blots may be reused with other probes for several times (up to 10 times in our hands). The former probe is stripped off either by washing in 5 mM EDTA at 60°C (2-4× 15 min), or by repeating the

denaturation/neutralization step. For blots, the stripping temperature can also be elevated to 95°C and SDS can be added to 0.1 % final concentration.

4. Several alternative enzymatic substrates are available for chemiluminescent detection (e.g. AMPPD, CSPD®, Lumigen PPD™, CDP-Star™, etc.; Höltke et al. 1995). Colorigenic substrates used in earlier studies (Zischler et al. 1989a, Pena et al. 1991, Bierwerth et al. 1992) suffer from several drawbacks: (i) re-use of membranes is limited by insoluble dye precipitates, which are only incompletely removed even by the use of organic solvents; and (ii) the dye precipitate is formed on the membrane itself, thus allowing only one hard-copy record per experiment.

5. Nonradioactive in-gel hybridization has repeatedly been reported using colorigenic enzymatic substrates (Zischler et al. 1989b, Pena et al. 1991,Yavachev 1991). In our hands, attempts to create chemiluminescent signals after digoxigenin-based in-gel hybridization were not successful.

6. Oligonucleotide probes can also be directly linked to alkaline phosphatase, thus reducing the number of steps required (Klevan et al. 1995). Though being more expensive, direct labeling is a useful alternative if only few different probes are routinely used.

7. If desired, the sensitivity of the chemiluminescence reaction can be further increased by the addition of enhancers (Klevan et al. 1995).

Hybridization of microsatellites to RAPD fragments

Recently, a novel strategy was developed that combines arbitrarily or microsatellite-primed PCR with microsatellite hybridization (Richardson et al. 1995, Cifarelli et al. 1995, Ender et al. 1996). In the first step of this procedure, genomic DNA is amplified with either a single arbitrary 10-mer primer (as is usually used in RAPD analysis; Williams et al. 1990) or a microsatellite-complementary 15-mer or 10-mer primer (Gupta et al. 1994, Weising et al. 1995b). PCR products are then electrophoresed, photographed, blotted, and hybridized to a ^{32}P- or digoxigenin-labeled mono-, di-, tri- or tetranucleotide repeat probe such as $[GA]_8$, $[CA]_8$, $[CC]_{12}$ or $[CAA]_5$. Subsequent autoradiography reveals reproducible, probe-dependent fingerprints which are completely different from the ethidium bromide staining patterns, and polymorphic at an intraspecific level. This method was coined RAMPO (for random amplified microsatellite polymorphisms; Richardson et al. 1995), RAHM (for random amplified hybridization microsatellites; Cifarelli et al. 1995) or RAMS (for randomly amplified microsatellites; Ender et al. 1996).

A) Materials
   Same as described in the previous sections, with the following additions:
   1. Thermocycler.
   2. *Taq* DNA polymerase: 5 U/µL (e.g. Amplitaq, Perkin-Elmer).
   3. 10× PCR buffer: 200 mM Tris-HCl pH 8.3, 500 mM KCl, 20 mM MgCl$_2$ (10× PCR buffer is often supplied by the manufacturer of the *Taq* DNA polymerase and may or may not contain MgCl$_2$).
   4. PCR primers (5 pmol/µL): random 10-mer oligonucleotides (e.g. Operon) or micro-satellite-complementary oligonucleotides (e.g. [GATA]$_4$).

5. dNTP stock: 2 mM dATP, 2 mM dCTP, 2 mM dGTP, 2 mM dTTP.
6. Template DNA: 5-20 ng/μL.

B) Experimental protocol

1. Use thin-walled tubes to set up a PCR with 25 μL reaction volumes containing 20 mM Tris-HCl pH 8.4, 50 mM KCl, 2 mM $MgCl_2$, 0.2 mM each of dATP, dCTP, dGTP and dTTP, 0.4 μM primer (10 pmoles per reaction), 1 unit *Taq* DNA polymerase and 15-100 ng of template DNA. Pipetting errors are minimized by preparing "master mixes" sufficient for all samples. A master mix includes the enzyme, 10× buffer, $MgCl_2$, and dNTPs; primer and template are added separately to each tube.
2. Mix the contents, centrifuge the tubes briefly, and overlay the reaction solution with two drops of mineral oil to prevent evaporation (mineral oil is not required in some thermocyclers equipped with heatable lids). Insert the tubes into the thermocycler.
3. For random 10-mer primers, use the following program: 1 cycle of 3 min at 94°C (initial denaturing), 39 cycles of 15 s at 94°C (denaturing), 30 s at 35°C (annealing) and 90 s at 72°C (elongation), and 1 cycle of 3 min at 72°C (final elongation step). For microsatellite-complementary primers, use the following touch-down program (modified from Don et al. 1991): 1 cycle of 3 min at 94°C (initial denaturing step), 39 cycles of 20 s at 94°C (denaturing), 30 s at 57°C, 53°C or 50°C (annealing; *see* below), and 40 s at 72°C (elongation), and 1 cycle of 4 min at 72°C (final step of elongation). Annealing is performed at stepwise reduced temperatures (cycle 1: 57°C; cycle 2: 53°C; cycles 3-39: 50°C). The annealing temperature range has to be adapted for oligonucleotides of different GC content.
4. Mix half of the sample with loading buffer, electrophorese in 1.5% agarose (see section on agarose gel electrophoresis), stain with ethidium bromide and document banding patterns by photography. Store the remainder of the sample at 4°C.
5. Denature, neutralize, and dry or blot the gel as described in a previous section.
6. Hybridize the gel or membrane to a [32]P- or digoxigenin-labeled microsatellite-complementary oligonucleotide probe as described in a previous section.
7. Document the results by autoradiography or chemiluminescence and photography.

C) Comments

1. The occurrence of RAMPO bands may be explained as follows. Any RAPD reaction or MP-PCR probably creates many thousand different products of various abundance. The majority of minor fragments will remain below the detection level of ethidium bromide staining. However, the ubiquitous presence of microsatellites in eukaryotic genomes provides a means of visualizing a subset of such minor amplification products by hybridization. The signal intensity of fragments harboring a certain microsatellite motif will depend both on the length of this motif and on the abundance of the fragment.
2. The RAHM (Cifarelli et al. 1995) and RAMPO (Richardson et al. 1995) approach were extensively tested with different species of yams *(Dioscorea)*, olive, sugarbeet, and sunflower, but to our experience works with all plant species tested so far. Since a similar strategy (RAMS; Ender et al. 1996) was shown to yield

polymorphic fragments among different *Daphnia* species (Crustacea), the technique is probably generally applicable in eukaryotes.
3. The reproducibility of weak RAPD fragments is sometimes questioned. Though RAMPO visualizes such small fragments, patterns were generally highly reproducible. An example of a reproducibility check of RAMPO patterns obtained with different yam samples is shown in Figure 4.
4. Hybridizing bands may be isolated, cloned, sequenced and used as a probe on RAPD or RFLP gels (Cifarelli et al. 1995, Ender et al. 1996). Cloned RAMPO fragments were generally shown to contain microsatellites, which are often polymorphic within and between species.

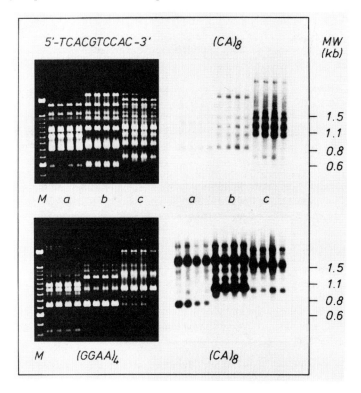

FIGURE 4. Random amplified microsatellite polymorphisms (RAMPOs) between three different yam species (a: *Dioscorea cayenensis*, b: *D. rotundata*, c: *D. bulbifera*). To test the reproducibility of RAMPO analysis, four replicate experiments were performed with each of the samples. Upper panel, left: RAPD profiles obtained with primer OPG-08 (5'-TCACGTCCAC-3'). Upper panel, right: OPG-08 products probed with (CA)$_8$. Lower panel, left: PCR profiles obtained with the microsatellite-complementary primer (GGAA)$_4$. Lower panel, right: (GGAA)$_4$ products probed with (CA)$_8$. Positions of molecular weight markers (lane M) are given in kilobases (kb).

Evaluation of fingerprint patterns

A fingerprint pattern obtained from a particular DNA sample is rarely informative on its own. Instead, patterns originating from different samples have to be compared to each other. To that end, individual bands are assigned to particular positions, and different lanes are screened for comigrating (i.e. "matching") bands. Mobility of DNA fragments is often somewhat uneven across a gel, and the resulting "band shifts" can lead to misinterpretations of band matching. Such artifactual variation can be counteracted by including internal molecular weight standards within each lane: a small amount of marker (about 700 pg) is added to each sample, and visualized by hybridization and autoradiography (Taggart and Ferguson 1994). However, even if highly efficient standardization procedures are employed, the decision of what is a "match" may still be somewhat problematic. It is advisable to analyze only the easily scorable part of a fingerprint pattern and to have the patterns interpreted by several independent researchers.

Multilocus DNA profiles are usually translated into discrete characters, i.e. "0" for absence and "1" for presence of a band at a particular position. These "0/1 matrices" can then be exploited for different types of studies. The most frequent applications are paternity testing, genotype identification, estimation of genetic diversity and relatedness, and linkage analysis. Some remarks concerning the evaluation of fingerprint bands are given below and in the "Comments, limitations and overview" section. However, a detailed discussion of the (mainly statistical) problems associated with the application of multilocus DNA fingerprints in different areas of research is beyond the scope of this Chapter (*see* Lynch 1991, Reeve et al. 1992, Brookfield 1992, Brookfield et al. 1993, Chakraborty and Jin 1993, Pena and Chakraborty 1994, Danforth and Freeman-Gallant 1996).

Since bands are generally inherited in a Mendelian fashion, an obvious application of RFLP fingerprinting is the estimation of relatedness between parents and offspring (reviewed by Pena and Chakraborty 1994). For paternity testing, the fingerprint patterns of mother, father and offspring are compared. All offspring bands must either be derived from mother or father. A large number of non-paternal bands will exclude paternity with a high level of probability, while the occasional occurrence of such bands may be a consequence of mutation (*see* Jeffreys et al. 1991 for a discussion).

For the evaluation of higher-order relationships, a "similarity index" (F) can be calculated from band sharing data of each pair of fingerprints according to the formula:

$$F = \frac{2m[ab]}{m[a] + m[b]}$$

where $m[a]$ and $m[b]$ represent the total number of bands present in lanes [a] and [b], respectively, and $m[ab]$ the number of bands which both lanes share (Wetton et al. 1987). F can acquire any value between 0 and 1, where 0 means 'no bands in common', and 1 means 'patterns are identical'. A distance matrix can be prepared from the similarity values, which can be further transformed into a dendrogram by phenetic methods (for details see Swofford and Olsson 1990).

Finally, DNA fingerprints can be used for genetic mapping. Bands which are polymorphic between two parents of a cross can be analyzed for linkage to each other as well as to other types of markers. A variety of computer programs are available that calculate recombination values and map distances, give a most likely order of markers within linkage groups, and thus create genetic maps (e.g. LINKAGE-1, Suiter et al. 1983; MAPMAKER, Lander et al. 1987; JOINMAP, Stam 1993).

Comments, limitations and overview

In recent years, the development of molecular markers based on DNA polymorphisms has greatly facilitated research in many biological disciplines. A large number of marker techniques have been developed and are now available to the scientific community. In view of the diversity of methods an important question arises: which marker technique is appropriate for which areas of research? Basically, there are four main application areas for molecular markers: (i) genotype *identification*, (ii) the analysis of genetic *diversity*, (iii) the estimation of genetic *relatedness*, and (iv) genetic *mapping*. Here we will discuss some advantages and disadvantages of RFLP fingerprinting in general, and oligonucleotide fingerprinting in particular, as compared to other marker techniques. We will also give some recommendations as to where the technique appears appropriate, and where not.

Since its introduction to genome analysis of plants (Weising et al. 1989) and fungi (Meyer et al. 1991), the potential of oligonucleotide fingerprinting to generate informative markers has been amply demonstrated in a large number of plant and fungal species (e.g. Meyer et al. 1991, Weising et al. 1991a,b, 1992, Vosman et al. 1992, Kaemmer et al. 1992, 1995, DeScenzo and Harrington 1994, Rogstad 1994, Ramakrishna et al. 1994, Morjane et al. 1994, Sharma et al. 1995, Hamann et al. 1995, Depeiges et al. 1995, Sharon et al. 1995, Sastry et al. 1995, Piquot et al. 1996, Wanner et al. 1996). This obvious success is based on several advantages of the technique itself, most importantly its high discriminatory potential, its versatility with regard to DNA of various complexity and origin (microsatellites are present in each eukaryotic genome), the rapid screening of the genome with few multilocus probes, and the excellent reproducibility. Moreover, neither is there any sequence information about the target genome necessary nor any laborious cloning involved, as is the case for RFLP analysis. Obvious disadvantages of the method include a relatively complex laboratory protocol with several experimental traps, and the requirement of higher amounts (and probably higher purity) of DNA as compared to PCR-based techniques. A general disadvantage shared with other multilocus approaches is the insufficient allelic information provided by multilocus banding patterns. Thus, fingerprint bands have to be considered as dominant markers. Transformation into codominant markers is principally possible, but requires tedious cloning steps.

RAMPO is a relatively new technique (Richardson et al. 1995; Cifarelli et al. 1995; Ender et al. 1996) which combines several advantages of oligonucleotide fingerprinting,

RAPD-PCR and microsatellite-primed PCR, i.e. the speed of the assay, the high sensitivity, high level of variability detected, and no requirement for *a priori* DNA sequence information. Since a single gel yields several independent polymorphic fingerprints, RAPD gels can be exploited more efficiently. By replacing restriction of genomic DNA with RAPD-PCR or MP-PCR to generate the pool of DNA fragments for microsatellite probing, the RAMPO procedure is also applicable to studies with only traces of DNA. On the other hand, RAMPO also suffers from the disadvantages inherent to RAPD analysis and microsatellite-primed PCR, i.e. the high sensitivity to a change in reaction conditions and limited reproducibility across laboratories (Penner et al. 1993, Weising et al. 1995b).

The high versatility of microsatellite-complementary oligonucleotide probes for RFLP fingerprinting in plants has been demonstrated by several authors. Sharma et al. (1995) studied the abundance and polymorphism of different simple sequence repeat motifs in 4 accessions of cultivated chickpea *(Cicer arietinum)* using 14 different restriction enzymes. Among 38 tested probes, 35 produced detectable hybridization signals. The abundance and level of polymorphism of the target sequences varied considerably, and no obvious correlation existed between abundance, fingerprint quality, and sequence characteristics of a particular motif. Depeiges et al. (1995) used oligonucleotide probes complementary to all possible microsatellites of the mono-, di-, tri- and tetranucleotide repeat type (49 in total) to screen *Eco*RI and *Hind*III-digested *Arabidopsis thaliana* and yeast DNA by Southern analysis. Interestingly, most probes did not reveal any signal or only a few bands. Only nine probes resulted in clear fingerprints with at least one of the two species, most notably $[A]_{18}$, $[ATG]_6$, $[AAG]_6$, $[GTG]_6$, $[GGAT]_4$ and $[AAAC]_4$. Similar observations were made by Sharon et al. (1995). The different results obtained with chickpea vs. *Arabidopsis thaliana* are most probably caused by differences in the size and complexity of the two genomes.

In general, pattern complexity as well as the level of polymorphism revealed by DNA fingerprinting with both mini- and microsatellite-complementary probes mainly depends on (i) the investigated species or population and its reproductive biology (i.e. selfing, outcrossing, apomixis, vegetative propagation, inbreeding; Wolff et al. 1994, Kraft et al. 1996) and (ii) the sequence motif used for hybridization. The influence of the restriction enzyme is comparatively low. It is still not clear which features (if any) are essential for a fingerprint probe to reveal high or low levels of polymorphism. Consequently, the optimal combination of species, probe and restriction enzyme has to be determined empirically. Depending on the probe, species-, variety- or individual-specific patterns may be observed within the same genus (*see* Weising et al. 1992, Sharma et al. 1995).

The high discriminatory potential of minisatellite and microsatellite-based RFLP fingerprints immediately suggested their use in *genotype identification* in humans and other species. Not only unrelated individuals, but also first-order relatives could be discriminated by an individual-specific fingerprint (Jeffreys et al. 1985a,b, 1991). Consequently, forensic analysis and paternity testing became the first application areas for RFLP fingerprinting (reviewed by Pena and Chakraborty 1994). Since somatic mutations of fingerprint patterns are comparatively rare, clonally related organisms generally yielded identical fingerprints.

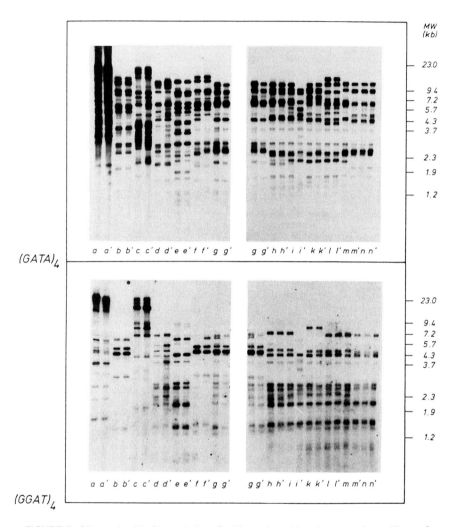

FIGURE 5. Oligonucleotide fingerprinting of wild species, old and present-day cultivars of tomato. Total leaf DNA of two individual plants per cultivar was digested with *Hinf*l, electrophoresed in 1.2% agarose gels (5 μg DNA per lane), and in-gel-hybridized to the ^{32}P-labeled oligonucleotide probes [GATA]₄ and [GGAT]₄ a,a': *Lycopersicon hirsutum;* b,b': *L. pimpinellifolium;* c,c': *L. peruvianum;* d,d' to n,n': different varieties and cultivars of *L. esculentum.* Positions of size markers are given in kilobases. The data obtained by seven oligonucleotide probes were analyzed phenetically and yielded a dendrogram compatible with the known history of cultivated tomato (Kaemmer et al. 1995).

This somatic stability of RFLP fingerprints made the technique highly useful for the unequivocal identification and discrimination of plant cultivars (e.g. banana: Kaemmer et al. 1992), micropropagated plants (e.g. *Achillea:* Wallner et al. 1996), apomictic plants (e.g. several *Rubus* species; Kraft et al. 1996),clonally propagating wild plants (e.g. *Sparganium erectum*; Piquot et al. 1996) and asexually propagating phytopathogenic fungi (Weising et al. 1991b, Morjane et al. 1994, Sastry et al. 1995). For vegetatively propagated plant cultivars, RFLP fingerprints largely fulfill the criteria considered essential for cultivar identification by Bailey (1983): (i) maximum variation between cultivars, (ii) minimum variation within cultivars, (ii) stability over time, and (iv) independence from environmental factors. For sexually propagating plants, however, criterion (ii) is usually not fulfilled, since fingerprints are often variable within cultivars.

Other application areas of multilocus fingerprinting include the estimation of higher order *genetic relatedness* between organisms (e.g. Kaemmer et al. 1995) and of *genetic diversity* among and within populations (e.g. Piquot et al. 1996). Such analyses are usually based on the extent of band sharing between individuals. The statistical significance of the obtained results is influenced by several assumptions, most importantly the homology of comigrating bands. The validity of these assumptions is controversially discussed, not only for RFLP fingerprinting (Lynch 1991, Danforth and Freeman-Gallant 1996), but also for RAPD data (Rieseberg 1996). Nevertheless, our own results obtained with banana (Kaemmer et al. 1992) and tomato (Kaemmer et al. 1995; Figure 5), the analysis of apomictic *Rubus* species (Kraft et al. 1996) and cultivated rice accessions (Zhou and Gustafson 1995) as well as numerous other studies (reviewed by Weising et al. 1994) suggest that multilocus fingerprints may well serve as one similarity criterion among others for estimating intraspecific relatedness. Though genetic relationships may sometimes be overestimated due to elevated background band-sharing (Zhou and Gustafson 1995), fingerprint data are particularly valuable in species where genetic variability is limited, e.g. as a consequence of predominantly asexual reproduction (Kraft et al. 1996) or genetic bottlenecks (Kaemmer et al. 1995).

Its high sensivity and the possibility of screening the genome rapidly with few probes has prompted the use of multilocus fingerprinting for linkage analysis and genome mapping, e.g. in humans (Wells et al. 1989) and fungi (Romao and Hamer 1992). Lack of allelic information is less of a problem in fungi, where a haploid lifestyle prevails. In our own studies on the phytopathogenic fungus *Ascochyta rabiei,* we found that oligonucleotide fingerprint bands were inherited in a Mendelian fashion and segregated in the progeny of a cross between two mating types (Figure 6). A genetic linkage map of *Ascochyta rabiei* based on oligonucleotide fingerprint data is currently being developed in our laboratory (Geistlinger, Weising and Kahl, in preparation).

Additional problems have to be considered when fingerprinting data are applied to genetic mapping. One problem concerns mutation rate. In human minisatellites, mutation to new length alleles usually occurs at rates from 0.5-1 % per gamete and generation (Jeffreys et al. 1991), but can be as high as 5% (Jeffreys et al. 1988). Considerable proportions of nonparental fingerprint bands were also observed in plants (Rogstad 1994; Hüttel, Weising and Kahl, unpublished) and fungi (DeScenzo and Harrington 1994), when microsatellite-complementary probes were used. A second, maybe even more serious problem for mapping, is the tendency of mini- and microsatellite-derived fingerprint bands to occur in clusters, unlike the much shorter PCR-detected microsatellites (*see* Bell and Ecker 1994).

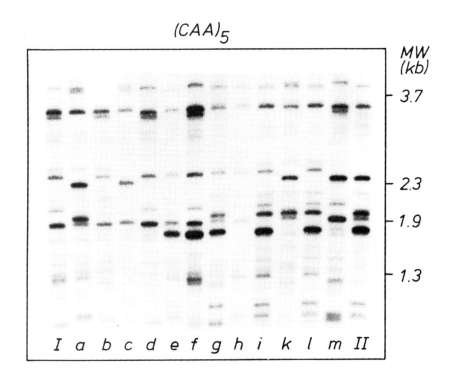

FIGURE 6. Mendelian segregation of oligonucleotide fingerprint bands in a core population (lanes a-m) resulting from a cross between mating types I (Mat I) and II (S6) of the chickpea blight fungus, *Ascochyta rabiei* (teleomorph *Didymella rabiei*). Genomic DNA was purified from mycelia, digested with *Hinf*I, electrophoresed in 1.6 % agarose gels (5 μg DNA per lane), and in-gel-hybridized to the ^{32}P-labeled oligonucleotide probe [CAA]$_5$. Positions of size markers are given in kilobases. Each band of the progeny can be traced back to one of the parental mating types.

Clustering of RFLP fingerprint bands was observed in several organisms, including dogs (Jeffreys and Morton 1987), swans (Meng et al. 1990), pea (Dirlewanger et al. 1994) and tomato (Arens et al. 1995; Broun and Tanksley 1996). In pea, four polymorphic, [GAA]$_5$-detected fingerprint bands mapped to the same linkage group within 42 cM of each other

(Dirlewanger et al. 1994), while in tomato, $[GATA]_4$-detected fingerprint bands formed several clusters in certain chromosomal regions (Arens et al. 1995).

Perspectives

Taken together, the unequivocal *identification and discrimination* of genotypes is probably the most appropriate application area of hybridization-based multilocus fingerprinting of plants and fungi (and other eukaryotic organisms as well). Here, the inherent advantages of the technique such as high sensitivity and reproducibility are very important, while difficulties with allelic designation, high mutation rate, clustering of bands and nonhomology of comigrating fragments are of minor significance only. Nevertheless, oligonucleotide fingerprinting and RAMPO will also prove valuable as two techniques among others for the establishment of maps, and the estimation of genetic diversity and intraspecific relatedness in plant and fungal species with limited levels of genetic variation, especially if no sequence information is available.

Acknowledgements

Research of the authors is supported by BMZ (grant 89.7860.3-01.130) and DFG (grants Ka 332/14-1, 14-2 and 14-3). We thank Siggi Kost, Juliane Ramser, Jörg Geistlinger and Bruno Hüttel for preparation of figures.

References

Ali S, Müller CR and Epplen JT (1986) *Hum Genet* 74:239-243.
Arens P, Odinot P, Van Heusden AW, Lindhout P and Vosman B (1995) *Genome* 38:84-90.
Bailey DC (1983) In: Tanksley SD and Orton TJ (eds), *Isozymes in Plant Genetics and Breeding,* Part A. Elsevier, Amsterdam, pp. 425-441.
Bell CJ and Ecker JR (1994) *Genomics* 19:137-144.
Bierwerth S, Kahl G, Weigand F and Weising K (1992) *Electrophoresis* 13:115-122.
Blackwell M and Chapman RL (1993) *Meth Enzymol* 224:65-77.
Bronstein I and McGrath P (1989) *Nature* 338:599-600.
Brookfield JFY (1992) *Mol Ecol* 1:21-26.
Brookfield JFY, Carter RE, Mair GC and Skibinski DOF (1993) *Mol Ecol* 2:209-218.
Broun P and Tanksley SD (1993) *Plant Mol Biol* 23:231-242.
Broun P and Tanksley SD (1996) *Mol Gen Genet* 250:39-49.
Bult C, Kallersjo M and Suh Y (1992) *Plant Mol Biol Rep* 10:273-284.
Carter RE, Wetton JH and Parkin DT (1989) *Nucleic Acids Res* 17:5867.
Chakraborty R and Jin L (1993) In: Pena SDJ, Chakraborty R. Epplen JT and Jeffreys AJ (eds), *DNA Fingerprinting: State of the Science.* Birkhäuser, Basel. pp. 153-175.
Cifarelli RA, Gallitelli M and Cellini F (1995) *Nucleic Acids Res* 23:3802-3803.
Couch JA and Fritz PJ (1990) *Plant Mol Biol Rep* 8:8-12.
Danforth BN and Freeman-Gallant CR (1996) *Mol Ecol* 5:221-227.
Depeiges A, Goubely C, Lenoir A, Cocherel S. Picard G. Raynal M, Grellet F and Delseny M (1995) *TheorAppl Genet* 91:160-168.
DeScenzo RA and Harrington TC (1994) *Phytopathology* 84:534-540.
Dirlewanger E, Isaac PG, Ranade S, Belajouza M, Cousin R and De Vienne D (1994) *Theor Appl Genet* 88:17-27.
Do N and Adams RP (1991) *Biotechniques* 10:162-166.
Don RH, Cox PT, Wainwright BJ, Baker K and Mattick JS (1991) *Nucleic Acids Res* 19:4008.
Doyle JJ and Doyle JN (1987) *Phytochem Bull* 19:11-15.

Ender A, Schwenk K, Städler T, Streit B and Schierwater B (1996) *Mol Ecol* 5:437-441.

Epplen JT (1992) In: Chrambach A, Dunn M and Radola BJ (eds), *Advances in Electrophoresis,* Vol. 5. VCH, Weinheim, Germany. pp. 59-112.

Fang G, Hammar S and Grumet R (1992) *Biotechniques* 13:52-55.

Fulton TM, Chunwongse J and Tanksley SD (1995) *Plant Mol Biol Rep* 13:207-209.

Grady DL, Ratliff RL, Robinson DL, McCanlies EC, Meyne J and Moyzis RK (1992) *Proc Natl Acad Sci USA* 89:1695-1699.

Gupta M, Chyi Y-S, Romero-Severson J and Owen JL (1994) *Theor Appl Genet* 89:998-1006.

Hamann A, Zink D and Nagl W (1995) *Genome* 38:507-515.

Höltke HJ, Ankenbauer W, Mühlegger K, Rein R, Sagner G, Seibl R and Walter T (1995) *Cell Mol Biol* 41:883-905.

Jeffreys AJ, Wilson V and Thein SL (1985a) *Nature* 314:67-73.

Jeffreys AJ, Wilson V and Thein SL (1985b) *Nature* 316:76-79.

Jeffreys AJ and Morton DB (1987) *Animal Genetics* 18: 1-15.

Jeffreys AJ, Royle NJ, Wilson V and Wong Z (1988) *Nature* 332:278-281.

Jeffreys AJ, Turner M and Debenham, P (1991) *Am J Hum Genet* 48:824-840.

Kaemmer D, Afza R, Weising K, Kahl G and Novak FJ (1992) *Bio/Technology* 10:1030-1035.

Kaemmer D, Weising K, Beyermann B, Börner T, Epplen JT and Kahl G (1995) *Plant Breeding* 114:12-17.

Kanazawa A and Tsutsumi N (1992) *Plant Mol Biol Rep* 10:316-318.

Klevan L, Horton L, Carlson DP and Eisenberg AJ (1995) *Electrophoresis* 16:1553-1558.

Kraft T, Nybom H and Werlemark G (1996) *Plant Syst Evol* 199:93-108.

Lander ES, Green P, Abrahamson J, Barlow A, Daly A, Lincoln SE and Newbury L (1987) *Genomics* 1:174-181.

Ledwith BJ, Manam S. Nichols WW and Bradley MO (1990) *Biotechniques* 9:149-152.

Li Q-B, Cai Q and Guy CL (1994) *Plant Mol Biol Rep* 12:215-220.

Lin J-Z and Ritland K (1995) *Plant Mol Biol Rep* 13:210-213.

Liston A and Rieseberg LH (1990) *Ann Missouri Bot Gard* 77:859-863.

Lohe AR and Brutlag DL (1986) *Proc Natl Acad Sci USA* 83:696-700.

Lynch M (1991) In: Burke T, Dolf G, Jeffreys AJ and Wolff R (eds), *DNA Fingerprinting: Approaches and Applications.* Birkhäuser, Basel. pp. 113-126.

Meng A, Carter RE and Parkin DT (1990) *Heredity* 64:73-80.

Meyer W, Koch A, Niemann C, Beyermann B. Epplen JT and Börner T (1991) *Curr Genet* 19:239-242.

Michaels SD, John MC and Amasino RM (1994) *Biotechniques* 17:274-276

Milligan BG (1992) In: Hoelzel AR (ed), *Molecular Genetic Analysis of Populations,* IRL Press, Oxford. pp. 59-88.

Miyada CG and Wallace RB (1987) *Meth Enzymol* 154:94-107.

Morjane H, Geistlinger J, Harrabi M, Weising K and Kahl G (1994) *Curr Genet* 26:191-197.

Murray MG and Thompson WF (1980) *Nucleic Acids Res.* 8:4321-4325.

Neuhaus-Url G and Neuhaus G (1993) *Transgenic Res* 2:115-120.

Paterson AH, Brubaker CL and Wendel JF (1993) *Plant Mol Biol Rep* 11:122-127.

Pena SDJ, Macedo AM, Gontijo NF, Medeiros AM and Ribeiro JCC (1991) *Electrophoresis* 12: 146-152.

Pena SDJ and Chakraborty R (1994) *Trends Genet* 10:204-209.

Penner GA, Bush A, Wise R, Kim W, Domier L, Kasha K, Laroche A, Scoles G, Molnar SJ and Fedak G (1993) *PCR Methods Applic* 2:341-345.

Phillips WJ, Chapman CGD and Jack PL (1994) *Theor Appl Genet* 88:845-851.

Piquot Y, Saumitou-Laprade P, Petit D, Vernet P and Epplen JT (1996) *Mol Ecol* 5:251-258.

Ramakrishna W, Lagu MD, Gupta VS and Ranjekar PK (1994) *Theor Appl Genet* 88:402-406.

Rassmann K and Tautz D (1992) *Mol Ecol* 1:135.

Reeve HK, Westneat DF and Queller DC (1992) *Mol Ecol* 1:223-232.

Rether B, Delmas G and Laouedj A (1993) *Plant Mol Biol Rep* 11:333-337.

Richardson T, Cato S, Ramser J, Kahl G and Weising K (1995) *Nucleic Acids Res* 23:3798-3799.

Rieseberg LH (1996) *Mol Ecol* 5:99-105.

Rogstad SH (1993) *Meth Enzymol* 224:278-294.

Rogstad SH (1994) *Theor Appl Genet* 89:824-830.

Romao J and Hamer JE (1992) *Proc Natl Acad Sci USA* 89:5316-5320.

Rosewich UL and McDonald BA (1994) *Meth Mol Cell Biol* 5:41-48.

Rowland LJ and Nguyen B (1993) *Biotechniques* 14:735-736.

Sambrook J, Fritsch EF and Maniatis T (1989) *Molecular Cloning, a Laboratory Manual*, 2nd Edition. Cold Spring Harbor Laboratory Press, Cold Spring Harbor, New York.

Sastry JG, Ramakrishna W, Sivaramakrishnan S, Thakur RP, Gupta VS and Ranjekar PK (1995) *Theor Appl Genet* 91:856-891.

Schäfer R, Zischler H, Birsner U, Becker A and Epplen JT (1988) *Electrophoresis* 9:369-374.

Sharma PC, Winter P, Bünger T, Hüttel B, Weigand F, Weising K and Kahl G (1995) *Theor Appl Genet* 90:90-96.

Sharon D, Adato A, Mhameed S, Lavi U, Hillel J, Gomolka M, Epplen C and Epplen JT (1995) *HortScience* 30:109-112.

Stam P (1993) *Plant J* 3:739-744.

Suiter KA, Wendel JF and Case JS (1983) *J Hered* 74:203-204.

Swofford, DL and Olsen, GJ (1990) In: Hillis DM and Moritz C (eds), *Molecular Systematics*. Sinauer Associates, Sunderland. pp. 411-501.

Sytsma KJ, Givnish TJ, Smith JF and Hahn WJ (1993) *Meth Enzymol* 224:23-37.

Taggart JB and Ferguson A (1994) *Mol Ecol* 3:271-272.

Tizard R, Cate RL, Ramachandran KL, Wysk M, Voyta JC, Murphy OJ and Bronstein I (1990) *Proc Natl Acad Sci USA* 87:4514-4518.

Tourmente S, Deragon JM, Lafleuriel J, Tutois S, Cuvillier C, Espagnol MC and Picard G (1994) *Nucleic Acids Res* 22:3317-3321.

Vergnaud G (1989) *Nucleic Acids Res* 17:7623-7630.

Vosman B, Arens P, Rus-Kortekaas W and Smulders MJM (1992) *Theor Appl Genet* 85:239-244.

Wallner E, Weising K, Rompf R, Kahl G and Kopp B (1996) *Plant Cell Reports* 15:647-652.

Weising K, Weigand F, Driesel AJ, Kahl G, Zischler H and Epplen JT (1989) *Nucleic Acids Res* 17:10128.

Weising K, Beyermann B, Ramser J and Kahl G (1991a) *Electrophoresis* 12: 159-169.

Weising K, Kaemmer D, Epplen JT, Weigand F, Saxena M and Kahl G (1991b) *Curr Genet* 19:483-489.

Weising K, Kaemmer D, Weigand F. Epplen JT and Kahl G (1992) *Genome* 35:436-442.

Weising K, Ramser J. Kaemmer D and Kahl G (1994) In: Schierwater B. Streit B. Wagner GP and DeSalle R (eds) *Molecular Ecology and Evolution: Approaches and Applications*. Birkhäuser. Basel. pp. 45-59.

Weising K, Nybom H, Wolff K and Meyer W (1995a) *DNA Fingerprinting in Plants and Fungi*. CRC Press, Boca Raton, USA.

Weising K, Atkinson RG and Gardner RC (1995b) *PCR Methods Applic* 4:249-255.

Wells RA, Green P and Reeders ST (1989) *Genomics* 5:761-772.

Westneat DF, Noon WA, Reeve HK and Aquadro CF (1988) *Nucleic Acids Res* 16:4161.

Wetton JH, Carter RE, Parkin DT and Walters D (1987) *Nature* 327:147-149.

Williams JGK, Kubelik AR, Livak KJ, Rafalski JA and Tingey SV (1990) *Nucleic Acids Res* 18:6531-6535.

Winberg BC, Zhou Z. Dallas JF, McIntyre CL and Gustafson JP (1993) *Genome* 36:978-983.

Wolff K, Rogstad SH and Schaal BA (1994) *Theor Appl Genet* 87:733-740.

Yavachev L (1991) *Nucleic Acids Res* 19:186.

Zhou Z and Gustafson JP (1995) *Theor Appl Genet* 91:481-488.

Zischler H, Nanda I, Schäfer R, Schrnid M and Epplen JT (1989a) *Hum Genet* 82:227-233

Zischler H, Schäfer R and Epplen JT (1989b) *Nucleic Acids Res* 17:4411.

Zischler H, Kammerbauer C, Studer R, Grzeschik KH and Epplen JT (1992) *Genomics* 13:983-990.

Fingerprinting of DNA and RNA using arbitrarily primed PCR

Thomas Vogt, Françoise Mathieu-Daude, Frank Kullmann, John Welsh and Michael McClelland
Sidney Kimmel Cancer Center, La Jolla, CA 92037, USA

Introduction

Arbitrarily primed PCR (AP-PCR) (Welsh and McClelland 1990, Williams et al. 1990, Caetano-Anollés et al. 1992) is a versatile method that generates fingerprints of genomes using arbitrarily selected primers under conditions where the primers will initiate synthesis on DNA, even when the match with the template is imperfect. Some of these priming events occur on opposite strands. The most efficient of these pairs of priming events compete with each other during PCR amplification to produce a fingerprint of a few to over 100 prominent PCR products. Such products have often been referred to as random amplified polymorphic DNA (RAPD; Williams et al. 1990). AP-PCR fingerprints have been exploited in ways largely analogous to the uses of restriction fragment length polymorphisms, including genetic mapping, taxonomy, phylogenetics, clinical epidemiology, and the detection of mutations in cancers.

AP-PCR and RAPD have most frequently been used to find polymorphisms as taxonomic markers in population studies of a wide variety of organisms. In this context, the method is directly applicable to clinical epidemiology and historical ecology. For example, *Borrelia burgdorferi*, *sensu lato*, the cause of Lyme disease in humans, was subdivided into different species (Welsh et al. 1992b). As a result, the complex manifestations of this disease can be more easily understood (Lebech et al. 1994, Cinco et al. 1993). Asikainen et al. (1995) have found relationships between periodontal disease and genetic variants of *Actinobacillus* species. Similarly, another group defined the genetic variability of *Pseudomonas cepacia* in patients with cystic fibrosis (Bingen et al. 1993). AP-PCR is an ideal tool for the study of nosocomial infections, e.g. outbreaks of *Proteus mirabilis* infections in pediatric hospitals (Bingen et al. 1993) or of methicillin resistant *Staphylococcus aureus* (Fang et al. 1993). Hundreds of papers in this field have appeared over the past few years (e.g. Makino et al. 1994, Giesendorf and Quint 1995, Menard and Mouton 1995, VanCouwenberghe et al. 1995, van Belkum et al. 1995, Odinot et al. 1995, Yao et al. 1995).

Aside from the epidemiology and taxonomy of microorganisms, AP-PCR has been used to detect dominant polymorphic markers for genetic mapping. Using DNA from a family of recombinant inbred mice, we and others have placed several hundred anonymous markers derived from AP-PCR on the mouse genetic map (Welsh et al. 1991). AP- PCR and RAPD have also found wide use in the molecular genetics of plants. For example, Sobral and colleagues have placed 200 markers on the sugarcane genetic map (al-Janabi et al. 1993), while similar projects are proceeding in several laboratories on many other plants (Pooler and Hartung 1995, Prince et al. 1995, Salimath et al. 1995, Schena et al. 1995, Sharma et al. 1995, Shaw and But 1995, Stammers et al. 1995, Tanhuanpaa et al. 1995, Wachira et al. 1995, Yeh et al. 1995, Yi et al. 1995).

AP-PCR has found applications in cancer research, since it is possible to detect somatic genomic changes that occur in neoplasms. Such alterations include allelic losses or gains and deletion or insertion mutations (Perucho et al. 1995). Their occurrence has potential to serve as biomarker of tumor development and progression and might provide clues for the understanding of carcinogenesis as well. Amplified bands usually originate from unique sequences, although a small fraction arise from repetitive elements. Amplification is semi-quantitative, in that the differences between samples in a particular template sequence are largely preserved during amplification. Therefore, information on the overall allelic composition of a genome can be estimated by careful control of AP-PCR. Such investigations require careful adjustment of the template DNA concentration. It is also recommended that each genomic DNA be fingerprinted at several concentrations to control for template quality (Welsh and McClelland 1990).

Thus DNA fingerprinting by AP-PCR provides a complementary molecular approach to the cytogenetics of cancer. It allows an unbiased examination of genetic alterations of the cancer cell genome and has led to a better understanding of the genomic instability of colon cancer cells (Peinado et al. 1992), malignant melanomas, and glioblastomas (Vogt et al. 1996). Tumor-specific somatic genetic alterations can be easily detected by comparing fingerprints of DNA from normal and tumor tissue isolated from the same individual. Fingerprint bands reflecting these somatic mutations have been cloned and characterized.

The extension of AP-PCR fingerprinting to RNA (Liang and Pardee 1992, Welsh et al. 1992a) has resulted in a tool with exciting potential for detecting differential gene expression. One variation of AP-PCR applied to RNA uses a 5' anchor primer such as oligo(dT)CA for reverse transcription and an arbitrary primer for priming the second-strand cDNA (Liang and Pardee 1992). At the arbitrary priming sites, the 3' seven or eight nucleotides of the primer usually match well with the template, but the nucleotides toward the 5' end of the primer also influence which sequences amplify. Shortly after our invention of AP-PCR, we began to explore its use in the detection of differentially regulated genes (Welsh et al. 1992a). In our approach, RNA arbitrarily primed PCR (RAP-PCR), we use arbitrary priming in both directions so that we could sample coding regions with greater frequency and so that we could apply the method to both prokaryotes and eukaryotes. The term "differential display" was coined by Liang and Pardee and the name has sometimes been applied to all variants of RNA fingerprinting using arbitrary primers. Using these methods, it is possible to obtain a partly abundance-normalized sample of cDNA produced in a single tube in a few hours.

RNA fingerprinting can provide a complex molecular phenotype reflecting changes in the abundances of hundreds of RNA species under various conditions. There are many biological situations where differential gene expression results in distinguishable phenotypes, for example, different types of tissues and cells, and cells responding to hormones, growth factors, stress, or the heterologous expression of certain genes. Comparison of RNA fingerprints from different treatment groups allows one to draw inferences regarding gene regulation. Hypotheses regarding signal transduction can be tested, and new hypotheses can be generated using this information. There are many possible applications in biomedical and biological investigations. Fragments of differentially expressed genes can be cloned directly from PCR-amplified products isolated from the gel. Now, there are many examples of the successful use of this technique (Welsh et al. 1995, Suzuki et al. 1995, Perucho et al. 1995, Mok et al. 1994, Wong and McClelland 1994, Ralph et al. 1993, McClelland et al. 1993, Welsh et al. 1992b, Wada-Kiyama et al. 1992, Adati et al. 1995).

Several other methods exist for detecting differential gene expression and cloning differentially expressed genes that do not rely on a biological assay. Most of these methods fall into two general categories: subtractive hybridization and differential screening. Each of these approaches has its strengths and weaknesses. *Subtractive hybridization* has been used successfully by some laboratories (Aasheim et al. 1994). The difficulty of subtractive hybridization methods probably derives from the nature of hybridization kinetics. Abundant genes hybridize faster and to a greater level of completion than low abundance genes, making them more amenable to subtractive methods. Unfortunately, many interesting regulatory genes are of the low abundance type, where hybridization is difficult to drive to completion. In contrast, RAP-PCR, especially the nested variant of it, lead to partial abundance normalization and thus rarer transcripts might be sampled. Another problem is that significant differences in expression that do not fall into the "all-or-nothing" category can easily be missed by subtractive techniques. Nevertheless, a dramatic improvement of subtractive hybridization is the so-called representational difference analysis (RDA; Lisitsyn et al. 1993), which has also been extended to the analysis of RNA (Hubank and Schatz 1994, Tsuchiya et al. 1994, Braun et al. 1995). *Differential screening* (Maser and Calvet 1995) suffers from similar drawbacks. In a typical differential screening experiment, radioactive probe is made from cDNA from two cell types, for example, and used to screen a cDNA library prepared from one of the two. Occasionally, clones from the library hybridize to one or the other but not to both probes. Unfortunately, low abundance messages do not yield sufficient probe mass to allow favorable hybridization kinetics. Nevertheless, perhaps the most exciting possible alternative to AP-PCR for detecting differentially expressed genes is based on differential hybridization of complex cDNA probes to dot blots (Bernard et al. 1996, Schena et al. 1995).

Over the past few years, a multitude of variants and improvements of AP-PCR have been developed. In this chapter, we describe our current preferred protocols for DNA and RNA fingerprinting as well as the related techniques which are required for isolation and identification of the genes and sequences involved. For additional and related techniques such as Southern or Northern blotting please also refer to standard manuals for molecular biology (Sambrook et al. 1989), since only the protocols for AP-PCR and RAP-PCR will be extensively discussed.

DNA fingerprinting protocols

Extraction of genomic DNA

Special attention should be paid to extraction of high quality genomic DNA for fingerprinting. The absence of degradation and contaminants is equally as important as the control of concentration in samples to be compared, because of the aforementioned semi-quantitative nature of AP-PCR. Generally, this requires optical density (OD) measurements, concentration adjustments, and inspecting the DNA after electrophoresis on an agarose gel before starting AP-PCR. For the extraction of DNA from cells, blood, and tissue samples (proteinase K, phenol/chloroform extraction, ethanol precipitation) please refer to standard protocols (Sambrook et al. 1989). For some tissues as well as material from plants, however, special procedures have to be applied. For sugarcane and its relatives a protocol was developed by Honeycutt et al. (1992), which should also work for plants other than sugarcane. For tissue applications, its sometimes more convenient and successful to take one of the commercially available kits for extraction, e.g. QIAamp (Qiagen).

CAUTION: When working with human or animal tissues be aware of the infectious risks (HIV, Hepatitis B, etc.) and wear gloves all the time when you are dealing with these sources.

Fingerprinting of genomic DNA with arbitrary primers

To carry out genomic fingerprinting using AP-PCR, the genomic DNA from sources to be compared (e.g. frozen normal and tumor tissue of the same individual, different microorganisms, different clinical microbiological specimens, etc.) is diluted in TE buffer or water to a stock of about 100 ng/μL. The measurement of DNA concentration is subject to several possible artifacts and is often not very accurate, so adjustments may be needed if the intensities of most of the arbitrarily primed PCR products are not uniform between DNA samples. This is one of the reasons why it is always necessary, in the first instance, to fingerprint in duplicate or triplicate concentrations per sample, using template DNA at concentrations differing by 2-fold or 4-fold. In the following protocol 50 ng and 100 ng of template DNA are used in duplicate reactions for this reason. Most of our applications worked at this scale of concentration or with even less template DNA (down to 20-40 ng).

If you are working with radionuclides you should know all safety rules which apply. Alternatively, silver staining can be used (Vogt et al. 1996). However, the flexibility of doing several exposures of one single radioactive labeled gel is a major advantage of the radioactive protocol.

A) Materials
1. Thermal cycler (e.g. model GenAmp 9600, Perkin-Elmer).
2. PCR tubes and caps.
3. Bases and racks fitting the thermal cycler used.
4. Sequencing gel electrophoresis apparatus (convenient dimensions: 40 cm long, 30 cm wide, 0.4 mm thick spacers; needs about 100 mL gel mixture to fill) and adequate power supply (2,000 V, 50 W).
5. Gel dryer.
6. X-ray films and exposure cassettes.
7. Multichannel pipettor.
8. Genomic DNA diluted to 100 ng/μL in TE or dH$_2$O.
9. TE buffer (10 mM Tris-HCl pH 8.0, 1 mM EDTA).
10. 10× PCR buffer (100 mM Tris-HCl, 100 mM KCl, pH 8.3 for Stoffel fragment; 100 mM Tris-HCl, 500 mM KCl, pH 8.0 for AmpliTaq polymerase).
11. dNTP mix (5 mM each dNTP).
12. 25 mM MgCl$_2$ solution.
13. Arbitrary 10- to 20-mer primers (used singly or in pairs).
14. *Taq* polymerase: e.g. AmpliTaq DNA polymerase (5U/μL), or AmpliTaq DNA polymerase Stoffel fragment (10 U/μL; Perkin-Elmer).
15. Autoradiogram markers preferably α-^{32}P-dCTP [>3000 Ci/mmol (1-2 μCi/reaction tube)] or α-^{35}S-dATP [>1000 Ci/mmol (5 μCi/reaction tube)].
16. 10× TBE buffer (0.089 M Tris-borate, 0.025 M disodium EDTA, pH 8.3).
17. 6% polyacrylamide-8 M urea gel (run in 1× TBE buffer).
18. Denaturing loading buffer (95% formamide, 0.1% bromophenol blue, 0.1% xylene cyanol, 10 mM EDTA).

b) Experimental protocol
1. Example: A reaction mixture for 20 tubes is prepared (30 μL total volume each) to compare 10 genomes (2 concentrations each). To control for pipetting errors add components sufficient for a 22 reaction master mix on ice in the following order:

Reagent	Stock conc.	Final conc.	μL/reaction	μL/22 reactions
dH$_2$O	–	–	10.42	229.2
PCR buffer	10×	1×	3	66
dNTPs (each)	5 mM	0.2 mM	1.2	26.4
Primer (1 or 2)	100 μM	0.6 μM	0.18	4
α[^{32}P]dCTP	10 mCi/mL	2 μCi/reaction	0.2	4.4
MgCl$_2$	25 mM	4 mM	4.8	105.6
AmpliTaq Stoffel	10 U/μL	2 U/reaction	0.2	4.4
Template			10	–
				440 (master mix)

2. Place 20 tubes into a PCR rack on ice.
3. If you have a stock of 100 ng/μL, prepare templates (first row of 10 tubes) by adding 1.5 μL to 13.5 μL of TE or dH$_2$O, and then mixing. Take 5 μL out of this dilution and add it to 5 μL of TE or dH$_2$O in a second tube (second row of tubes) to obtain 2 tubes per reaction with 100 ng/10 μL and 50 ng/10 μL template. This two-fold dilution can be conveniently done by a multipipetter.
3. Subsequently, distribute the master mix to all 20 tubes (20 μL each).
4. Perform the reaction in a thermal cycler:
   When using *18-mers or longer primers* the cycling protocol should be five low stringency cycles (95°C, 60 s; 40°C, 60 s; 72°C, 2 min) then 30 high stringency cycles (95°C, 30 s; 65°C, 30 s; 72°C, 2 min). When using *10-mers* the cycling protocol should be 30 low stringency cycles (95°C, 60 s; 35°C, 60 s; 72°C, 2 min). 10-mer primers work best with Stoffel fragment while *Taq* polymerase might give better results with longer primers.
5. Dilute 3 μL of the complete reaction mix with 12 μL of formamide-dye buffer and incubate at 68°C for 15 min or heat to 95°C for 3 min.
6. Immediately chill the solution on ice and load 2 μL onto an 6% polyacrylamide-8 M urea sequencing gel (for 100 mL: 42.6 g urea, 15 mL 40% acrylamide-bisacrylamide (19:1) solution, 10 mL of 10× TBE, 600 μL ammonium persulfate solution, 60 μL TEMED) and electrophorese for 4 h at 50 W in 1× TBE buffer. Both DNA concentrations of each sample should be loaded side by side to control for reproducibility.
   NOTE: Acrylamide is neurotoxic. Avoid skin contact and inhaling the non-polymerized solution.
7. Dry the gel under vacuum at 80°C and directly expose to an X-ray film at room temperature without an intensifier screen. If possible, attach three or more luminescent labels (autoradiogram markers) to the dried gel in order to localize the bands in the gel in case the isolation of one of the bands is desired. [32]S gels usually require an exposure of 2 or more days, while ^{32}P gels require less than 24 h.

RNA fingerprinting protocols

RNA purification

We use the RNeasy kit (Qiagen). However, similar to most purification methods, the purified RNA is often not entirely free of contaminating genomic DNA. This can be a problem because the genome is more than 10 times as complex as the RNA population, resulting in better matches with the primer. We therefore routinely treat the RNA with RNase-free DNase I prior to fingerprinting. This should be done in the presence of a RNase-inhibitor (both enzymes available from Boehringer-Mannheim) in a Tris-MgCl$_2$ buffer. The kit procedure is then repeated for cleaning the treated RNA. The RNA concentration is checked spectrophotometrically. Equal aliquots are then electrophoresed on a 1% agarose gel and ethidium bromide stained to compare large and small ribosomal RNA, qualitatively. When starting with fresh RNA, we further recommend to perform one RAP-PCR leaving out the reverse transcriptase as a control of DNA contamination. Thus, identification of DNA related bands is possible.

A) DNase treatment.
1. 1M MgCl$_2$ solution.
2. 1M Tris-HCl, pH 8.0.
3. dH$_2$O.
4. RNase inhibitor, 40 U/μL (Boehringer Mannheim).
5. RNase-free DNase I, 10 U/μL (Boehringer Mannheim).
6. Sterile tubes 0.5 mL.
7. Waterbath at 37°C.

B) Digestion protocol
1. For convenient pipetting prepare at least 70 μL of DNase digestion mix by adding:

Reagent	Amount (μL)	Final concn. after adding to RNA template as suggested
dH$_2$0	52.5	–
1M MgCl$_2$	7	0.01 M
1M Tris-HCl, pH 8.0	7	0.01 M
RNase inhibitor, 40U/μL	2.1	1.2 U/μL
DNase 10 U/μL	1.4	0.2 U/μL
	70 (total)	

2. Mix 5 μL of digestion mix with each 45 μL of untreated RNA.
3. Incubate for 45 min in a water bath at 37°C.

Fingerprinting of total cellular RNA

In RAP-PCR (Welsh et al. 1992a), the synthesis of first-strand cDNA is initiated from an arbitrary primer at those sites in the RNA that best match the primer. Second-strand synthesis is achieved by arbitrary priming using a thermostable DNA polymerase, at sites where the primer finds the best matches. Poorer matches at one end of the amplified sequence can be compensated for by very good matches at the other end. These two steps result in a collection of molecules that are flanked at their 3' and 5' ends by the exact sequence (and complement) of the arbitrary primer. These serve as templates for high-stringency PCR amplification and result in fingerprints similar in appearance to those generated from genomic DNA. Because both primers are internal to the transcript, open reading frames are found in about 30% of products using this variation of RNA fingerprinting. The ratio of the intensities of RAP-PCR products between samples correlates with the ratio of abundances of the corresponding RNA. Although the intensities of different bands within the same fingerprint vary independently, the intensity of a band between fingerprints appears to be proportional to the concentration of its corresponding template sequence.
NOTE: Before you start, remember always to wear gloves, use sterile RNase-free equipment, and prepare reactions on ice!

A) Materials
In addition to the reagents in "DNA fingerprinting protocol" you will need the following items:
1. 100 mM DTT (dithiothreitol).
2. Murine leukemia virus reverse transcriptase (MuLV-RT, 75 U/μL).

3. 10× RT buffer (500 mM Tris-HCl pH 8.3, 500 mM KCl, 40 mM $MgCl_2$).
4. RNase inhibitor, 40 U/ μL (Boehringer Mannheim).
5. First (25 μM) and second (100 μM) strand synthesis primers.
6. 100 mM $MgCl_2$.
7. RNA stock solutions with about 400 ng/μL (preferably for the fingerprinting of two 2-fold dilutions, i.e. about 400, 200 and 100 ng/5 μL template volume).

B) Experimental protocol
1. Example: 8 RNA samples are to be compared at 3 concentrations each. First place 3 rows of PCR tubes into a rack on ice. As described for DNA fingerprinting, dilute your stocks with water so that you have, for instance, 400, 200 and 100 ng/5 μL in each set of three tubes for one sample.
2. Prepare first strand reaction mixture (for 26 reactions to compensate for pipetting errors) by adding:

Reagent	Stock concn.	Final concn. after adding to template	μL/reaction	μL/26 reactions
RNase inhibitor	40 U/μL	2 U/μL	0.5	13
RT buffer	10×	1×	1	26
DTT	100 mM	10 mM	1	26
dNTP	5 mM	0.2 mM	0.4	10.4
dH_2O	–	–	1.05	27.3
primer	25 μM	2 μM	0.8	20.8
MuLV-RT	75 U/μL	18.75 U/reaction	0.25	6.5
			5 (total)	130 (total)

2. Distribute 5.0 μL of this reaction mix to all samples. Final volume will be 10 μL.
3. Place reactions into cycler and ramp over 5 min to 37°C and hold at 37°C for 60 min. Subsequently heat samples to 94°C for 5 min to "kill" the reverse transcriptase.
4. Add 30 μL of dH_2O to all samples. The cDNA made is sufficient for at least four different PCR second strand synthesis reactions. You now have 40 μL of first strand cDNA solution (12.5 mM Tris-HCl, 12.5 mM KCl, 1.0 mM $MgCl_2$, 0.5 μM first strand synthesis primer, 0.05 mM dNTP)
5. Prepare second primer PCR master mix:

Reagent	Stock concn.	Final concn. in master mix	μL/reaction	μL/26 reactions
dH_2O	–	–	5.07	131.8
10× PCR buffer	10×	1×	2	52
dNTPs	5 mM each	0.4 mM each	0.8	20.8
primer	100 μM	8 μM	0.8	20.8
α-^{32}P-dCTP	10 mCi/mL	2 μCi/reaction	0.13	3.38
$MgCl_2$	100 mM	8 mM	0.8	20.8
Taq pol Stoffel	10 U/μL	4 U/reaction	0.4	10.4
Template			10	–
			20 (total)	260 (master mix)

6. Distribute 10 μL of the master mix to each of the 24 tubes.
7. Cycling conditions: When using primers longer than 18-mers cycle through one low stringency step (94°C, 5 min; 40°C, 5 min; 72°C 5 min) followed by 30 high

stringency steps (94°C, 1 min; 55°C, 1 min; 72°C, 2 min). When using 10-mers, cycle through 30 low stringency cycles (94°C, 1 min; 35°C, 1 min; 72°C, 2 min). While 10-mers work best with Stoffel fragment, *Taq* polymerase might give better results with long primers.

8. Mix 3 μL of the complete reaction mix with 12 μL of formamide-dye buffer and incubate at 68°C for 15 min.
9. Immediately chill the solution on ice and load 2 μL onto an 6% polyacrylamide-8 M urea sequencing gel (for 100 mL: 42.6 g urea, 15 mL 40% acrylamide-bisacrylamide (19:1) solution, 10 mL of 10× TBE, 600 μL ammonium persulfate solution, 60 μL TEMED). Electrophorese for 4 h at 50 W in 1× TBE buffer. All three concentrations of each sample should be loaded side by side.
   NOTE: Acrylamide is neurotoxic. Avoid skin contact and inhaling the non-polymerized solution.
10. Transfer gel to Whatman paper.
11. Dry the gel under vacuum at 80°C and directly expose to an X-ray film at room temperature without an intensifier screen. Place several luminescent labels (autoradiogram markers) on the dried gel to facilitate alignment of the autoradiogram with the gel in case the isolation of one of the bands is desired. The ^{32}P gels require less than 24 h for a first assessment of the result. Multiple exposures for variable time can follow depending on the intensity of interesting bands.

Isolation and purification of DNA from fingerprints

Isolation of DNA sequences from acrylamide gels and reamplification.

Once a fingerprint has been generated, it is often necessary to purify and clone bands that display polymorphisms or that indicate differential gene expression for further characterization or use as biomarkers.

A) Materials
   1. Disposable scalpels.
   2. PCR and electrophoresis material as described above.

B) Experimental protocol
   1. Align the autoradiogram luminescent labels on the gel with their exposed images. Use a needle to mark the exact position of the band in the dried gel, then excise with the scalpel.
   2. Place the excised portion of the gel (approx. 0.5-1×2-3 mm) in 50 μL of TE and incubate at 60°C for 3 h to elute the DNA.
   3. Dilute eluted DNA with water 100-fold. Take 10 μL as a template and follow precisely the AP-PCR protocol but using ordinary *Taq* polymerase. Perform 25 low stringency cycles (95°C, 30 sec; 35°C, 30 sec; 72°C, 2 min) for 10-mers or using 60°C annealing temperature for 18-mers or longer primers. The final volume of reaction can be chosen according to subsequent procedures planned. If you want a "SSCP" cleaning of your product before cloning then a 20 μL total volume is sufficient. If you plan to extract the product from an agarose gel for direct cloning after a cold PCR then you are better off taking a volume of 40 μL.

4. Analyze the PCR product by gel electrophoresis running the sample next to the original fingerprint to verify its size and purity. Re-exposure of the gel will confirm the accuracy of excision of the band.

Purification of the desired product by native "Single Stranded Conformation Polymorphism" gels

After a differentially amplified RAP product or a polymorphic band in fingerprints of genomic DNA is detected, eluted and reamplified, it appears that this is most often a mixture of several different products of similar size (Mathieu-Daude et al. 1996). To identify the desired band, we use a native polyacrylamide gel to separate these sequences of the reamplified mixture based on single stranded conformation polymorphisms. For one SSCP gel of the aforementioned size mix: 22.5 mL Hydrolink MDE™ gel (Baker), 5.5 mL 10× TBE, 4.5 mL glycerol, 57.5 water, 360 μL 10% ammonium persulfate solution, 36 μL TEMED. Run in 0.6× TBE. We recommend reamplification of both the region carrying the desired product and a corresponding region from an adjacent lane where the product is less prominent or not visible. Denaturation of the reamplified DNA followed by side-by-side comparison on an SSCP gel allows the classification of reamplified material into: (i) those that can be directly cloned because the amplified product is relatively pure (two bands appear), (ii) those that need to be reamplified from the SSCP gel before cloning, and (iii) those that are too complex for further study. This screen should save considerable effort now wasted on directly cloning unsuitable products from fingerprinting experiments. The main drawback of this strategy is that the stochiometry of the PCR products in the excised bands can change during reamplification. If reamplification is not used, the product can be detected in two weeks with an intensifying screen.

To avoid amplification but shorten the exposure time our most recent strategy is to reload the lanes of interest in a number of adjacent lanes and elute the bands of interest from the resulting swathe of lanes. The material eluted is now hot enough to detect in an SSCP gel by exposure for less than one day using an intensifying screen.

Cloning of a reamplified PCR product

When the correct band is identified, it makes sense to clone and sequence the product in order to place it into the context of known sequences and biological phenomena, or alternatively define entirely new markers or differentially expressed genes within the experimental setting. For this purpose, the product should be reamplified without hot dCTP and run on an agarose gel to check size and estimate concentration. Arbitrarily chosen oligonucleotide primers can be chosen that contain recognition sites for restriction endonucleases that digest DNA to produce DNA fragments with staggered ends compatible with the cloning sites of commercially available plasmid or phagemid vectors. In this case, one can digest both the vector and reamplified band DNA with the appropriate restriction enzyme(s), ligate, and transform using standard protocols (Sambrook et al. 1989). Frequently, however, arbitrarily chosen oligonucleotides do not contain recognition sites for restriction enzymes. Here, it helps that *Taq* polymerase often adds an adenine to the ends of a product. Several commercially available systems allow a direct cloning of these products. We had good results with the TA cloning kit (Invitrogen). It provides all important reagents for ligation (pCR^{TM}2.1 vector) and transformation as well as competent cells. It comes with a handbook.

You will additionally need:
1. Microcentrifuge.
2. Thermocycler and tubes for ligation at 14°C.
3. Water baths at 37°C and 42°C.
4. LB medium and agar dishes.
5. 20 mg/mL 5-bromo-4-chloro-3-indolyl-p-D-galactoside (X-Gal) in DMF.
6. 50 mg/mL ampicillin stock.
7. Freezers at -70°C and -20°C.
8. Incubator and shaker with a constant temperature of 37°C.

After blue-white screening of the clones:
1. Pick at least 10 white colonies per desired product.
2. Grow colonies overnight in 2-5 mL LB broth containing 50 μg/mL of ampicillin for plasmid isolation.

NOTE: You can save a lot of time if you determine in advance which of your growing white clones really contain the insert. For this purpose, when picking the colonies dip the pipet tip into 50 μL of water and pipet up and down. Take 35 μL of this suspension for clonal expansion in LB overnight and dilute the remaining solution with 15 μL of water. Take 5 μL of this dilution and set up a conventional PCR with the supplied universal primers. Use high stringent conditions and run all your products on an agarose gel. Next day you will be able to extract plasmid DNA only from those clones which proved to have the insert of the expected size. It is wise to freeze aliquots of those clones with 15% DMSO at -80°C for later use.

DNA sequencing according to Sanger

Plasmid DNA extraction

For plasmid preparation from expanded clones please refer to standard protocols (Sambrook et al. 1989). Alternatively, the commercially available Miniprep kits work very well and fast, e.g. Qiaprep (Qiagen).

Sequencing with Sequenase

For sequencing of the double stranded vector insert, we use the Sequenase DNA sequencing kit (USB/Amersham). Before use. the DNA must be denatured as follows: add 0.1 volumes of 2 M NaOH, 2 mM EDTA and incubate 30 min at 37°C. Neutralize the solution with 3 M sodium acetate and precipitate the DNA with 2-4 volumes of ethanol at -70°C for 15 min. Pellet using a 15 min spin in a tabletop centrifuge and resuspend pellet in 7-9 μL water.

For details of the sequencing reaction protocol please refer to the vendor's information. After sequencing the reactions are directly loaded on a denaturing polyacrylamide gel similar to fingerprints. For autoradiography ^{35}S-dATP is preferentially used.

Database Query

After the sequence of a specific DNA marker or a differentially expressed RNA is determined, the internet provides convenient web sites to check the databases of the National Center for Biological Information. Here you can fill out forms and let the computer do the job of calculating the probabilities of hitting a sequence already deposited in the databases. If you do not have web access you can use the e-mail service of the NCBI. If you want to obtain the documentation files, send a message consisting of just the word "help" (without the quotes) to: blast@ncbi.nlm.nih.gov. For a free subscription to "NCBI News", the NCBI newsletter, send a request along with your name and postal mailing address to: info@ncbi.nlm.nih.gov. For submissions there is also a new tool available called "BankIt" through the NCBI's home page on the World Wide Web. The URL is http://www.ncbi.nlm.nih.gov.

Southern hybridization

In many cases, hybridization analysis may be necessary to confirm that the cloned band corresponds to the band visualized in the AP-PCR fingerprint. A dried AP-PCR gel can be transferred to a nitrocellulose or nylon blotting membrane. The transfer is performed by capillary action overnight using standard conditions (buffer 20× SSC: dissolve 175.3 g NaCl and 88.2 g sodium citrate in 800 mL dH_2O, adjust to pH 7, and add water to 1 L) (Sambrook et al. 1989). Before use, the piece of dried polyacrylamide gel is soaked for a few minutes in 20× SSC. After DNA-transfer, the gel is treated with UV C to crosslink the DNA to the membrane.

Using this procedure 10-50% of the radioactive material is usually transferred to the blotting membrane. The transfer efficiency is not the same for all the bands but depends on size: bands larger than about 1500 bp will transfer with lower efficiency.

The probe for the hybridization step is prepared directly from the reamplified and purified PCR product or from a plasmid obtained by a miniprep. We recommend that the product is run on an agarose gel and extracted from the gel with the Qiaquick kit (Qiagen). About 25 ng of purified PCR product needs to be labeled (e.g. random primed DNA labeling kit, Boehringer-Mannheim). Unincorporated nucleotides are removed by using reagents such as Oligoclean (Amersham). Hybridization is performed using standard methods (Sambrook et al. 1989). Specific hybridization by the probe to the band of the correct size and distribution among lanes is evidence of successful cloning. The presence of the background ^{32}P-bands from the original fingerprint often facilitates the identification of new hybridizing bands. Cross-hybridization with other bands amplified with the same or different arbitrary primers can also yield interesting information.

Chromosomal localization of DNA markers

If human systems are in the scope of the investigation, the commercially available rodent-human hybrids can be used for chromosomal localization of DNA markers. This is particularly useful in cancer research applications. The approach works out as follows.

Once the sequence of a band is determined, specific primers can be designed. DNA panels of rodent/human somatic cell hybrids are available, e.g. PCR amplifiable mapping panel no. 2 from NIGMS (Coriell Cell Repositories, Camden, NJ). PCR is performed with 50-100 ng of genomic DNA from the different hybrids using the specific PCR primers. The PCR products are analyzed by non-denaturing 6% polyacrylamide gel electrophoresis. Because each cell line contains only one or a few human chromosomes, successful amplification of the specific product will indicate that the sequence of interest is located in one of the chromosomes present in this/these cell line(s).

Northern vs. quantitation with specific primers of potentially differentially expressed genes

If RAP-PCR has generated candidate RNA species, which might be regulated under certain experimental conditions, subsequent to all the aforementioned cloning procedures and Southern confirmation of successful cloning, a second RNA quantitation tool is required to confirm differential expression.

The classical tool for this is the Northern blot. Several standard protocols are available for this (Sambrook et al. 1989). The probe is made analogous to the Southern probes mentioned above. There are many alternative ways to quantitate RNA. All of them need some practice and have their limitations. To mention some of them, there is the RNase protection assay (Eriksson et al. 1995), the primer elongation technique (Sambrook et al. 1989), and the PCR based approach of quantitative RT-PCR (Foley et al. 1993). A concept which we are fostering is the run of a AP-PCR protocol with specific primers and low stringency conditions. Here, the best matching target should be the DNA from which the sequence for the specific primers was derived. This should be the predominant product of the PCR. Aside from this, however, a lot of the other AP-PCR products will appear on the fingerprint and serve as an internal standard for amplification to compare different sources of cDNA. This concept would overcome the problem of RT-PCR approaches, which generally suffer from the lack of a decent internal control with the same primers. Further investigations have to be done to come to a definite evaluation of this recent development in our lab.

Comments

How to interpret fingerprints

In AP-PCR, the comparison of one particular genomic DNA fingerprint with its matching samples will provide four theoretical types of changes: (i) new bands in a particular sample or losses of bands; (ii) changes in molecular weight of amplified fragments, reflected by changes in the mobility of bands; (iii) increases in the intensity of a band; and (iv) decreases in the intensity of a band.

These changes can represent polymorphic markers for mapping or somatic mutations when tumor and normal tissue is compared. Only changes that are apparent in *both* concentrations tested should be taken as potentially real. For examples of gels and their interpretation refer to the publications referenced above.

M   N   NS   STA   BIS   TPA   DPP

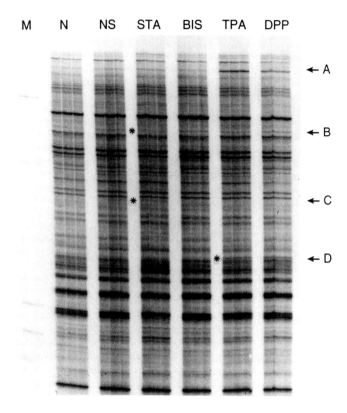

← A

← B

← C

← D

FIGURE 1. The figure shows a typical RAP-PCR fingerprint. The primers were "OPN24" 5'-
AGGGGCACCA-3' for first strand synthesis, and "KinaseA1+" 5'-GAGGGTGCCTT-3' for
second strand synthesis. This is an example of one of the most recent applications of this
technique to questions of drug-design and development of new cancer treatments. A human
melanoma cell line (WM35) was treated with a panel of protein kinase C active drugs, which
have different profiles regarding their isoform-specific activities and thus different potential to
stop melanomas from growing and to cause differentiation (*STA*, Staurosporine; *BIS*,
Bisindolylmaleimide I; *TPA*, Phorbol-12-myristate-13-acetate; *DPP*, 12-Deoxyphorbol-13-
Phenylacetate; *N* is control; *NS* is control plus DMSO which was used as a solvent). Each
sample is fingerprinted at three different RNA masses, 450 ng, 300 ng, and 150 ng, to ensure
reproducibility of any differences between samples. It can be seen that DPP and TPA appear
to be the most closely related drugs. They share the up-regulation of two transcripts marked
with "A" and "D". The less toxic indolcarbazole *BIS* does not show this phenomenon, thus it
might act as anticancer-drug by affecting a different subset of PKC-isoforms with less toxic
side-effects. The non-specific kinase inhibitor *STA*, which is also too toxic for any therapeutic
use, shows a markedly different RNA-phenotype: "B" marks an almost complete loss of a
band and "C" a new prominent product. Thus, when designing new drugs, the comparison of
these RNA-phenotypes or RT-PCR using products derived from such gels have the potential
to direct research before any costly applications are done in animals or humans.

In RAP-PCR differences in band intensity can represent differentially expressed genes in the chosen treatment groups. Only changes that are apparent in all three concentrations tested should be interpreted as real. For an example of a gel and its interpretation refer to Figure 1.

It should be clear that, given two or more RNA populations, differences in the fingerprints will result when corresponding templates are represented in different amounts. There are, however, other intrinsic limitations that should be considered in the design of an experiment based on RAP-PCR. While these fingerprints contain anywhere from 40-150 bands, typical RNA populations for eukaryotic cells have complexities in the tens of thousands of molecules. Also, the influence of the abundance of mRNA on the fingerprint must be considered. The fingerprints are not totally "abundance normalized" in that more abundant RNA species will be more likely to produce a visible AP-PCR product. Therefore, searching for one particular differentially expressed gene will often be futile if the RNA is rare. Rather, the method is more appropriate for problems where several differentially expressed genes are anticipated. This is not so great a limitation as it might initially appear. First of all, many developmental and pathological phenomena are accompanied by many dozens or even hundreds of alterations in gene expression. Secondly, much of the technological development associated with sequencing, such as fluorescent-tagged primers and automated gel reading, and capillary electrophoresis, is readily adaptable to RAP-PCR. Therefore, a large fraction of the genes expressed in many situations can be surveyed given existing technology and this capability will be greatly enhanced by further technological developments.

Choice of primers for AP-PCR

Though the particular sequence of a primer does not play an important role for AP-PCR and RAP-PCR, some rules do apply and the length of the primer as well as the type of enzyme used have considerable effects on fingerprints. First, the primers should not have stable secondary structure. Second, the sequence should be chosen such that the 3' end is not extensively complementary to any other sequence in the primer. In particular, palindromes should be avoided. Third, primers of 10-20 nucleotides in length can be used. Longer primers have some small theoretical advantages. In particular longer primers can contain more sequence information designed to aid in subsequent steps in the experiment, such as cloning, sequencing, etc. Primers 10 nucleotides in length are cheaper and can be obtained in convenient kits from several companies (e.g. Genosys).

If a larger number of fragments, i.e. an information-rich fingerprint is intended to be generated and resolved on an acrylamide gel, this can be achieved by several strategies: (i) select, in a preliminary screen, 10-mers that yield a large number of fragments using AmpliTaq holoenzyme; (ii) use 10-mers with Stoffel fragment which generates more PCR products; (iii) use longer primers, such as 18-mers; (iv) use primers that are biased toward more common sequences in the genome by performing a statistical analysis; and (v) use very short primers in very large amounts to generate extremely complex fingerprints (Caetano-Anollés et al. 1992).

Our current protocols presented here, prefer arbitrary 10-mer or 11-mer primers in combination with the AmpliTaq Stoffel fragment (Perkin-Elmer) or KlenTaq (Ab peptides). These enzymes generate almost twice as many easily visible PCR products as does AmpliTaq holoenzyme using the same primer and template DNA. Furthermore, this enzyme increases the number of arbitrary 10-mers that give productive fingerprints from about 75% to over 90%. At least for primers of 10 bases or longer, the more complex the PCR fingerprint the more reproducible it seems to be. However, too many products make the pattern difficult to interpret. Thus, reliable fingerprints can be obtained by screening arbitrary primers for moderately complex patterns and then using these on the population under study.

If a simple pattern is desired which might be analyzed on agarose gels as well, then it is important to be aware of the "context effect". This situation arises for very simple but not for complex patterns. The fingerprint is the result of competition between many PCR products. The fewer the winners are, the greater is the effect of their presence and absence on the probability of other PCR products being amplified. For example, if two similar but nonidentical genomes give very simple fingerprints with the same primer and differ by one or more of the major PCR products, then one must be concerned that the difference(s) may affect the probability of amplification of other PCR products. One symptom of such a context effect in a mapping population would be nonparental products that appear due to the absence of a prominent polymorphism from one parent in some offspring. This phenomenon may be a good reason to avoid using simple fingerprint patterns that contain fewer than 10 prominent PCR products. For the purposes of fingerprinting, it is often best to use arbitrary primers in pairwise combinations (Welsh and McClelland 1991). A few primers can be used in a very large number of pairwise combinations. For example, 20 primers can be used in 380 (19×20) combinations. One potential disadvantage of using pairwise combinations of primers is that some of the products will have the same primer at both ends, and these products may be shared between fingerprints using that primer.

One advantage of using primers in pairwise combinations is that the products with different primers at each end can be directly sequenced using conventional PCR sequencing kits. However, these sequences are not always of the highest quality, perhaps because no part of a fingerprinting gel is entirely free of other products. Sometimes it is best to clean the products first.

There is no requirement that the primers be entirely arbitrary and indeed we have designed primers to take advantage of the hypermutability of CpG and dinucleotide repeats (Welsh et al. 1991). Recently, others have taken this idea even further (Sharma et al. 1995, van Soolingen et al. 1993). For RAP-PCR or differential display, primers may encode transcribed repeats such as zinc fingers or kinase motifs (McClelland et al. 1994) or a conserved exon (McClelland et al. 1995). However, there is no evidence for significant enrichment of the corresponding genes except in one report (Stone and Wharton 1994).

Kinetics of AP-PCR and the "C_0t effect": How many cycles?

A simple working model for each of the steps in RAP-PCR says that if the initial priming event resulting in first strand synthesis is represented by the probability p_1, and the second priming event is represented by the probability p_2, the exponential growth of that product in the subsequent PCR can be approximately modeled as:

$$m = p_1 p_2 c \, X^n$$

where X is a number between 1 (no amplification) and 2 (perfect amplification), c is the concentration of the template molecule, n is the number of effective exponential growth cycles, and m is the mass of the product. By effective exponential growth cycles, we refer to only those cycles that occur before any of the reaction components become limiting. Usually the limiting component is the polymerase, which is eventually titrated by the primed sites to the extent that only linear amplification can occur. The mass of each of the products can, in principle, be characterized by specific values for these variables. Naturally, there will occur instances in which any one of the variables may dominate. For example, given a normal distribution for the product, $p_1 p_2 c \, X^n$, there will occasionally be high values for c that compensate for low values of $p_1 p_2 \, X^n$. Simply put, the fingerprint cannot be expected to be abundance-normalized. The over-representation of abundant messages can be a serious problem when the genes of interest give rise to rare messages. We have developed a partial solution to this problem, termed "nested" RAP-PCR (Ralph et al. 1993, McClelland et al. 1995).

More recently, however, more subtle analyses of reaction kinetics revealed that the rate of amplification of abundant PCR products generally declines faster than that of less abundant products in the same tube in the later cycles of PCR (Mathieu-Daude et al. 1996). As a consequence, differences in product abundance diminish as the number of PCR cycles increases. Rehybridization of PCR products, which may interfere with primer binding, can explain this significant feature in late cycles. Rehybridization occurs with a half-time dependent on the reciprocal of the DNA concentration. Thus, if multiple PCR products are amplified in the same tube, reannealing occurs faster for the more abundant PCR products. In RT-PCR using an internal control, this results in a systematic bias against the more abundant of the two PCR products. In RNA fingerprinting by RAP-PCR or differentially display of cDNA, very large or absolute differences in the expression of a transcript between samples are preserved but smaller real differences may be gradually erased as the PCR reaction proceeds. Thus, this "C_0t effect" may systematically cause an underestimate of the true difference in starting template concentrations. However, differences in starting template concentrations will be better preserved in the less abundant PCR products. Furthermore, the slow down in amplification of abundant products will allow these rarer products to become more visible in the fingerprint which may, in turn, allow rarer cDNA to be sampled more efficiently. In some applications, where the object is to stochiometrically amplify a mixture of nucleic acids, the bias against abundant PCR products can be partly overcome by limiting the number of PCR cycles and, thus, the concentration of the products. In other cases, abundance normalization at later cycles may be useful, such as in the production of normalized libraries (Mathieu-Daude et al. 1996).

Comparison of RNA Fingerprinting protocols and further improvements

What are the advantages and disadvantages of the RAP-PCR protocol (McClelland et al. 1995) compared to that of Liang and Pardee (1992)? In contrast to an arbitrary selection step in both directions, the protocol of Liang and Pardee (1992) uses an oligo(dT)CA, or similar oligo(dT)-XM, primer for reverse transcription followed by an arbitrary primer in the other direction. The method samples 3' ends that are mostly noncoding. In contrast, arbitrary priming from both ends can sample open reading frames (ORFs) in about 30% of mRNA products. Because the authentic reading frame often occurs in arbitrarily selected stretches of a few hundred bases in a typical mRNA sequence. Our strategy has the advantage that conserved ORFs between species or protein family members can occasionally be observed in database searches without further cloning (Ralph et al. 1993).

One variant of the protocol we use can be designed to prime arbitrarily only once after denaturation before switching to high stringency annealing (e.g. 60°C for an 18-mer primer); any contaminating dsDNA is usually primed only once and is therefore not amplified during PCR. The protocols employing 10-mers or an anchored oligo(dT)-NN primer use 35°C annealing steps throughout and can sample contaminating dsDNA in the classic manner of RAPD (Williams et al. 1990). Finally, the oligo(dT)-XM primers are as highly promiscuous as other primers, if not more so, and must result in many products that are sampled from hnRNA or inside mRNA and other products in which the oligo(dT)CA primer occurs at both ends. Regardless of the arbitrary primed strategy used, to ensure sampling of only mRNA and not hnRNA or residual genomic DNA, the RNA must first be poly(A)-selected.

Each of these approaches has strengths and weaknesses. Although the protocol using an anchored oligo(dT)-NN primer is excellent, arbitrary priming from both directions, followed by nesting, can have significant advantages, not least of which is the sampling of ORFs.

The RAP-PCR method is amenable to various possible improvements and, given the previously demonstrated utility of RAP-PCR, it is worth developing some of these possibilities. For example, in principle, rTth DNA polymerase (Perkin-Elmer) could be used as both the reverse transcriptase and the DNA polymerase (Myers and Gelfand 1991), reducing the number of experimental steps. Conditions exist that are compatible with both properties of the enzyme (Young et al. 1993). Alternatively, a viral reverse transcriptase and Taq DNA polymerase can be mixed from the beginning. These enzymes also have different optima, but it should be possible to find a compatible buffer system.

The power of examining many RNA samples in parallel

One of the features of RAP-PCR that distinguishes it from most other methods for detecting differential gene expression is the possibility to examine 100 and more anonymous RNAs under many conditions in the same experiment. Thus if, for example, a cell line is treated in eight different manners and then RNA is fingerprinted, it is possible to look for genes in a vast number of possible regulatories categories. In this example each RNA may be up-regulated, down-regulated or unchanged in any of the eight treatments, giving a total or 3^8 possible regulatory categories, or 6561 categories, in all. Thus, by judicious selection of treatments it is possible to parse genes into narrow regulatory categories of interest and to find transcripts that represent new or unexpected regulatory categories. With the advent of many other methods to find differentially expressed genes, it is perhaps this ability to examine RNA in parallel that will ensure that the method remains useful for some time to come.

References

Aasheim HC, Deggerdal A, Smeland EB and Hornes E (1994) *Biotechniques* 16:716-721.

Adati N, Ito T, Koga C, Kito K, Sakaki Y and Shiokawa K (1995) *Biochim Biophys Acta* 1262:43-51.

al-Janabi SM, Honeycutt RJ, McClelland M and Sobral BW (1993) *Genetics* 134:1249-1260.

Asikainen S, Chen C and Slots J (1995) *Oral Microbiol Immunol* 10:65-68.

Bernard K, Auphan N, Granjeaud S, Victorero G, Schmitt-Verhulst A-M, Jordan B and Nguyen C (1996) *Nucleic Acids Res* 24:1435-1442.

Bingen E, Weber M, Derelle J, Brahimi N, Lambert-Zechovsky NY, Vidailhet M, Navarro J and Elion J (1993) *J Clin Microbiol* 31:589-593.

Braun BS, Frieden R, Lessnick SL, May WA, Denny CT (1995) *Mol Cell Biol* 15:623-630.

Caetano-Anollés G, Bassam BJ and Gresshoff PM (1992) *Mol Gen Genet* 235:57-165.

Cinco M, De Giovannini R, Fattorini P, Florian F and Graziosi G (1993) *Microbiologica* 16:23-32.

Eriksson PS, Nilsson M and Matejka GL (1995) *Brain Res Dev Brain Res* 84:208-214.

Fang FC, McClelland M, Guiney DG, Jackson MM, Hartstein AI, Morthland VH, Davis CE, McPherson DC and Welsh J (1993) *Jama* 270:1323-1328.

Foley KP, Leonard MW and Engel JD (1993) *Trends Genet* 9:380-385.

Giesendorf BA and Quint WG (1995). *Cell Mol Biol (Noisy-le-grand)* 41:625-638.

Honeycutt R, Sobral BWS, Keim P and Irvine JE (1992). *Plant Mol Biol Rep* 10:66-72.

Hubank M. and Schatz D-G (1994) *Nucleic Acids Res* 22:5640-5648.

Lebech A M, Hansen K, Wilske B and Theisen M (1994) *Med Microbiol Immunol (Berl)* 183:325-341.

Liang P and Pardee A (1992) *Science* 257:967-971.

Lisitsyn N, Lisitsyn N and Wigler M. (1993) *Science* 259:946-951.

Makino S, Okada Y, Maruyama T, Kaneko S and Sasakawa C (1994) *J Clin Microbiol* 32:65-69.

Maser RL and Calvet JP (1995) *Semin Nephrol* 15:29-42.

Mathieu-Daude F, Cheng R, Welsh J and McClelland M (1996) *Nucleic Acids Res* 24:1504-1507.

Mathieu-Daude F, Welsh J, Vogt T and McClelland M (1996) *Nucleic Acids Res*, in press.

McClelland M., Chada K, Welsh J and Ralph D (1993) *Exs* 67:103-115.

McClelland M, Mathieu-Daude F and Welsh J (1995) *Trends Genet* 11:242-246.

McClelland M, Ralph D, Cheng R and Welsh J (1994) *Nucleic Acids Res* 22:4419-4431.

Menard C. and Mouton C (1995) *Infect Immun* 63:2522-2531.

Mok SC, Wong KK, Chan RK, Lau CC, Tsao SW, Knapp RC and Berkowitz R. S. (1994) *Gynecol Oncol* 52:247-252.
Myers TW and Gelfand DH (1991) *Biochem* 30:7661-7666.
Odinot PT, Meis JF, Van den Hurk PJ, Hoogkamp-Korstanje JA and Melchers WJ (1995) *Epidemiol Infect* 115:269-277.
Peinado MA, Malkhosyan S, Velazquez A and Perucho M (1992) *Proc Natl Acad Sci USA* 89:10065-10069.
Perucho M, Welsh J, Peinado MA, Ionov Y and McClelland M (1995) *Methods Enzymol* 254:275-90.
Pooler MR and Hartung JS (1995) *Curr Microbiol* 31:377-381.
Prince JP, Lackney VK, Angeles C, Blauth JR and Kyle MM (1995) *Genome* 38:224-31.
Ralph D, McClelland M and Welsh J (1993) *Proc Natl Acad Sci USA* 90:10710-10714.
Salimath SS, de Oliveira AC, Godwin ID and Bennetzen JL (1995) *Genome* 38:757-763.
Sambrook J, Fritsch EF and Maniatis T (1989) *Molecular cloning: A laboratory manual.* Cold Spring Harbor Laboratory Press, Cold Spring Harbor, New York.
Schena M, Shalon D, Davis RW and Brown PO (1995) *Science* 270:467-470.
Sharma PC, Huttel B, Winter P, Kahl G, Gardner RC and Weising K (1995) *Electrophoresis* 16:1755-1761.
Shaw PC and But PP (1995) *Planta Med* 61:466-469.
Stammers M, Harris J, Evans GM, Hayward MD and Forster JW (1995) *Heredity* 74:19-27.
Stone B and Wharton W (1994) *Nucleic Acids Res* 22:2612-2618.
Suzuki Y, Wanaka A, Tohyama M and Takagi T (1995) *Neurosci Res* 23:65-71.
Tanhuanpaa PK, Vilkki JP and Vilkki HJ (1995) *Genome* 38:414-416.
Tsuchiya H, Tsuchiya Y, Kobayashi T, Kikuchi Y, Hino O (1994) *Jpn J Cancer Res* 85:1099-1104.
van Belkum A, Kluytmans J, van Leeuwen W, Bax R, Quint W, Peters E, Fluit A, Vandenbroucke-Grauls C, van den Brule A, Koeleman H, et al. (1995) *J Clin Microbiol* 33:1537-1547.
van Soolingen D, de Haas PE, Hermans PW, Groenen PM and van Embden JD (1993) *J Clin Microbiol* 31:1987-1995.
VanCouwenberghe CJ, Cohen SH, Tang YJ, Gumerlock PH and Silva J Jr. (1995) *J Clin Microbiol* 33:1289-1291.
Vogt T, Stolz W, Landthaler M, Rüschoff J and Schlegel J (1996) *J Invest Dermatol* 106:194-197.
Wachira FN, Waugh R, Hackett CA and Powell W (1995) *Genome* 38:201-10.
Wada-Kiyama Y, Peters B and Noguchi CT (1992) *J Biol Chem* 267:11532-11538.
Welsh J, Chada K, Dalal SS, Cheng R, Ralph D and McClelland M (1992a) *Nucleic Acids Res* 20:4965-4970.
Welsh J and McClelland M (1990) *Nucleic Acids Res* 18:7213-7218.
Welsh J and McClelland M (1991) *Nucleic Acids Res* 19:5275-5279.
Welsh J, Petersen C and McClelland M (1991) *Nucleic Acids Res* 19:303-306.
Welsh J, Pretzman C, Postic D, Saint Girons I, Baranton G and McClelland M (1992b) *Int J Syst Bacteriol* 42:370-377.
Welsh J, Rampino N, McClelland M and Perucho M (1995) *Mutat Res* 338:215-229.
Williams JG, Kubelik AR, Livak KJ, Rafalski JA and Tingey SV (1990) *Nucleic Acids Res* 18:6531-6535.
Wong KK and McClelland M (1994) *Proc Natl Acad Sci USA* 91:639-43.
Yao, J. D., Conly, J. M., and Krajden, M. (1995) *J Clin Microbiol* 33, 2195-8.
Yeh FC, Chong DK and Yang RC (1995) *J Hered* 86:454-460.
Yi QM, Deng WG, Xia ZP and Pang HH (1995) *Hereditas* 122:135-141.
Young KK, Resnick RM. and Myers TW (1993) *J Clin Microbiol* 31:882-886.

Randomly amplified polymorphic DNA (RAPD) analysis

J. Antoni Rafalski
Biotechnology Research, Agricultural Products Department, E.I. du Pont de Nemours and Co. Inc., Wilmington, DE 19880-0402, USA

Introduction

Amplification of genomic DNA using at least one short oligonucleotide primer, in low stringency conditions, results in multiple amplification products from loci distributed throughout the genome (Williams et al. 1990, Welsh and McClelland 1990). This observation led to the development of genome mapping (Williams et al. 1990) and fingerprinting (Welsh and McClelland 1990) applications. Depending on the specific conditions of amplification or product separation and detection, the arbitrary primer amplification methods were termed RAPD (randomly amplified polymorphic DNA; Williams et al. 1990), AP-PCR (arbitrarily primed PCR; Welsh and McClelland 1990), or DAF (DNA amplification fingerprinting; Caetano-Anolles et al. 1991). These methods rapidly became popular because of their simplicity and applicability to any genome. In late 1996, a literature search revealed over 3,000 references to RAPDs. Obviously, it is not possible to give a comprehensive literature review here. However, many review articles and protocols are available (e.g. Caetano-Anolles 1993, Rafalski et al. 1991, Rafalski et al. 1994, Tingey and del Tufo 1993, Waugh and Powell 1992, Williams et al. 1993). Here we provide an up-to-date RAPD laboratory protocol and discuss some aspects of reproducibility and interpretation of the RAPD assay.

Materials

Equipment

1. Thermocycler (for example, Perkin Elmer PE9600). Thermocyclers capable of using 96-well plates or tube assemblies are very convenient and compatible with robotic or multichannel pipettors that may be used to assemble the RAPD reaction. Thermocyclers with heated lids do not require adding oil to the PCR reactions and are

therefore most convenient. Some state-of-the-art thermocyclers are compatible with 384-well plates, increasing sample throughput significantly.

2. Horizontal agarose gel apparatus.
3. Electrophoretic power supply capable of delivering voltages up to 200 V at up to 150 mA.
4. Ultraviolet light box (transilluminator), set-up for taking photographs of ethidium bromide-stained agarose gels or an electronic imaging set-up equipped with a CCD camera (e.g. Eagle Eye, Stratagene).
5. Spectrophotometer or fluorometer for the determination of DNA concentration.

Reagents

1. Water: Sterile, de-ionized or distilled water should be used for preparing all reagents and pre-mixes.
2. Genomic DNA from species of interest: Most frequently used DNA isolation protocols provide DNA of sufficient quality, providing DNA has been quantitated carefully (see comments). We use the CTAB procedure (Murray and Thompson 1980), or the "Dellaporta prep" (Chen and Dellaporta 1994, Dellaporta et al. 1985). If problems arise, phenol extraction followed by ethanol precipitation is frequently helpful.
3. PCR buffers: 10× PCR buffer II (100 mM Tris-HCl pH 8.3, 500 mM KCl) and 10× Stoffel buffer (100 mM Tris-HCl pH 8.3, 100 mM KCl). Store frozen. PCR and Stoffel buffers are available from Perkin-Elmer and other suppliers. 10× standard PCR buffer (100 mM Tris-HCl pH 8.3, 500 mM KCl, 15 mM $MgCl_2$, 0.01% [w/v] gelatin) may also be used with appropriate adjustment of Mg^{+2} concentration.
4. Deoxynucleotide triphosphate mixture (2 mM each, dGTP, dATP, dTTP, dCTP): Ready-made solutions of dNTPs are available from Perkin-Elmer, Pharmacia and other suppliers. If dNTP solutions are made from dry reagents, the solution should be adjusted to pH 7.5 with 0.1 M Tris (unbuffered) or 0.1 M NaOH, checking pH by spotting 1 μL aliquots on a strip of pH paper. Aliquot into plastic test tubes and store frozen.
5. Arbitrary sequence primers: Sets of suitable 10-mer primers are available from several suppliers (e.g. Operon). Dilute in sterile distilled water to 4 μM and store frozen at -20°C.
6. 25 mM $MgCl_2$ solution: Prepare by diluting sterile 1 M $MgCl_2$ with sterile distilled water, aliquot and store frozen. Some *Taq* polymerase brands are sold with 25 mM $MgCl_2$ solution included. The *Taq* polymerase buffer stock provided by some suppliers contains magnesium, necessitating appropriate adjustment of the amount of extra $MgCl_2$ added.
7. *Taq* DNA polymerase: AmpliTaq® (Perkin-Elmer) or other high quality *Taq* polymerase is preferable. Contamination with DNA seen with some brands of *Taq* polymerase may cause appearance of bands in the "no DNA" controls.
8. Stoffel fragment of *Taq* DNA polymerase (Perkin Elmer).
9. Agarose: SeaKem GTG® (FMC Bioproducts) or other supplier.

10. TBE buffer for agarose gel electrophoresis: 1 M Tris-HCl, 0.9 M boric acid, 0.01 M EDTA pH 8.3. The buffer can be made in the laboratory or purchased pre-made (Life Technologies or other suppliers).
11. Ethidium bromide DNA stain (Life Technologies or other suppliers).
12. Electrophoretic size standards, for example bacteriophage λ DNA HindIII digest and bacteriophage ϕX174 HaeIII digest (e.g. Life Technologies, New England BioLabs).
13. YO-PRO-1 iodide (Molecular Probes; cat. no. Y3603): 1 mM in DMSO.

Experimental protocol

RAPD reactions are assembled as described in the following tables when using either Stoffel fragment or the *Taq* polymerase holoenzyme (Table 1).

TABLE 1. RAPD conditions.

Reagent	Stock	Final conc.	Amount used
A) Stoffel fragment *Taq* polymerase			
DNA	25-125 ng/5 μL	5-25 ng	1-5 μL
Stoffel Buffer	10x	1x	2.5 μL
	100 mM KCl	10 mM KCl	
	100 mM Tris-HCl	10 mM Tris-HCl	
	pH 8.3	pH 8.3	
MgCl$_2$	25 mM	2.5 mM	2.5 μL
Primer	4 μM	0.4 μM	2.5 μL
dNTP mix	2 mM each	0.1 mM each	1.25 μL
Stoffel fragment	10 U/μL	2 U	0.2 μL
Sterile distilled H$_2$O			complete to 25 μL
B) *Taq* polymerase holoenzyme			
DNA	25-125 ng/5 μL	5-25 ng	1-5 μL
PCR Buffer II	10x	1x	2.5 μL
	500 mM KCl	10 mM KCl	
	100 mM Tris-HCl	10 mM Tris-HCl	
	pH 8.3	pH 8.3	
MgCl$_2$	25 mM	1.7 mM	1.7 μL
Primer	4 μM	0.4 μM	2.5 μL
dNTP mix	2 mM each	0.1 mM each	1.25 μL
Taq polymerase	5 U/μL	0.5-1 U	0.1-0.2 μL
Sterile distilled H$_2$O			complete to 25 μL

It is preferable to wear gloves throughout the RAPD reaction preparation procedure. PCR buffer, dNTPs, MgCl$_2$ solution and primer solutions are thawed from frozen stocks, mixed by vortexing and placed on ice. DNA, if stored frozen (see below) should also be thawed out and mixed gently.

If several DNA samples are amplified using a large number of primers, as in identifying polymorphisms between mapping parents, the primers are pipetted first into test tubes compatible with the thermocycler used. For each DNA being tested, a pre-mix is then prepared including, in the following order, water, buffer, magnesium, dNTPs and DNA. Polymerase is then added to the pre-mix, the pre-mix is vortexed briefly, then aliquoted into the tubes already containing primers. Oil is added if recommended by the thermocycler manufacturer. The tubes are sealed and placed in the thermocycler and the cycling is started immediately.

Alternatively, if many DNA samples are amplified using one or a few primers, as in the case of analyzing segregation of a RAPD locus in a population of individuals, the DNA samples are first added to the test tubes on ice. A pre-mix containing, in the following order, water, buffer, magnesium, dNTPs, and primer is prepared. Polymerase is then added to the pre-mix, the pre-mix is briefly vortexed, then aliquoted into the tubes already containing DNA. The tubes are capped and placed in the thermocycler and the cycling is started immediately.

Cycling conditions may be modified depending on the thermocycler used. The following conditions work well on the Perkin Elmer DNA Thermal Cycler or Model 480 Thermal Cycler: 40 to 45 cycles of 1 min at 94°C, 1 min at 36°C, 2 min at 72°C, followed by 1 cycle of 7 min at 72°C and a 4°C incubation. Faster thermocyclers, for example Perkin-Elmer model 9600, allow shortening of the protocol: denaturation for 3 min at 94°C, followed by 40-45 cycles of 15 sec at 94°C, 30 sec at 36°C and 1 min at 72°C, concluding with 7 min at 72°C and a 4°C incubation. In all cases, use the shortest available transitions between temperatures.

After the cycling is finished, add an appropriate volume of a gel loading solution (Sambrook et al. 1989), for example 2.5 μL of 50% glycerol and 0.2% bromophenol blue. Mix well and load approximately half of the volume of the reaction on a horizontal agarose gel cast in 1× TBE buffer (Sambrook et al. 1989). DNA size standards (for example, a mixture of bacteriophage λ DNA HindIII digest with bacteriophage ϕX174 HaeIII digest) are electrophoresed alongside the RAPD reactions. We prefer to include 0.5 μg/mL ethidium bromide in both gel and 1× TBE electrophoresis running buffer. One may also stain the gel with ethidium bromide after electrophoresis.

CAUTION: Ethidium bromide is mutagenic and should be handled with appropriate care. Gloves should be worn at all times. Materials Safety Data Sheets (MSDS), available from the reagent supplier, should be consulted as to the proper handling and disposal.

Gel concentration of 1.4% is usually used. Amplifications using Stoffel fragment of Taq polymerase produce amplification products in a lower molecular weight size range than amplifications using Taq polymerase holoenzyme, and therefore agarose concentration may be adjusted accordingly. Some investigators prefer to use higher concentrations, up to 2.5% agarose or up to 3-4% of NuSieve® agarose (FMC Bioproducts). The gel is electrophoresed under standard conditions, typically 5-10 V/cm gel length. After the bromophenol blue dye has reached three-fourths of the gel length, the electrophoresis is complete and the gel is examined and photographed under UV light.

CAUTION: a UV-protective face shield and protective gloves should be worn.

For a description of the gel photography see Sambrook et al. (1989), or consult manufacturer literature concerning the use of electronic gel image recording devices. Figure 1 shows a typical example of a RAPD pattern obtained using DNA from a population of recombinant inbred individuals.

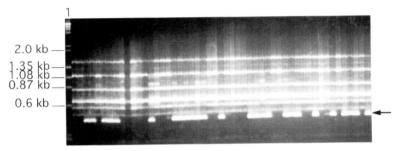

FIGURE 1. Image of an ethidium bromide-stained 1.4% agarose gel separation of RAPD reaction products obtained using the protocol described here and Stoffel fragment of Taq DNA polymerase. DNA samples (5 ng each) from a recombinant inbread population of soybean derived from a cross of Asgrow A3803 by PI437.654 were amplified using primer 11F5 (5'-CCTGCGGATG-3'). Lane 1 contains DNA size standards consisting of a mixture of bacteriophage lambda DNA *Hind*III digest and bacteriophage φX174 *Hae*III digest. Sizes of relevant bands are indicated. An arrow indicates a segregating band that was scored. Several other bands appeared to be segregating, but were judged not to be reliable enough to be scored.

As an alternative to agarose, other separation methods, including acrylamide gel or capillary electrophoresis may be used. Also, instead of ethidium bromide detection, other dyes (SYBR-green; Molecular Probes) may be used. Radioactively or fluorescently labeled primers may also be used.

Depending on the objective of the experiment, polymorphisms are noted, segregation of polymorphic bands scored, or the pattern compared with an existing database of RAPD fingerprints.

Comments

It is prudent to optimize all parameters of the RAPD assay, in particular DNA, magnesium and primer concentration, cycling conditions (especially annealing temperature), and the type and amount of thermostable DNA polymerase. The following factors influence reproducibility of RAPD profiles within and between laboratories:
• DNA concentration
• Reproducibility of thermocycler profiles
• Primer quality and concentration
• Magnesium concentration
• Choice of DNA polymerase
• Pipetting accuracy

DNA concentration

Optimize the amount of DNA used per RAPD reaction. Below a certain critical concentration of genomic DNA, RAPD amplification is no longer reproducible (Williams et al. 1993). It is essential to stay above this critical concentration. Excessive DNA concentration is likely to produce poor resolution or "smears". Usually 5-25 ng of DNA per 25 μL reaction is appropriate, but it is best to do a series of RAPD reactions using a couple of primers and a set of serial dilutions of each genomic DNA to identify empirically the useful range of DNA concentrations, for which reproducible RAPD patterns are obtained.

Most "quick" DNA preparations contain large amounts of RNA (sometimes 10 times in excess over DNA), so spectrophotometric DNA concentration measurements of such preparations (A_{260}) are likely to be in error. It is advisable to remove the RNA by digesting with DNase-free RNase A, followed by removal of UV-absorbing unincorporated nucleotides via gel filtration. Alternately, RNase digestion may be followed by DNA quantitation using fluorescent intercalating dyes. Such an assay may be conveniently performed in a 96-well plate with an appropriate fluorescence plate reader.

For example, 10 μL samples of genomic DNA are deposited in wells of a 96-well assay plate compatible with the fluorescence-measuring instrument to be used. A series of dilutions of genomic DNA of known concentration are placed in the same plate. Calf thymus or salmon sperm DNA is suitable, providing the concentration is carefully determined from A_{260} readings taken after thorough dialysis against 10 mM Tris-HCl pH 7.5 and 0.1 mM EDTA. A volume of 100 μL of a RNase-YoPro dye mix is added [10 mM Tris-HCl pH 7.5, 1 mM EDTA, 50 mM NaCl, 1 μM YoPro iodide and 10 μg/mL DNase-free RNase A prepared according to Sambrook et al. (1989)]. The plate is incubated at room temperature for 30 min to allow RNase A to digest contaminating RNA. Fluorescence intensities of individual wells are measured using a plate fluorometer (for example Millipore Cytofluor). Excitation wavelength of ca. 490 nm and emission wavelength of ca. 510 nm should be used. A calibration curve is constructed from the readings from standard DNA dilution series, and the genomic DNA samples are quantitated using the calibration curve.

If careful quantitation of genomic DNA is not practical, the amount used should be optimized by performing RAPD on a series of serial dilutions as indicated above.

Reproducibility of thermocycler profiles

RAPD amplifications are sensitive to the thermal profile of the thermocycler. Different thermocyclers have different thermal profiles, even if they are set to identical cycling parameters. This may be true even of different units of the same model from the same manufacturer. In some inferior machines there is considerable difference in temperatures between the internal wells and wells on the perimeter of the thermal block. As the thermocyclers age, their thermal profile may also change. This is especially true of the Peltier device machines. RAPD profiles obtained on different machines will be reproducible if the thermal profiles are at least approximately the same. This can be accomplished by measuring the actual thermal profile of the reference machine using a thermocouple and a data logger or a strip chart recorder, and then setting the other thermocycler to such cycling parameters that reproduce as exactly as possible the reference profile.

Primer quality and concentration

Most commercially-obtained primers are of satisfactory quality. Nevertheless, some failures of RAPD are due to a poor primer or low primer concentration (i.e. concentration lower than nominal). Primers stored as a frozen solution occasionally deteriorate especially when frequently thawed and re-frozen, and should be checked periodically. Many failures of RAPD reactions are due to inadequate mixing of the primer stock after thawing. The stock should be always thawed out completely and mixed thoroughly before use.

Magnesium concentration

The *Taq* polymerase and Stoffel fragment of *Taq* polymerase have different requirements for Mg^{+2} ions. Care should be taken that magnesium concentration in the amplification reaction is optimal. Low concentrations of Mg^{+2} produce few bands. Problems may arise from the presence of up to 1 mM of EDTA in TE buffer, in which genomic DNA is frequently dissolved. EDTA complexes magnesium and reduces the effective concentration of available Mg^{+2}. If significant volumes of EDTA-containing DNA solutions are added, the magnesium concentration in the reaction should be increased appropriately. The optimal concentration for a particular application should be determined empirically by performing a series of RAPD reactions at different Mg^{+2} concentrations in the range of 1-4 mM.

Pipetting accuracy

Pipetting small volumes, especially below 3 μL, could be a source of errors. Many multi-channel pipettors are not accurate below 5 μL. Therefore it is best to dilute various stock solutions to a concentration that would allow pipetting of 5 μL or more. This is especially significant if a large number of reactions are performed, making visual confirmation of each pipetting step impractical.
NOTE: Pipettes require regular servicing.

Choice of DNA polymerase

Taq polymerase was the first enzyme used for RAPD amplifications (Williams et al. 1990) and AP-PCR (Welsh and McClelland 1990). However, Sobral and Honeycutt (Sobral and Honeycutt 1993) pointed out that the Stoffel fragment of *Taq* polymerase (Lawler et al. 1993) appears to give better reproducibility of AP-PCR profiles, while at the same time changing the size distribution of amplification products towards lower molecular weights. This observation also holds for RAPD analysis. It is recommended to compare side-by-side RAPD profiles using *Taq* polymerase and Stoffel fragment and choose the enzyme that provides more satisfactory results. We preferentially use Stoffel fragment for RAPD. Due to the different amplification product size distribution (Rafalski et al. 1994), adjustment in the concentration of the separating gel matrix may be required for optimal results.

Limitations

Reproducibility of RAPD profiles has been a subject of considerable discussion (Penner et al. 1993, Skroch and Nienhuis 1995). The guidelines provided in the previous section always assure satisfactory reproducibility. In mapping applications it is recommended that a suitable statistical test be applied to ascertain that the segregation ratios conform to the expectation (for example 3:1 in an F_2 population or 1:1 in a backcross population).

Attention has to be paid to the design of mapping experiments involving dominant markers, including RAPD. When mapping dominant traits, only the RAPD amplification products in coupling with the trait are informative, i.e. amplification products derived from the mapping parent that carries the dominant form of the gene being mapped (Williams et al. 1993).

Because of the random nature of genome sampling, the RAPD assay is not an appropriate technique when the difference between the two genomes being compared is limited to an extremely small genomic fraction. For example, RAPD would generally not allow identification of single mutation or a very small deletion (0.001% of the genome or less). However, RAPD would efficiently identify localized or dispersed differences that constitute a significant fraction of the genomes. For example, donor genome segments in near-isogenic lines resulting from several backcrosses (1-10% of the donor genome) can be efficiently identified using RAPD. Another similar appropriate application of RAPD is bulk segregant analysis (Michelmore et al. 1991), used to identify markers closely linked to a trait or a locus of interest.

Recent research indicates that, relative to other methods, RAPD markers tend to underestimate genetic distances between more distantly related individuals, for example in inter-specific comparisons (Powell et al. 1996). It is, therefore, appropriate to exercise caution when using RAPD for taxonomic studies above the species level. Errors could also result from incorrect allele assignment based on mobility similarities. It is advisable to use Southern blotting to verify that bands of similar mobility are, in fact, allelic (Williams et al. 1993).

Overview

Despite the limitations discussed above, RAPD analysis offers the following advantages:
- non-radioactive detection
- no prior DNA sequence information for a genome is required
- universal primers work in any genome
- very small amounts of genomic DNA are sufficient
- multiplex detection of polymorphism
- experimental simplicity
- no need for expensive equipment beyond a thermocycler and a transilluminator

While I cannot provide here a comprehensive review of RAPD applications, a list of some typical applications may be useful:
- generating genetic maps
- mapping traits in segregating populations
- mapping traits in near-isogenic lines
- mapping traits using bulk segregant analysis
- saturating regions of a genome with markers
- fingerprinting individuals
- germplasm analysis
- measurement of genetic distances between individuals
- estimating relative parental contributions in backcrosses

While no one marker system can be considered ideal for all applications, the RAPD method provides a valuable tool in the repertoire of a molecular geneticist, allowing relatively easy entry into mapping and fingerprinting applications.

Acknowledgments

I would like to thank my colleagues at DuPont for the refinement of the RAPD protocols in the course of performing literally millions of RAPD reactions. I am grateful to Joe del Tufo, Julie Vogel, Renato Tarchini and Kunsheng Wu for their comments on the manuscript. Joe del Tufo provided Figure 1.

References

Caetano-Anollés G (1993) *PCR Methods Applic* 3:85-94.
Caetano-Anollés G, Bassam BJ and Gresshoff PM (1991) *Biotechnology* 9:553-557.
Chen J and Dellaporta S (1994) Urea-based Plant DNA Miniprep. In: Freeling M, Walbot V (eds). Maize Handbook. Springer-Verlag, New York, NY. pp. 526-527.
Dellaporta SL, Wood J and Hicks JB (1985) Maize DNA miniprep. In: Molecular Biology of Plants. 1st ed. Cold Spring Harbor Laboratory, Cold Spring Harbor, NY. pp. 36-37.
Lawler FC et al. (1993) *PCR Methods Applic* 2:275-287.
Michelmore RW, Paran I and Kesseli RV (1991) *Proc Natl Acad Sci USA* 88:9828-9832.
Murray MG and Thompson WF (1980) *Nucleic Acids Res* 8:4321-4325.
Penner GA et al. (1993) *PCR Methods Applic* 2:341-345.
Powell W et al. (1996) *Molecular Breeding,* in press.
Rafalski A, Tingey S and Williams JGK (1994) Random amplified polymorphic DNA (RAPD) markers. In: Gelvin SB, Schilperoort RA (eds). Plant Molecular Biology Manual. Kluwer Academic Publishers, Dordrecht, vol H4. pp. 1-8.
Rafalski JA, V.Tingey S and Williams JGK (1991) *AgBiotech News & Information* 3:645-648.
Sambrook J, Fritsch EF and Maniatis T (1989) Molecular cloning. A laboratory manual. Second Edition, (2nd Ed. ed.) Cold Spring Harbor Laboratory, Cold Spring Harbor, N.Y.
Skroch P and Nienhuis J (1995) *Theor Appl Genet* 91:1086-1091.
Sobral BWS and Honeycutt R (1993) *Theor Appl Genet* 86:105-112.
Tingey SV and del Tufo JP (1993) *Plant Physiol* 101:349-352.
Waugh R and Powell W (1992) *Trends Biotechnol* 10:186-191.
Welsh J and McClelland M (1990) *Nucleic Acids Res* 18:7213-7218.
Williams JGK, Kubelik AR, Livak KJ, Rafalski JA and Tingey SV (1990) *Nucleic Acids Res* 18:6531-6535.
Williams JGK, Rafalski JA and Tingey SV (1993) In: Wu R (ed), Methods in Enzymology. Academic Press, Orlando, FL., vol 218. pp. 704-740.

RAPD-DGGE: applications to pedigree assessment, cultivar identification and mapping

Ismail M. Dweikat and Sally A. Mackenzie
Department of Agronomy, Purdue University, West Lafayette, IN 47907, USA

Introduction

The advent and modification of PCR-based approaches for the mapping and fingerprinting of genetic materials has produced a myriad of strategies capable of almost unlimited resolution of large-sized genomes with reasonable resource input. Much of this resolution is the result of an increasing number of reproducible data points that can be collected per experiment. What has remained somewhat problematic in many mapping and fingerprinting efforts is the limited amount of DNA polymorphism available within some species and population structures. One effective means of dealing with limited DNA polymorphism has been the application of denaturing gradient gel electrophoresis (DGGE) to the fractionation of PCR products.

DNA fragments subjected to DGGE migrate according to melting properties of the fragments as well as fragment size (Myers et al. 1985). This is by virtue of an increasing concentration gradient of denaturant (generally formamide and urea) established within the gel. A DNA fragment, when subjected to a gradient running from low denaturant concentration (0-40%) at the origin, to high (40-80%) at the gel base, will migrate at a uniform rate until it reaches a concentration at which melting is initiated. The primary melting domain within the fragment forms a "bubble" or "fork"-like configuration to produce a partially single-stranded molecule that greatly slows in migration. Two like-sized fragments differing in sequence will demonstrate distinct melting properties, reflected in their migration to different points within the denaturing gel. Thus, fragment polymorphisms are observed not only as a consequence of differences in fragment size, but DNA sequence as well. Several studies have demonstrated that as little as a single base pair difference can be resolved using this methodology (Abrams et al. 1990, Myers et al. 1985), allowing the monitoring of allelic variants in human populations, as well as pedigree assessments in highly selected populations (Dweikat et al. 1993, He et al. 1992), and gene mapping in species that are particularly low in DNA fragment length polymorphisms by conventional methodologies (Dweikat et al. 1994, 1996, Procunier et al. 1995). Consequently, this approach offers the investigator an opportunity to test for sequence variation amongst a large number of individuals without undertaking a much more labor-intensive and, in some cases, nonfeasible DNA sequencing approach. Resolution of polymorphism resulting from sequence variation can be further enhanced by the development of GC-rich primers that act as "clamps" to stabilize the bubble structure (Abrams et al. 1990, Sheffield et al. 1989)

The DGGE procedure is well-suited for the detection of DNA polymorphisms in situations where restriction fragment length polymorphism is low but DNA sequence variations are expected. Over 75% of the reported uses of DGGE involve the detection of allelic variation in humans. Generally this is accomplished using PCR amplification products produced from gene-specific oligonucleotide primers. The monitoring of intragenic sequence variation using DGGE has also facilitated population studies in complex microbial (Muyzer et al. 1993, Wawer and Muyzer 1995), and domestic (Johnston and Fernando 1995) and marine (Norman et al. 1994) animal populations. In higher plant systems, the power of genetic analysis is immeasurably enhanced by the availability of reproductive systems that are amenable to the development of large-scale F_2-derived recombinant inbred populations and backcross-derived near-isogenic populations. These, combined with PCR-based random amplified polymorphic DNA (RAPD) or amplified fragment length polymorphism (AFLP) analyses, provide the tools necessary for high-resolution genomic investigation. In crops such as wheat (*Triticum aestivum* L.), lack of sufficient DNA polymorphism amongst breeding lines has greatly complicated the application of RAPD analysis. The use of DGGE to resolve RAPD products in wheat and other self-pollinating grain crops can dramatically increase resolution of DNA polymorphism (He et al. 1992, Procunier et al. 1994).

Materials

Gel apparatus and supplies

1. Heat-treated glass gel plates.
2. Gel spacers (0.75-1.0 mm thick).
3. Waterbath circulator with a range of 10-100°C.
4. Power supply capable of providing 200 V and 150 mA.
5. Peristaltic pump.
6. Stir plate.
7. Gradient maker (10 mL minimum capacity).
8. Hamilton syringe.
9. Gel electrophoresis system. We routinely use the SE600 vertical PAGE unit (Hoeffer) or the Protein II apparatus (BioRad). We find that the SE600 vertical unit allows more uniform heating of the buffer.

Buffer and stock solutions

1. 20× TAE: 800 mM Tris, 400 mM sodium acetate, 20 mM Na$_2$EDTA; adjust pH to 7.6 with glacial acetic acid.
2. DGGE stock solutions: 11.68 g acrylamide, 0.32 g bis-acrylamide, and 5 mL 20× TAE in 100 mL of distilled water. Add 4 mL formamide and 4.2 g urea or 20 mL formamide and 21 g urea for 10% or 50% denaturant gels (100% denaturant gels contain 7 M urea and 40% formamide).
3. 10% ammonium persulfate (prepare fresh daily).
4. N,N,N',N'-tetramethylethylenediamine (TEMED).
5. Loading buffer: 20% sucrose or ficoll and 0.1% bromphenol blue in 1× TAE buffer.
6. Ethidium bromide (10 mg/mL) in water.
7. 1 M KOH in 50% ethanol.

Experimental protocol

This protocol is based on the procedure originally described by Myers et al. (1985,1987).

1. Clean glass plates by soaking in 1 M KOH dissolved in 50% ethanol for at least 1 h and rinsing in hot water followed by deionized water. Allow to air dry.
2. Assemble the plates with spacers and brackets for casting.
3. Place a small (3 mm) stir bar in the high density solution chamber of the gradient maker (chamber proximal to gel apparatus) and attach gradient maker tubing to a gel loading or 0.23 gauge needle.
4. Prepare two small beakers. Add 10 mL of 10% denaturant stock solution and 100 μL of 10% ammonium persulfate to a first beaker, mix well, and then add 3 μL TEMED. Mix well. Add 10 mL of 50% denaturant stock solution and 100 μL of 10% ammonium persulfate to a second beaker and mix well. Add 3 μL TEMED and mix well again. No degassing is necessary.
5. Pour the solution of the first beaker (10% denaturant) into the gradient maker low density chamber (distal to the gel apparatus) and open the inter-chamber valve briefly, allowing a very small amount of the solution to pass between the chambers in order to eliminate bubbles.
6. Pour the solution of the second beaker (50% denaturant) into the gradient maker high density chamber. Turn on the stir plate to allow mixing in the 50% denaturant chamber and simultaneously open the outlet valve.
7. Immediately open the inter-chamber valve to allow mixing between chambers and turn on the peristaltic pump for a pump rate of approximately 3-4 mL/min.
8. During gel pouring, hold needle still to avoid disruption of the gradient and fill plates to approximately 1 cm from the rim.
9. Insert comb at an angle to avoid introduction of air bubbles.
10. Allow gel polymerization for a minimum of 2 h at room temperature.
11. Prepare 4.5-5 L of 1× TAE running buffer and fill buffer chamber of gel apparatus.
12. Set circulating water bath to 65°C, attach tubing to heat exchange unit of the gel apparatus, and begin water circulation at least 30 min prior to gel loading to pre-warm the running buffer.
13. Gently remove gel comb and fill wells with pre-warmed 1× TAE running buffer.

14. Add loading buffer to the DNA samples in a ratio of 1:5 (loading buffer:sample) and underlay DNA samples into the gel wells using a Hamilton syringe.
15. Fill upper buffer chamber with warmed 1× TAE buffer and attach electrodes to power supply. Gel should be run at a constant voltage of 150 V for 7 h.
16. After electrophoresis, gel can be stained in ethidium bromide (1 mL stock solution per 10 mL 1× TAE buffer). Stain with gentle shaking for 30 min, followed by destaining in deionized water for at least 2 h.

Important tips

1. Glass plates, spacers and clamps must be very clean.
2. Standard glass plates are sometimes susceptible to cracking during a DGGE gel run. Consequently, it is advisable to use heat-treated glass plates.
3. A minimum of 2 h are required for complete gel polymerization. Incomplete gel polymerization will significantly affect gel resolution. After polymerization, a gel can be stored overnight at 4°C prior to running.
4. When the gel clamps are not properly tightened, buffer seepage will result in distortion of the outer gel lanes.
5. The pH of running buffer and gel must be equal. When this is not the case, glass cracking or poor gel resolution can occur.
6. The running buffer can be reused 2-3 times without affecting gel quality.
7. Greater resolution of DNA bands is achieved with smaller DNA sample volumes.
8. Once the gel run is complete, or if it is unexpectedly interrupted due to power failure, the undisturbed gel can be retained in the gel apparatus for as long as 48 h without significant loss of resolution.
9. The usefulness of DGGE for various analyses has led to commercialization of the procedure. An apparatus designed for DGGE is now marketed by BioRad as 'D-GENE'.

Limitations

An obvious limitation that accompanies the incorporation of DGGE to the analysis of RAPD products is the labor intensity of the procedure. DGGE is conducted using an polyacrylamide gel system, more technically demanding and time-consuming to prepare than the standard agarose gel. Furthermore, staining of the gel with ethidium bromide, followed by a period of destaining, is required for best results. The procedure requires more specialized equipment, including a waterbath circulator, a gradient maker, and a peristaltic pump. However, we have incorporated what should be some useful tips to allow for efficient preparation, storage, and rapid processing of the gels. Although additional time and skill are required to conduct RAPD-DGGE, the enhanced resolution should, we believe, merit the effort.

Overview

DGGE has been successfully applied to gene mapping studies in wheat by our laboratory and others (Dweikat et al. 1994,1996, Procunier et al. 1995). Wheat is a crop of unusually

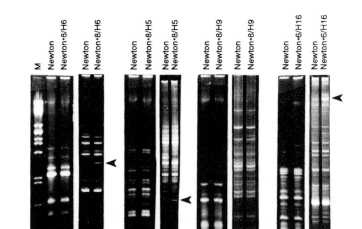

FIGURE 1. A comparison of resolution achieved by fractionating RAPD products using agarose gel electrophoresis versus DGGE. DNA markers associated with Hessian fly resistance gene H5, H6, H9 or H16 in wheat are presented in contrast to a near-isogenic line, "Newton", that is susceptible to all known biotypes of Hessian fly. Panels A contain RAPD products fractionated by agarose gel electrophoresis; panels B are the corresponding samples fractionated in 12% polyacrylamide by DGGE using a 10-50% denaturant gradient. Panels 1-4 represent RAPD products derived from four different primer reactions. Arrows indicate DNA polymorphisms that have been demonstrated to cosegregate with resistance genes H5, H6, H9 or H16. Lane M contains *Pst*I-digested lambda DNA for molecular weight estimation of the agarose fractionated samples.

large genome size and low DNA polymorphism, significantly enhanced by the RAPD-DGGE procedure (He et al. 1992, Procunier et al. 1994). An example of DNA markers associated with Hessian fly resistance loci that were resolved using RAPD-DGGE is shown in Figure 1. We have also applied DGGE to the discrimination of highly related germplasm in oat, wheat and barley (Dweikat et al. 1993). An example of the polymorphism resolved amongst related lines of barley is demonstrated in Figure 2.

Other examples of successful application of RAPD-DGGE to crop species or their pathogens include the identification of DNA markers linked to the S-locus in rye (*Secale cereale* L.) (Wehling et al. 1994), identification of DNA markers linked to a resistance locus in sorghum (Lee et al. 1996), cultivar identification in sunflower (Brunel 1994), and DNA fingerprint assessment of genetic heterogeneity in inbred populations of soybean cyst nematode (Clayton and Mackenzie, unpublished data).

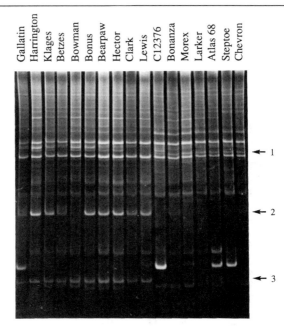

FIGURE 2. PCR amplification of genomic DNA from 17 lines of barley using RAPD primer combination OpA16/OpA17. The DNA samples were fractionated in 12% polyacrylamide and DGGE using a 10-50% denaturant gradient. The gel was run in 1X TAE buffer, pH 7.6 at 60C and stained with ethidium bromide. Arrows 1, 2, and 3 indicate polymorphisms that distinguish 2-rowed barley types (Gallatin-Lewis) from 6-rowed types (C12376-Chevron) (from Dweikat et al. 1993).

References

Abrams ES, Murdaugh SE and Lerman LS (1990) *Genomics* 7:463-475.

Brunel D (1994) *Seed Sci Technol* 22:185-194.

Dweikat IM, Ohm H, Mackenzie S, Patterson F, Cambron S and Ratcliffe R (1994) *Theor Appl Genet* 89:964-968.

Dweikat I, Ohm H, Patterson F and Cambron S (1996) *Theor Appl Genet*, in press.

Dweikat IM, Mackenzie S, Levy M and Ohm H (1993). *Theor Appl Genet* 85:497-505.

He S, Ohm H and Mackenzie S (1992 *Theor Appl Genet* 84:573-578.

Johnston DA and Fernando MA (1995). *Parasitology Res* 81:91-97.

Lee T-C, Dweikat IM and Dunkle LD (1996) *Plant Genome VI*. Abstract P164.

Muyzer G, DeWaal ED and Witterlinden AG (1993) *Appl Environ Microbiol* 59:695-700.

Myers RM, Lumelsky N, Lerman LS and Maniatis T (1985) *Nature* 313:495-497.

Myers RM, Maniatis T and Lerman LS (1987) *Methods Enzymol* 155:501-527.

Norman JA, Moritz C and Limpus CJ (1994) *Mol Ecol* 3:363-373.

Procunier JD, Townley-Smith TF, Fox S, Proshar S, Gray M, Kim WK, Czarnecki E and Dyck PL (1995) *J Genet Breeding* 49:87-91.

Procunier JD, Wolf M and Howes NK (1994). *Biotechnol Techniques* 8:707-710.

Sheffield VC, Cox DR, Lerman LS and Myers RM (1989) *Proc Natl Acad Sci USA* 86:232-236.

Wawer C and Muyzer G (1995) *Appl Environ Microbiol* 61:2203-2210.

Wehling P, Nackauf B and Wricke G (1994) *Plant J* 5:891-893.

Nucleic acid scanning by amplification with mini-hairpin and microsatellite oligonucleotide primers

Gustavo Caetano-Anollés
Plant Molecular Genetics, The University of Tennessee, Knoxville, TN 37901-1071, USA

Introduction

Nucleic acid templates can be amplified with at least one arbitrary oligonucleotide primer to produce specific signatures with which to characterize the sequence of virtually any nucleic acid, even anonymous in nature (Livak et al. 1992, Bassam et al. 1995). This nucleic acid "scanning" strategy can be used in a wide range of comparative and experimental applications, including DNA and RNA fingerprinting, genome mapping, marker-assisted breeding, molecular evolution and systematics, and population biology (reviewed in Caetano-Anollés 1993, 1994, 1996, Rafalski and Tingey 1993, McClelland et al. 1995). Nucleic acid scanning can also be applied to the fingerprinting of cDNA populations, PCR products, extrachromosomal nucleic acids, and cloned DNA. Three techniques were originally described, randomly amplified polymorphic DNA (RAPD; Williams et al. 1990), arbitrarily primed PCR (AP-PCR; Welsh and McClelland 1990) and DNA amplification fingerprinting (DAF; Caetano-Anollés et al. 1991a). These techniques vary in the length of primers used, the primer-to-template ratios, and the way amplification products are resolved and detected. They also produce markedly different amplification profiles, varying from simple (RAPD) to complex (DAF), and detect different levels of polymorphic DNA.

The versatile nature of nucleic acid scanning can improve the performance of the amplification reaction and meet the needs of particular applications. For example, there is sometimes a requirement to increase or decrease the ability to distinguish organisms at different taxonomic levels. Generally, this is done by using more than one nucleic acid typing technique capable of resolving genomes, for example, at the species or subspecies level. However, amplification fingerprints can be "tailored" directly in the number and range of amplification products, their multiplex ratio, the level of detection of

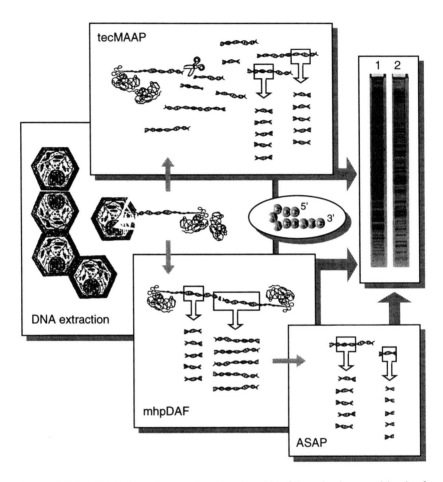

FIGURE 1. Three fingerprint tailoring strategies capable of detecting increased levels of polymorphic DNA in nucleic acid scanning applications. The strategies are based on DNA polymerase-driven amplification of defined loci (boxed DNA segments), generally using mini-hairpin primers (depicted in the center of the figure).

polymorphic DNA, and even the nature of amplified sites (Caetano-Anollés et al. 1991b). Fingerprint tailoring can be accomplished using several variations of the original nucleic acid scanning strategy. For example, selective restriction fragment amplification (SRFA) (Zabeau and Vos 1993), generally known by AFLP analysis, produces highly complex profiles by arbitrary amplification of restriction fragments ligated to adaptors with hemi-specific primers harboring adaptor-complementary 5' termini (Vos et al. 1995). Other strategies increase profile complexity and polymorphic DNA levels by prior digestion of template DNA with one or more type II restriction endonucleases that bind to 4 bp recognition sequences (Caetano-Anollés et al. 1993), when using mini-hairpin primers (Caetano-Anollés and Gresshoff 1994), or when producing arbitrary signatures from amplification products (ASAP) (Caetano-Anollés and Gresshoff 1996). Finally, primers can be designed to be complementary to consensus elements representing interspersed repetitive sequences (IRS) (such as the human *Alu* sequences), satellite sequences (including mini- and microsatellites), transposable elements, telomere associated sequences (TAS), and coding sequences such as histones, rRNA and tRNA genes.

This chapter describes the use and application of three nucleic acid scanning techniques that have the ability to tailor fingerprint pattern: (i) mini-hairpin primed DAF (mhpDAF), (ii) ASAP analysis, and (iii) template endonuclease cleaved multiple arbitrary amplicon profiling (tecMAAP) (Figure 1). DAF with primers containing hairpin-turn structures at their 5' termini can be used to obtain reliable fingerprints from almost any template nucleic acid, regardless of its complexity (Caetano-Anollés and Gresshoff 1994). mhpDAF minimizes some of the inherent limitations of nucleic acid scanning reactions. In ASAP, initial fingerprints generated by nucleic acid scanning with arbitrary primers are again fingerprinted by reamplification with new primers (Caetano-Anollés and Gresshoff 1996). These "fingerprints of fingerprints" are generally produced using mini-hairpin primers or primers complementary to interspersed repetitive sequences, provided their sequence differs significantly from that of primers used to generate the original amplification reactions. Finally, tecMAAP is based on a dual-step reaction where the template nucleic acid is first subjected to endonuclease digestion and then amplified with one or more arbitrary oligonucleotide primer (Caetano-Anollés et al. 1993). These nucleic acid scanning techniques are capable of fingerprinting a wide variety of templates, including whole genomes, plasmids, cloned DNA, and PCR products, and of detecting increased levels of polymorphic DNA. The protocols provided include the separation of the generated amplification products by polyacrylamide gel electrophoresis (PAGE) in vertical denaturing 0.45 mm-thick gels using a 8×10 cm format (Caetano-Anollés and Bassam 1993), open-faced discontinuous denaturing or nondenaturing slab gels (Allen et al. 1989, Doktycz 1993), and by using a semi-automated miniaturized electrophoretic and staining device (PhastSystem, Pharmacia) (Caetano-Anollés et al. 1996). Gels are backed on polyester film, silver stained using the procedure of Bassam et al. (1991) and preserved by drying at room temperature.

Materials

DNA amplification

1. Nucleic acid template: 1-50 ng/μL stock solutions.
2. AmpliTaq Stoffel fragment DNA polymerase (10 units/μL) (Perkin-Elmer).
3. Stoffel (STF) buffer (10× stock): 100 mM Tris-HCl, 100 mM KCl, pH8.3.
4. TTNK10 buffer (10× stock): 200 mM Tris-HCl, 1% Triton X-100, 40 mM $(NH_4)_2SO_4$, 100 mM KCl, pH 8.6.
5. Deoxynucleoside triphosphate (dNTP) stock: 2 mM of each dNTP.

6. Magnesium solution: 25 mM $MgCl_2$ (for use with STF buffer) or 100 mM $MgSO_4$ (with TTNK10 buffer).
7. Oligonucleotide primers: 30 μM or 300 μM stock solutions.

Electrophoresis

1. Tris-borate-EDTA (TBE) buffer (10× stock): 1 M Tris-HCl, 0.83 M boric acid, 10 mM $Na_2EDTA \cdot H_2O$, pH 8.3.
2. Leading ion buffer: 50 mM formate and 130 mM Tris-HCl, pH 8.5.
3. Trailing ion buffer: 200 mM of an appropriate trailing acid species such as Proline, Glycine or Tricine (*see* Doktycz 1993) in 100 mM NaOH.
4. Acrylamide-crosslinker-urea solution: 0.2% piperazine diacrylamide, 9.8% acrylamide and 10% urea in 1× TBE.
5. 10% ammonium persulfate.
6. N,N,N',N'-tetramethylethylenediamine (TEMED).
7. Loading buffer A: 30% urea, 0.08 % xylene cyanol FF.
8. Loading buffer B: 40% urea, 3% Ficoll, 0.02% xylene cyanol, 0.02% bromophenol blue, in 1× TBE).
9. Polyester gel-backing film (e.g. GelBond PAG; FMC Bioproducts), buffer strips made of stacked filter paper, membrane syringe filters, vertical, isothermally controlled horizontal (e.g. EC1001, E-C Apparatus), or semi-automated electrophoretic unit (PhastSystem).

Silver staining

1. Fixative-stop solution: 7.5% [v/v] glacial acetic acid.
2. Silver solution: 1g/L silver nitrate, 1.5 mL/L 37% formaldehyde.
3. Developer: 30 g/L sodium carbonate, 2 mg/L sodium thiosulfate, 3 mL/L formaldehyde.
4. Anti-cracking solution: 35% [v/v] ethanol, 10% [v/v] glacial acetic acid, 1% [v/v] glycerol.
5. Staining trays with flat bottoms and straight sides, and shaker. Clear plastic lids from 1000-μL pipette-tip racks can be used as staining trays.

Handling of reagents

Reagents used to separate the nucleic acids must be electrophoresis grade and those used in silver staining of high purity analytical grade. Use deionized water with conductivities of less than 10 MΩ/cm. Acrylamide solutions are light sensitive and should be stored in dark bottles at 4°C. For silver staining, use relatively fresh formaldehyde and never store it in the cold. Prepare impregnation and developer solutions in advance and store at room temperature. Add formaldehyde and freshly prepared (at least weekly) sodium thiosulfate prior to staining. Acrylamide is a potent neurotoxin and silver is toxic. Handle these solutions with care and dispose of them appropriately. Recycle the used silver by precipitation with NaCl.

Experimental protocol

DNA amplification

A) DAF analysis with mini-hairpin primers
  1. Assemble an amplification cocktail (20 μL) by adding components in the following order: 10.6 μL water, 2 μL dNTP stock, 0.8 μL of 100 mM MgSO$_4$ stock, 2 μL of TTNK10 buffer, 0.6 μL of Stoffel DNA polymerase, 2 μL of each primer (3-30 μM final concentration), and 2 μL template DNA stock.
  2. If necessary, cover the amplification mix with 1-2 drops of heavy mineral oil.
  3. Amplify in temperature cycler, usually 35 cycles of 30 s at 96°C, 30 s at 30°C and 30 s at 72°C (when using an oven thermocycler).
  4. Retrieve each amplification mixture from the tubes by adding 100-200 μL of chloroform and pipetting out the aqueous droplet formed, or directly by using a long pipette tip or by rolling the contents of each tube over Parafilm.

B) Arbitrary signatures from amplification profiles (ASAP)
  1. Dilute DAF reactions to about 1 ng/μL template stock solution. Typically, DAF reactions contain 100-200 ng/μL of double stranded DNA.
  2. Assemble the amplification cocktail (20 μL) by adding the following components: 10.6 μL water, 2 μL dNTP stock, 0.8 μL of 100 mM MgSO$_4$ stock, 2 μL of TTNK10 buffer, 0.6 μL of Stoffel DNA polymerase, 2 μL of each primer (9 μM final concentration), and 2 μL diluted amplification products.
  3. Cover the amplification mixture with 1-2 drops of heavy mineral oil.
  4. Amplify in 35 cycles of 30 s at 96°C, 30 s at 30°C and 30 s at 72°C in an oven thermocycler.
  5. Retrieve the aqueous phase from reaction tubes.

C) Template endonuclease cleavage MAAP (tecMAAP)
  1. Digest template DNA with one or more restriction endonucleases. Add 2 μL of enzyme (usually 2 units) and 2 μL of appropriate restriction enzyme buffer to 2 μg of DNA in a total 10 μL reaction volume. Incubate at the recommended temperature for 1 h. Use preferably blunt-end or staggered-cutter enzymes that recognize 4 bp motifs.
  2. Confirm complete digestion by electrophoresis in agarose or polyacrylamide gels.
  3. Assemble the amplification mixture (20 μL) by adding the following components: 10.2 μL water, 2 μL dNTP stock, 1.2 μL of 25 mM MgCl$_2$ stock, 2 μL of STF buffer, 0.6 μL of Stoffel DNA polymerase, 2 μL of each primer (3-30 μM final concentration), and 2 μL digested template. Use MgSO$_4$ and TTNK10 buffer when using mini-hairpin primers.
  4. Cover the amplification mixture with 1-2 drops of heavy mineral oil, when required.
  5. Amplify in 35 cycles of 30 s at 96°C, 30 s at 30°C and 30 s at 72°C in an oven thermocycler.
  6. Retrieve the aqueous phase from reaction tubes.

Separation of amplification products

A) Vertical PAGE
1. Assemble electrophoretic rigs by placing the following components on the clamp assembly: large glass plate, polyester backing film, 0.45 mm spacers, and small glass plate. The backing sheet must be placed in tight apposition to the large glass plate, with its hydrophilic side up. Assemblage is better done under running distilled water to avoid dust particles. Make sure all rig components are flush against the bottom before tightening assembly screws. Then dry overnight in a dust-free area in the dark (backing sheets are light-sensitive). Rigs can be dried in an oven at 40-50°C to speed the process.
2. Place gel rigs in casting stand, and prepare to cast 4-15%T and 2%C polyacrylamide-urea gels. Mix 10 mL of the acrylamide-crosslinker-urea stock solution with 150 μL ammonium persulfate and 15 μL TEMED.
3. Deliver the gel mix between glass plate and backing sheet using a syringe coupled to a 0.45 μm-pore size membrane filter. Insert Teflon comb and allow to set (2 min) and. fully polymerize (30 min).
4. Place gels rigs attached to electrode core into buffer tank containing 1 $\times$ TBE.
5. Remove combs and rinse wells with buffer using a fine-gauge syringe needle.
6. Pre-electrophorese gels at 150 V for at least 5 min.
7. Rinse and load wells with 3 μL of a 5-10 fold dilution of each amplification reaction mixed with 3 μL of loading buffer.
8. Electrophorese at 150-300 V for about 30-90 min.
9. Disassemble gel rig and remove backed gels.

B) Open-faced discontinuous PAGE
1. Pour the gel mix by capillary action between glass plate and polyester backing film separated by 0.2-0.4 mm spacers.
2. Wash the gel with abundant water and air dry.
2. Before electrophoresis, rehydrate the gel with leading ion buffer.
3. Place the gel on the horizontal platen of an isothermally controlled electrophoresis apparatus maintained at 10°C.
4. Carefully place electrodes on top of the cathodal buffer strip containing trailing ion buffer and the anodal buffer strip containing leading ion buffer.
5. Apply DNA samples directly or diluted 3-fold in loading buffer to the gel surface, 1 cm from the cathode, using a sample applicator. Alternatively, the samples can be appied on the surface of squares made of polypropylene filter mesh.
6. Electrophorese at 300 V for about 1 h or until the bromophenol blue marker reaches the anode.

C) Miniaturized semi-automated PAGE
1. Place non-denaturing 5-15% Phastgels on the horizontal platen of a PhastSystem apparatus using a forceps.
2. Place buffer strips on the top the gel and sample applicator containing 0.5 μL of amplification reactions on amplicator holding arm.

3. Separate amplification products at 15°C using the following steps: pre-electrophoresis for 100 Vh (400 V, 10 mA, 2.5 W), sample application for 2 Vh (400 V, 1 mA, 2.5 W); DNA sample electrophoresis for 100 Vh (200 V, 10 mA, 2.5 W), and slow hold for 100 Vh (50 V, 1 mA, 0.5 W).
4. Remove gels at the holding phase (about 200 Vh total).

Silver staining

1. Fix gels in fixative-stop solution for 10 min.
2. Wash the gels 3 times with distilled water, 2 min each time.
3. Impregnate with silver solution for 20 min.
4. Rinse with distilled water for 5-20 s.
5. Develop the image with developer solution at 8°C (usually about 4 min).
6. Stop image development in fixative-stop solution for at least 1 min, and wash extensively with water.
7. If required, wash gels with anti-cracking solution for 5 min.
8. Dry stained gels at room temperature and store in photographic albums.

NOTE: The stain detects about 1 pg/mm^2 band cross-section of DNA (Bassam et al. 1991), and is commercially available (Promega). Silver staining of Phastgels can be automated using the PhastSystem automated silver staining unit and the following program routine: Step 1, 10 min fixation; Steps 2-4, 2 min washes; Step 5, 20 min silver impregnation; Step 6, 0.1 min wash; Step 7, 3.5 min image development; Step 8, stop reaction (1 min); and Step 9, 4 min final wash. The four reagents (water, fixative-stop, impregnation, and developer solutions) are delivered through 6 in-ports and discarded through 2 out-ports (silver discarded separately). Steps 7 and 8 use ice-cold solutions.

Comments

Principles

The scanning amplification reaction is generally driven by the annealing of a single arbitrary primer to short and complementary inverted repeats that are in near proximity but scattered throughout the template, and the successful strand-extension of the annealed oligonucleotides. Targeting of template sites occurs under a non-stringent reaction environment that is both adequate for annealing of short primers and specific enough to provide discrimination of legitimate and illegitimate amplicons. In nucleic acid scanning, a multiplicity of arbitrary sites is targeted, and specificity is therefore expressed as the ability to produce a "most parsimonious" reproducible fingerprint. The single most important variable in the amplification reaction is the mass ratio between primer and template concentrations, capable of adjusting the overall stringency of the amplification reaction. Operationally, this ratio defines the different nucleic acid scanning techniques. For example, while RAPD uses primer-to-template ratios of less than 1, DAF requires ratios ranging 5-50,000. A second important parameter is the length of the arbitrary primer, influencing the extent of mismatch priming (mispriming) in the reaction.

A model to explain the amplification of DNA with arbitrary primers was proposed by Caetano-Anollés et al. (1992) [see also Caetano-Anollés 1993, 1994, Caetano-Anollés and Gresshoff 1994]. The model is based on the competitive effects of primer-template and template-template interactions established predominantly during the first few cycles of the amplification process. Table 1 describes some factors probably involved in the amplification process. During the first few temperature cycles a group of cognate amplicons is selected for amplification during a "template screening" phase driven by primer-template-enzyme interactions that can accomodate primer-template mismatching events. First-round amplification products are initially single-stranded but have palindromic termini that can establish template-template interactions, forming hairpin-loops and duplexes. The primer will have to recognize and displace these structures to allow enzyme anchoring and primer extension. In subsequent rounds of amplification the different species of the reaction tend to establish and equilibrium, while the rare primer-template duplexes are enzymatically transformed into accumulating amplification products.

TABLE 1. Factors that determine the outcome of a genome scanning reaction

Factor		Timing[a]
1.	Mismatching between primer and template sequences	E,L
2.	Interaction of palindromic termini and internal sequences of an amplicon leading to the stabilization of hairpin structures in first-round amplification products	E
3.	Displacement of template hairpin loop complexes by the primer during subsequent rounds of amplification	E
4.	Number and nature of competing amplification sites	E
5.	Exposure or masking of annealing sites by secondary nucleic acid structure	E,L
6.	Enzyme recognition and anchoring to stable primer-template duplexes	E,L
7.	Formation of template-template duplexes	L

[a]The factor is proposed relevant during: E, early amplification including the initial template "screening" phase where cognate amplicons are selected for amplification. L, late amplification including entrance into linear amplification (*plateau* process).

Fingerprint tailoring is based on the improvement of at least three components of the nucleic acid scanning reaction, individually or in combination: (i) analysis of amplification products, (ii) primer design, and (iii) amplification strategy (Caetano-Anollés 1996). Detection of amplification products with sensitive techniques, such as capillary electrophoresis (CE), can increase throughput and polymorphic DNA (Caetano-Anollés et al. 1995). CE can resolve DAF products in about 30 min, following the sieving of high molecular weight products. Alternatively, fluorophore-labelled DAF products can be sized by laser scanning with automated DNA sequencers (Caetano-Anollés et al. 1992). For example, DAF with fluorophore-labeled primers can allow clear distinction of two closely related *Streptococcus uberis* isolates when amplification products are separated using a Gene Scanner ABI362 fluorescent fragment analyzer (Figure 2A). Several other techniques for fragment or sequence identification can be used, including matrix-assisted laser desorption/ionization (MALDI) mass spectrometry (Doktycz et al.

stringency of amplification. For example in tecMAAP, many sites are targeted but only few are preferentially amplified in a dynamic reaction where kinetics and primer-template interaction determine the outcome of a particular fingerprint. While endonuclease digestion with one or more 4 bp cutters eliminates many of possible amplicons, therefore reducing the effective length of template DNA, it also increases the nucleotide sequence being probed. Sequence variation within restriction sites will either directly eliminate some of preferentially amplified products, or will indirectly change the overall reaction kinetics creating new products or eliminating ones that were previously amplified. Monte Carlo simulation showed that restriction forces arbitrary octamers to mismatch with the template in the last one or two 5' terminal nucleotides (Caetano-Anollés 1994). Ultimately, differential cleavage of the template is responsible of enhancing the detection of polymorphic DNA without an appreciably change in the number of DNA fragments amplified. tecMAAP enhances up to 100-fold the detection of polymorphic DNA levels, allowing identification of closely related cultivars, plant accesions and near isogenic lines (Caetano-Anollés et al. 1993).

DNA amplification with mini-hairpin primers

The design of primer sequence can take advantage of how primer length, sequence and secondary structure influence amplification. For example, very short primers (5-6 nt) produce relatively simple DAF profiles that resemble those obtained in RAPD analysis. In this case, amplification with short primers is hampered by the existence of palindromic termini in amplification products capable of forming hairpin loops, and an inherent decrease of primer annealing efficiency with decreasing primer length (Caetano-Anollés et al. 1992). Such complicating effects can be minimized with the introduction of an extraordinarily stable and compact hairpin structure at the 5' end of the primer consisting of a loop of 3-4 nt and a 2 nt stem (Caetano-Anollés and Gresshoff 1994). These extraordinarily stable "mini-hairpin" sequences have high melting temperatures, unusually rapid mobilities during electrophoresis in polyacrylamide gels, and cause band compression during Maxam-Gilbert DNA sequencing (Hirao et al. 1988, 1989, 1992). Their extraordinary stability depends on the existence of a hairpin-turn region determined by the helical motif of the stem region, the loop-closing base pair, and stacking of a loop B form structure (Hirao et al. 1994). Mini-hairpins have been observed in natural DNA, such as in the replication origin of phage G4 or in rRNA genes (Hirao et al. 1989, 1990, Antao and Tinoco 1991a,b). Mini-hairpin primers are designed by attaching an arbitrary "core" sequence at the 3' end of the mini-hairpin. These mini-hairpin primers interfere with the formation of hairpin loops in amplification products (Caetano-Anollés and Gresshoff 1994). They can therefore be used to produce reliable "sequence signatures" from small template molecules such as plasmids, cloned DNA, and PCR products.

A number of closed sets of short mini-hairpin primers are possible by attaching arbitrary 3 nt cores containing one of 64 possible nucleotide sequences to the 3' terminus of a constant mini-hairpin. Mini-hairpins must belong to a group of stable oligomers that exhibit extraordinarily high melting temperatures and contain loops of 3-4 nt (T_m =71-76.5°C). Examples of these mini-hairpins include the oligonucleotide GCGAAGC (HP$_7$) and the GCGNAAGC series (where N = A, G, C or T) (Hirao et al. 1989). In particular,

1995) or the hybridization to oligonucleotide arrays (Salazar and Caetano-Anollés 1996). Figure 2B shows that a single arbitrary array can provide characteristic hybridization patterns for a wide range of organisms. Obviously, the improved analysis of fingerprints is confined to the actual originating amplification process. In contrast, tailoring of both primer design and amplification strategy is not limited by such constraints. DNA fingerprints of varying complexity can be obtained by altering primer sequence and therefore the specificity with which primers anneal to their targeted sites. Primers can be arbitrarily selected, or can be designed to: (i) recognize particular sequence motifs within the genome representing dispersed sequences, structural chromosomal domains, or consensus sequences complementary to gene families, (ii) contain stretches of secondary structure within certain primer domains, or (iii) include degenerate sequence. Ultimately, their design can be biased by choosing those that work best. Finally, the strategy of amplification itself can be modified to increase or decrease the complexity of fingerprint pattern or the percentage of amplification products that are polymorphic within a particular group of organisms or templates under study.

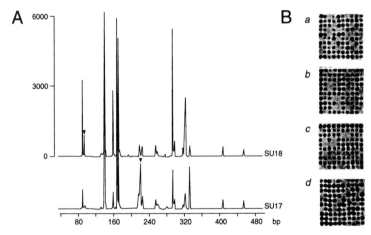

FIGURE 2. A. Automated separation and analysis of DAF profiles generated from *Streptococcus uberis* isolates of bovine origin. A 5'-end FAM-fluorescently labeled decamer (GTGACGTAGG) was used to amplify the closely related *S. uberis* strains SU17 and SU18, isolated from cow no. K1449, 7 d before calving and when developing clinical mastitis (4 d later). Amplification products were separated using a Gene Scanner ABI 362 fluorescent fragment analyzer using a 6% polyacrylamide gel. Arrowheads indicate two DNA polymorphisms. These two strains could not be distinguished in a previous study (Jayarao et al. 1992). The scale of the vertical axis indicates amount of fluorescence, and is linearly proportional to product amount. B. Nucleic acid scanning-by-hybridization (NASBH) of *Escherichia coli* (a), *Candida albicans* (b), *Acacia floribunda* (c) and human (d) DNA using an array of terminally-degenerate undecamers generates a similar number of positive signals.

Two mechanisms are mainly responsible for fingerprint tailoring: (i) an increase of the number of sites being probed in the template, during primer annealing or due to endonuclease restriction or post-amplification manipulations, and (ii) a change in the kinetics of the reaction due to novel primer-template interactions or variation in the

the decamer primer series HP$_7$-NNN was used to find a correlation between abnormal electrophoretic mobility behavior of the oligomers in denaturing polyacrylamide gels and amplification efficiency in DAF analysis of soybean cv. Bragg and bermudagrass cv. Tifway (Caetano-Anollés and Gresshoff 1996). The mini-hairpin HP$_7$ was chosen for this study because of it being the smallest and most stable (T$_m$=76.5) of described hairpins (Hirao et al. 1989, 1994). Previously, we found that mini-hairpin primers with arbitrary cores of different lengths exhibited abnormal electrophoretic mobilities (Caetano-Anollés and Gresshoff 1994, 1996). As originally described (Hirao et al. 1989), the mobilities of the mini-hairpin structures *per se* were appreciably higher than those of their 5' one-nucleotide lacking fragments. Similarly, the HP$_7$-NNN oligomers exhibited abnormal electrophoretic behavior. Oligomers migrated as hexamer or heptamer oligonucleotides (Figure 3), their mobilities (expressed as apparent oligonucleotide lengths) ranging 5.72-7.84. Mobilities varied with the sequence of the arbitrary core, and increased in the order A>C>>T=G, A=C>T=G, and G>>A>T=C for the first, second and third nucleotide from the 3' terminus, respectively. These fluctuations paralleled melting temperatures and were therefore indicative of differences in the stability of terminal mini-hairpins (Hirao et al. 1989). Similar tendencies were seen in part with octamer primers belonging to the GTCCANNN series, suggesting mobility is partially influenced by the sequence of the arbitrary core *per se*. However, no correlation between electrophoretic mobility and efficiency in DAF analysis was observed. The majority of primers produced complex fingerprints with a balanced number of amplification products (ranging 15-43 in the 50-700 bp interval) of high and low intensity. Only 5 out of the 64 primers tested failed to amplify soybean or bermudagrass DNA. However, it should be noted that the HP$_7$-CNN series contained several primers with very high amplification efficiencies, that perhaps target specific sequences particularly abundant in tested genomes.

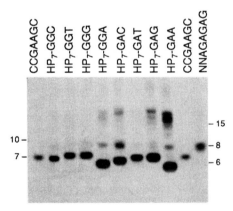

FIGURE 3. Abnormal electrophoretic mobilities of selected mini-hairpin decamers in denaturing polyacrylamide gels. Oligonucleotides (about 0.9 μmol/lane) were electrophoresed on a 20% polyacrylamide gel containing 7 M urea at room temperature and stained with silver. Controls include the 5' one-base-lacking mini-hairpin CGAAGC (6), and the unstructured oligomers CCGAAGC (7), NNAGAGAG (8) and CCGAAGCNNN (10). As reference, marginal lanes include the heptamer and octamer controls.

Silver staining detected alternative molecular species present at room temperature. Of the 64 decamer mini-hairpins, only 7 produced alternative conformers. In particular, 4 of these belonged to the series HP_7-GRN, where R=G or A (Figure 3). There is no explanation as to why these particular oligonucleotides can establish alternative conformations. In a previous study some mini-hairpin primers also exhibited various conformations including alternative hairpin structures, single strand unimolecular species, and in some cases a bimolecular duplex form, a structure with an internal loop that usually appears when the palindrome is 3 nt long (Caetano-Anollés and Gresshoff 1994). Alternative forms were eliminated when samples were heated at 90°C for 5 min before loading suggesting that they result from slow interconversion of oligonucleotide forms (Xodo et al. 1991). Furthermore, formation of single strands and duplex forms was also dependent on relatively high DNA concentrations and ionic strengths (Xodo et al. 1991), minimizing the role of alternative molecular species during amplification.

Importance of the terminal symmetry of amplification products

The importance of the mini-hairpin and arbitrary core sequences during amplification was shown by reamplification of DAF products engineered to have terminal symmetries of between 7-15 nt (Caetano-Anollés and Gresshoff 1994, 1996). Mini-hairpin primers with cores of varying length derived from the sequence GCGAGCTG generated DAF profiles that were then used as templates and re-amplified using these same primer sets. For example, primer CCGAGCTG and HP_7-CCGAGCTG successfully re-amplified products with terminal symmetries homologous to (and originally produced by) these primers, ultimately generating analogous fingerprints. In contrast, derived primers with cores of 3-7 nt reamplified only few of products generated with the octamer core, despite having perfect complementarity to the 3'-terminal nucleotides. Furthermore, all primers (with cores of $\geq$ 3 nt) were able to reamplify all engineered templates. However, the original amplification patterns were only reproduced when primer and product terminal sequence were complementary. The absence of the hairpin or a shorter or longer core sequence resulted in variant fingerprints. Results emphasize the importance of both hairpin and arbitrary core sequence in primer-template interaction, and highlight a crucial role for the previously proposed minimal 8 nt 3'-terminal primer domain in amplicon selection (cf. Caetano-Anollés et al. 1992).

To further examine the role of the arbitrary core sequence and the hairpin structure during amplicon screening, DNA from the circular plasmids pUC18 (2686 bp) and pBR322 (4363 bp) were amplified with the mini-hairpin HP_7-CTG (Caetano-Anollés and Gresshoff 1994). Those amplicons that were to be amplified with high probability from the template nucleotide sequence were predicted, and actual amplification products assigned to the predicted amplicons. The expected number of annealing sites within the pUC18 sequence that were complementary to the overall sequence of the mini-hairpin primer or of its arbitrary core were 0 and 111 in number, respectively. In theory, $2 \cdot 10^{-6}$ and 607 products were expected from an anonymous template corresponding to the size and GC content of this plasmid. To make the analysis possible, simulation studies used a mixture of 3 possible primer annealing species containing the core sequence and each of three regions of the hairpin (the loop, the 3' and the 5' palindromic sequence of the stem).

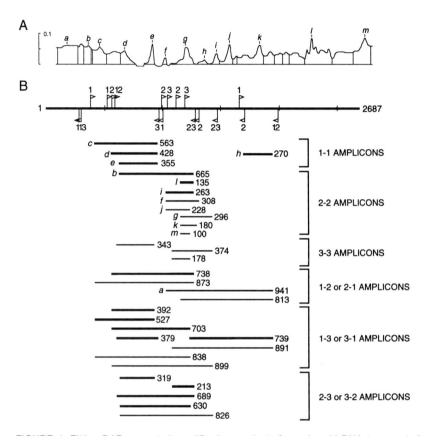

FIGURE 4. Fitting DAF generated amplification products from plasmid DNA to expected amplicons. A. Laser densitometric scan of a DNA pattern generated from plasmid pUC18 with mini-hairpin primer HP$_7$-CTG. Visual examination of the silver stained fingerprints confirmed assignment of numerical integration values to peaks (a-m). Band mobilities were reproducible with SE<2%. B. DAF simulation study and amplicon assignment of DAF products (identified in A) using the Lightspeed Pascal program *Amplify* (W. Engels, Genetic Dept., Univ. Wisconsin, Madison, WI). Three alternative primer species involving the annealing of the arbitrary core and either the 3' palindromic sequence (1, N$_5$GCCTG), the hairpin loop (2, N$_2$GAAN$_2$CTG) or the 5'-terminal palindrome sequence (3, GCN$_5$CTG). Primer annealing sites are shown located by flags along a linearized pUC18 sequence, and were identified using P=70, S=35, and P match weights of 15,12,10,1,1... (from 3' to 5'). Candidate amplicons are shown below the target sequence as bars, with heavier lines corresponding to more probable amplifications. Amplicon assigment was confirmed by restriction endonuclease digestion of amplification products with *BgI*I (identifying a,h), *Alw*NI (a,b,f,g,i-m), *Hae*II (a-f,i,j) and *Pvu*II (b-e) (Caetano-Anollés and Gresshoff 1994).

In this study all 13 observed major amplification products were assigned to predicted amplicons, and assignments confirmed by restriction endonuclease digestion of amplification products (Figure 4). Amplicons were defined by 13 of 18 possible annealing sites determined by 5-8 nt of primer sequence. Almost all products resulted from the annealing of the same primer annealing species and involved the arbitrary core and either the 3' palindrome or the loop. While only one tertiary product (*a*) originated from the annealing of two primer species (involving the loop and the 3' terminal palindrome), one other (*g*) involved the 5' terminal palindrome but together with the loop. Some sites allowed annealing of the core and the three hairpin regions but with considerable mismatching. Unassigned 19 predicted products always originated from different primer annealing species or involved the 5' palindromic region. Similar overall results were obtained in the fingerprinting of plasmid pBR322.

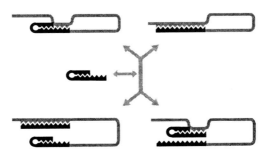

FIGURE 5. Model of possible interactions between mini-hairpin primer and first-round amplification products. Regions of the template that contain the mini-hairpin primer sequence are indicated in black. Note that mini-hairpin complementary sequences are incapable of forming hairpin-turn structures.

Results suggest that amplification driven by mini-hairpin primers is determined by the core region and either the 3' terminal palindrome or the loop of the mini-hairpin structure. The 5' palindromic sequence appears not involved in the amplification process, suggesting that the hairpin structure remains in tight conformation and does not form a duplex with the template during the amplicon screening phase. The preferential amplification of products with terminal symmetries extended to recognize the mini-hairpin sequence indicates the existence of annealing interactions at the amplicon termini. Direct physical interaction between the annealing sites of each amplicon is only possible when first-round amplification products loop to form hairpin structures. These hairpin-looped template strands then serve as preferential sites for annealing during the second round of amplification. The crucial amplicon screening of first-round (primer extended) long DNA strands obviously requires the active displacement of annealed template duplexes by the primer. Since both the 5' termini of the template and the primer contain mini-hairpin structures (the complementary sequence of the mini-hairpin is unstructured), displacement of hairpin-looped template may be favoured when the primer acquires a linear conformation and establishes duplex interactions. It should be noted that mini-hairpins oligonucleotides can form stable duplexes under denaturing conditions (Figure 3). However, the exact role of the hairpin-turn structure in this process still remains to be ascertained. A model of possible interactions during nucleic scanning of first-round amplification products is proposed in Figure 5.

Further evidence supporting the interaction between amplicon termini comes from the cloning and sequencing of amplification products (G. Caetano-Anollés, unpublished results). A high proportion of amplicons examined contained terminal regions of extended homology, suggesting stabilization and preferential amplification of corresponding first-round amplification products.

Mini-hairpin primers detected increased levels of polymorphic DNA in several plants, including centipedegrass (Caetano-Anollés and Gresshoff 1994), bermudagrass (Caetano-Anollés et al. 1995), and soybean (Caetano-Anollés and Gresshoff 1996). For example, mini-hairpin primers distinguished a γ-irradiation bermudagrass mutant from its parent, cultivar 'Tifway', at almost 5-fold higher levels than did standard primers (Caetano-Anollés et al. 1995). The enhanced resolving power of mini-hairpin primers may result from an increase in the size of the genome being probed during annealing. As discussed above, annealing of mini-hairpin primers appears influenced by secondary structure of DNA and interactions between amplicon termini. Generally, the scanning of extended annealing sites increases the chances of finding sequence variation. Mini-hairpin primers may accomplish this by targeting a high number of anealing sites with their short arbitrary cores and then selecting amplicons with extended annealing sites at their termini during scanning of first-round amplification products.

ASAP: fingerprints of fingerprints

ASAP analysis is based on the generation of fingerprints from fingerprints by re-amplification of genetic profiles with mini-hairpin or standard arbitrary primers (Caetano-Anollés and Gresshoff 1996). By definition, ASAP analysis is a dual-step amplification strategy that provides additional scanning of primary sequence within preselected amplicons. However, ASAP can generate allelic signatures from any nucleic acid fragment obtained by cloning, nucleic acid scanning, or PCR amplification. If entirely based on nucleic acid amplification, ASAP uses one or more primers in each amplification step, allowing a combinatorial use of oligomers in fingerprinting. Provided the sequence of the primers used in each step differs substantially from each other, unique fingerprints can be generated in each particular combination. For example a set of 10 oligomers can be used in 100 different pairwise combinations if used singly, or 10,000 if two primers are used during the first amplification step, 90% of which should produce unique fingerprints. However, the first amplification step determines the initial set of amplification products. Depending on the primer, the second amplification step will only provide variant signatures within the originally amplified fragments.

Primers in ASAP analysis can be designed to recognize particular sequence motifs or interspersed repetitive sequences in the template. For example, primers complementary to simple sequence repeats (SSR) present in microsatellite loci can be used to generate very simple ASAPs by re-amplification of DAF fingerprints (Caetano-Anollés and Gresshoff 1996). The advantage of targeting these loci is that they represent highly polymorphic regions, and in some cases, co-dominant markers with many allelic forms. The high informativeness of SSR loci has made them the target of a number of PCR-based techniques, where primers complementary to the sequence repeats are used directly or with arbitrary 5' or 3' terminal anchors (Meyer et al. 1993, Perring et al. 1993, Zietkiewicz et al. 1994, Wu et al. 1994). Unfortunately, these techniques amplify both the SSR motif and unrelated arbitrary sequences, and generate fingerprints with relatively high multiplex ratios where co-dominance may be difficult to interpret (Weising et al. 1995). The simple profiles of SSR-ASAP analysis mitigates the problem by generating very simple profiles containing in 50% of the cases only 1-2 prevalent amplification products. These products represent SSR loci and in many cases are clearly co-dominant. In order to target consistently about 10-15 kb of DAF amplified sequence which contains only few SSR annealing sites, the SSR primers have to be anchored at their 5' termini with ambiguous (ie. degenerate) nucleotides, and the amplification reaction must occur under stringent conditions to avoid mismatch priming.

Optimization of the amplification reaction

DNA amplification can be appropriately characterized by three parameters: *specificity*, *efficiency* (i.e. yield) and *fidelity*. These parameters are strongly influenced by the different components of the reaction and by thermal cycling. The judicious design of arbitrary primer sequence and the careful optimization of amplification parameters will ultimately result in reproducible and efficient amplification. Optimization of reaction components, concentrations, and thermal cycling depends on the identification of "reproducibility windows" (Bassam and Bentley 1994). These windows are defined as range of values within which amplification parameters exhibit little or no variation, and provide a measure of central tendency with which to define, for example, an experimental concentration, temperature or cycle number, and avoid borderline experimental conditions.

The optimization of the amplification reaction is a laborious process as a large group of interacting factors have profound effects on fingerprint, product number and efficiency of amplification. It relies on the sequential investigation of each variable and the design of large experiments. It should be noted that in reality optimum conditions are seldomly identified. Optimization has preceded several RAPD studies and rendered widely different optima (e.g. Akopyanz et al. 1992, Carlson et al. 1991, Devos and Gale 1992, Fekete et al. 1992, Munthali et al. 1992, Nadeau et al. 1992). This is in contrast to the few reports of optimization using DAF, where similar conditions were found optimal for the amplification of different organims such as centipedegrass (Weaver et al. 1995), bermudagrass (Caetano-Anollés et al. 1995), *Chrysanthemum* (Scott et al. 1996), *Petunia* (Cerny et al. 1996), and soybean (Prabhu and Gresshoff 1994). This consistency only holds true if genomes of similar complexity are examined, and appears related to the tolerance of the DAF reaction to wide changes in amplification conditions (Caetano-Anollés 1994). Table 2 shows recommended conditions for amplification of genomes of low and high complexity. These conditions should serve as a start for a DAF optimization exercise.

TABLE 2. Reproducibility windows for important DAF reaction components[a]

Reagents	DNA amplification: Plants and animals		Fungi and bacteria	
	Optimal	Recommended	Optimal	Recommended
Primer (μM)	2-10	3-6	3-9	3-6
Template (ng/μL)	0.01-2	0.1	0.1-10	1
MgCl$_2$ (mM)	1-8	1.5	4-8	6
dNTPs (μM)	50-300	200	50-300	200
Enzyme(units/μL)[b]	0.2-0.8	0.3	0.2-2	0.3

[a] Using buffer containing 10 mM Tris-HCl and 10 mM KCl (pH8.3).
[b] AmpliTaq Stoffel DNA polymerase.

The influence of different reaction components on the amplification reaction can be determined using an iterative process that simplifies an otherwise overwhelming full matrix analysis (Bassam et al. 1992, Weaver et al. 1995). This optimization strategy is based on simple matrix analysis in which several values for those experimental variables determined *a priori* to be most important are tested in combination with other variables. Similarly, a fractional factorial design has been used to study the influence of reaction components in RAPD analysis of *Chrysanthemum* (Wolff et al. 1993). Finally, the Taguchi method (Taguchi 1986) has also been applied to the study of interactions between specific reaction components in PCR and RAPD (Cobb and Clarkson 1994). Because of simplicity and proved success in industrial process design, the Taguchi method offers a simple optimization alternative (Cobb and Clarkson 1994). A number of reaction components at three different concentrations spanning expected reproducibility windows are laid in an orthogonal array that defines only a limited number of amplification reactions (Table 3). The measurement of an amplification parameter (such as product yield) is then used to estimate a quadratic loss function (termed signal-to-noise ratio, S). For each component the optimal conditions are those that give the largest S values. Polynomial regression generates S curves whose maximum represent reaction optima. Furthermore, a number of progressive trials can be used to better define those initially relevant factors.

Table 3 shows an experimental example of Taguchi optimization of bacterial DAF analysis for product yield. Optima coincided fairly well with optimal conditions obtained by iterative analysis based on reproducibility, maximum multiplex ratio and high efficiency (Bassam et al. 1992). The results of Taguchi optimization rely on the assumption that optima lie within the range tested and do not originate from tightly defined or vey relaxed peaks. Results also depict the way an amplification parameter is quantitated. An optimization of bacterial DNA fingerprinting such as that described by Bassam et al. (1992) will therefore require Taguchi optimization for depictors of amplification efficiency (product yield and range of products amplified) and specificity (number of products, range of products amplified, departure from a consensus fingerprint,

and reproducibility), and a series of optimization experiments that will progressively
define the different optima.

TABLE 3. DAF optimization using the Taguchi method.

A) Orthogonal array for 4 variables at 3 concentration levels
(low, medium and high) and calculation of amplification yields
in the amplification of *Escherichia coli* strain Smith 92 with
primer GTAACGCC.

Reaction tube	Primer (μM)	Template (ng/μL)	MgCl$_2$ (mM)	Enzyme (U/μL)	p
1	1	0.01	1	0.1	1
2	1	0.1	3	0.3	67
3	1	10	9	0.9	42
4	3	0.01	3	0.9	58
5	3	0.1	9	0.1	18
6	3	10	1	0.3	2
7	9	0.01	9	0.3	55
8	9	0.1	1	0.9	2
9	9	10	3	0.1	5

Quantitative estimates of amplification efficiency (amplification yield, *p*)
were obtained from each fingerprint by integrating relative peak areas from
density profiles generated using the *Image* analysis software.

B) Optimization using signal-to-noise ratios (S) calculated for
each reaction component using pooled amplification yields.

Component	Level Low	S_1	Medium	S_m	High	S_h	Optimal level
Primer (μM)	1 2 3	4.77	4 5 6	10.73	7 8 9	10.14	5.5
Template (ng/μL)	1 4 7	4.77	2 5 8	10.74	3 6 9	10.14	5.0
MgCl$_2$ (mM)	1 6 8	3.01	2 4 9	18.69	3 5 7	28.77	7.0
Enzyme (U/μL)	1 5 9	4.59	2 6 7	10.78	3 4 8	10.78	0.55

Values of *p* corresponding to the three listed reaction tubes (described in A)
for each concentration level (pooled amplification yields) were used to
calculate S using the equation:

$$S = -10 \log\left[\frac{1}{n}\sum_{i=1}^{n}\frac{1}{p_i^2}\right]$$

where *n* is the number of concentration levels. The optimal concentration level
for each component was then calculated by choosing levels which maximize S
after polynomial regression ($p = mx^2 + mx + c$).

Important parameters in the amplification reaction

Amplification with mini-hairpin primers has been optimized (Caetano-Anollés and Gresshoff 1994) by using a buffer containing Triton X-100 and ammonium sulfate (Caetano-Anollés et al. 1994). Amplification requires 2-12 mM $MgSO_4$ concentrations, with an optimal concentration range of 3-6 m M. Primers with long core regions are less tolerant of high magnesium levels. A minimum primer concentration of 1.5 mM is generally sufficient for reproducible amplification. However, higher primer concentrations (up to 30 μM) increase the efficiency of amplification but do not alter profile composition. In the study of PCR fragments, plasmids or viral DNA, reproducible patterns can only be obtained using more than 30 μM primer concentration. Template DNA levels should be comparable to those used in standard DAF (0.1-5 ng/μL). In ASAP analysis, primer concentration was found particularly important for reproducibility (Caetano-Anollés and Gresshoff 1996). Minimal primer concentration ranged 6-9 μM when using mini-hairpin decamers, and at 9 μM when using standard octamers. In contrast, ASAP primers with sequences partially complementary to DAF product termini tolerated lower primer concentrations (about 3 μM). The reproducibility window for template concentration ranges 0.001-1 ng/μL of DAF products.

Fingerprint evaluation

The comparison of fingerprint patterns involves the identification of matching components. Individual bands within each lane have to be *sized* relative to molecular weigh standards and then *matched* in the search for co-migrating bands. Band scoring is dependent on the type and conditions used during electrophoresis, band detection, and quality of DNA and amplification reagents. For example, during electrophoresis mobility of DNA fragments can shift due to irregularities of the electric field or protein or polysaccharide impurities in the samples. These problems can be usually corrected by running molecular weight standards every few lanes or by including them together with the samples, by using monomorphic products as internal references, and by running samples at least in duplicate. Scanned images of gels can also be manipulated to correct artifactual electrophoretic irregularities, such as the common "smiling" effect on PAGE in which centrally located bands exhibit higher mobilities, and warping due to variation in electric field.

During electrophoresis some bands are not well resolved as others. It is good practice to score bands within a region of the gel where fingerprint patterns are well discernable. For example, using a standard PAGE protocol for DAF analysis, only those products that are <500-700 bp in length are generally scored. Bands should be scored consistently throught the different samples to be compared. In some cases, co-migrating bands exhibit differences in intensity which could represent the dosage of the amplification products (perhaps indicative of the homozygous or heterozygous state) or existence of more than one co-migrating band (probably representing more than one locus). Exclude from the analysis those products that exhibit wide but unexplainable variations in intensity or lend themselves to ambiguous interpretation. Besides manual band scoring, bands can be sized and compared automatically with appropriate hardware and software tools. The fingerprint image can be recorded via a high-resolution video camera, scanning device, or phosphorimage analyzer and the mobility and intensity of the bands analyzed directly.

The dataset generated following band scoring and analysis of fingerprints needs to be interpreted according to each particular application. Applications include the assesment of genetic diversity, the identification of genotype, and segregation and linkage analysis of molecular markers. DAF patterns can be treated as multilocus fingerprint data, and are therefore subject to the advantages and limitations of the system (Caetano-Anollés 1994). For example, allelic pairs cannot be assigned to each other and allele frequencies cannot be calculated for each loci. Co-migrating bands can represent heterologous loci, and individual bands may not represent individual characters. Remember that DAF bands can be used as markers in genetic studies. They are largely dominant, and are therefore more accurate and better suited for genetic mapping in backcross, recombinant inbred or haploid populations than in F_2 populations. Be aware that some markers can exhibit atypical segregation. In soybean, 75% of DAF markers show true Mendelian inheritance while the rest segregate in a uniparental way, being either of maternal (probably cytoplasmic) or paternal origin (Prabhu and Gresshoff 1994).

Quantification of pairwise similarities or differences between fingerprints can be done using different algorithms. Most commonly, a "similarity index" is calculated as depictor of band sharing. For example, Jaccard's similarity coefficent (J) takes into consideration only those matches between bands that are present (Jaccard 1908). In turn, Nei and Li's coefficient (N) measures the proportion of bands shared as the result of being inherited from a common ancestor and represents the proportion of bands present and shared in both samples divided by the average of the proportion of bands present in each sample (Nei and Li 1979). These similarity coefficients are calculated by:

$$J = \frac{sp}{sp + (1 - s)} \qquad\qquad N = \frac{2sp}{2sp + (1 - s)}$$

where p is the proportion of shared bands present in both samples and s is the proportion of the total number of bands that are shared either present or absent in both samples. The N coefficient has been recommended for analysis of multilocus fingerprints (Lamboy 1994a). Please note that these two coefficients do not consider which of unshared bands belong to which sample. These coefficients can be corrected to accomodate the existence of artifactual false-negative or false-positive bands if they were present in the amplification reactions (Lamboy 1994a,b). These and other coefficients can be used in phenetic and phylogenetic analysis of fingerprints by using a number of statistical analysis software packages that use ordination techniques [such as *principal component analysis* (PCA) or *principal coordinate analysis* (PCO)], distance matrix or cluster analysis methods [such as *unweighted pair group method using arithmetic average* (UPGMA) or *neighbour-joining* (NJ) algorithms] or parsimony strategies [such as *phylogenetic analysis using parsimony* (PAUP)]. Examples of cluster analysis applications include PHYLIP (J. Felsenstein, Dept. of Genetics, Univ. of Washington, Seattle, WA) and NTSYS (Exeter Software, Setauket, NY), and of parsimony analysis applications PHYLIP, PAUP (D.L. Swofford, Illinois Natural History Survey,

Champaign, IL), McClade (W.P Madison and D.R. Maddison, Sinauer Assoc., Sunderland, MA), and Hennig86 (J.S. Harris, Port Jefferson Sta., New York, NY).

For the analysis of mapping populations in linkage analysis and genetic mapping, calculation of recombination values and map distances can be performed with a group of other computer programs, such as LINKAGE-1 (Suiter et al. 1983), MAPMAKER (Lander et al. 1987), GMENDEL (Liu and Knapp 1990), or JOINMAP (Stam 1993).

Cost and other practical considerations

The cost per fingerprint can be estimated to be US$1.27 for DAF and US$1.08 for RAPD analysis. On a per sample basis, this cost is comparable to that of RFLP and other PCR-based fingerprints, but does not include labor, laboratory equipment and space. About two thirds of these expenses correspond to the polymerase enzyme. Therefore, those users capable of preparing their own enzyme can decrease costs considerably. Several enzyme purification protocols are available (eg. Pluthero 1993, Harrell and Hart 1994). The cost of the primer is negligible (about 1%) but can be considerable when constructing a primer library for genetic mapping or fingerprinting purposes. Primer synthesis can now provide oligonucleotides at about US$0.01/base per nmol.

An important decision is that of a thermal cycler. A wide range of thermal cyclers is available with varied heat sources, and mechanisms of heat transfer and dissipation. Heat source can be electrical, electronic (Peltier) or irradiation-based (visible or IR light). Heat transfer can be via metallic blocks, air or liquids. Heat dissipation can be achieved by convection in air-driven (oven) units, thermal conductivity in block-based units (Peltier or water cooling), or can be based on robotic transfer of tubes to blocks preset at lower temperatures. Some thermal cyclers eliminate condensation by the uniform heating of the tubes and can avoid the use of oil overlays. Other thermal cyclers have sample temperature probes that accurately control cycling parameters. Finally, some units maximize heat transfer by using capillary tubes or thin-walled tubes; these units usually result in very fast cycling reactions. Depending on needs and performance, any of these units can be used. For example, the oven based thermal cyclers can accomodate a big number of tubes (500-1000), but are relatively slow. Light-driven units allow the use of very small samples, are fast, but can accomodate only few capillaries.

Acknowledgments

I will like to thank N.M. Salazar of Digital Diagnostica S.R.L. (Uruguay) for photographic material, and support from the International Atomic Energy Agency, Vienna (FAO/IAEA RCP580, 8151) and the Racheff Chair endowment.

References

Akopyanz N, Bukanov TU, Westblom TU, Kresovic S and Berg DE (1992) *Nucleic Acids Res* 20:5137-5142.
Allen RC, Graves G and Budowle B (1989) *Biotechniques* 7:736-744.
Antao VP and Tinoco Jr I (1991a) *Nucleic Acids Res* 20:819-824.
Antao VP, Lai SY and Tinoco Jr I (1991b) *Nucleic Acids Res* 19:5901-5905.

Bassam BJ and Bentley S (1994) *Australasian Biotechnol* 4:232-236.
Bassam BJ, Caetano-Anollés G and Gresshoff PM (1991) *Anal Biochem* 196:81-84.
Bassam BJ, Caetano-Anollés G and Gresshoff PM (1992) *Appl Microbiol Biotech* 38:70-76.
Bassam BJ, Caetano-Anollés G and Gresshoff PM (1995) Method for profiling nucleic acids of
    unknown sequence using arbitrary oligonucleotide primers. US Patent 5,413,909.
Caetano-Anollés G (1993) *PCR Methods Applic* 3:85-94.
Caetano-Anollés G (1994) *Plant Mol Biol* 25:1011-1026.
Caetano-Anollés G (1996) *Agro-Food-Industry High-Tech* 7:26-35
Caetano-Anollés G and Bassam BJ (1993) *Appl Biochem Biotechnol* 42:189-200.
Caetano-Anollés G and Gresshoff PM (1994) *Bio/Technology* 12:619-623.
Caetano-Anollés G and Gresshoff PM (1996) *Biotechniques* 20:1044-1056.
Caetano-Anollés G, Bassam BJ and Gresshoff PM (1991a) *Bio/technology* 9: 553-557.
Caetano-Anollés G, Bassam BJ and Gresshoff PM (1991b) *Plant Mol Biol Rep* 9:294-307.
Caetano-Anollés G, Bassam BJ and Gresshoff PM (1992) *Mol Gen Genet* 235: 157-165.
Caetano-Anollés G, Bassam BJ and Gresshoff PM (1993) *Mol Gen Genet* 241: 57-64.
Caetano-Anollés G, Bassam BJ and Gresshoff PM (1994) *PCR Methods Applic* 4:59-61.
Caetano-Anollés G, Bassam BJ, McCracken GF, Reddick B and Gresshoff PM (1996) *Pharmacia
    Applic Notes* 18-1115-51.
Caetano-Anollés G, Callahan LM, Williams PE, Weaver KR and Gresshoff PM (1995) *Theor
    Appl Genet* 91:228-235.
Carlson JE, Tulsieram LK, Glaubitz JC, Luk VWK, Kauffeldt C and Rutledge R (1991) *Theor
    Appl Genet* 83:194-200.
Cerny TA, Caetano-Anollés G, Trigiano RN and Starman TW (1996) *Theor Appl Genet* 92:1009-
    1016.
Cobb BD and Clarkson JM (1994) *Nucleic Acids Res* 22:3801-3805.
Devos KM and Gale MD (1992) *Theor Appl Genet* 84:567-572.
Doktycz MJ (1993) Discontinuous electrophoresis of DNA: adjusting DNA mobility by trailing
    ion net mobility. *Anal Biochem* 213:400-406.
Doktycz MJ, Hurst GB, Habibi-Goudarzi S, McLuckey SA, Tang K, Chen CH, Uziel M,
    Jacobson KB, Woychik RP and Buchanan MV (1995) *Anal Biochem* 230:205-214.
Fekete A, Bantle JA, Haling SM and Stich RW (1992) *J Bacteriol* 174:7778-7783.
Harrell II RA and Hart RP (1994) *PCR Methods Applic* 3:372-375.
Hirao I, Naraoka T, Kanamori S, Nakamura M and Miura K (1988) *Biochem Int* 16:157-162.
Hirao I, Nishimura Y, Naraoka Y, Watanabe K, Arata Y and Miura K (1989) *Nucleic Acids Res*
    17:2223-2231.
Hirao I, Ishida M, Watanabe K and Miura K (1990) *Biochem Biophys Acta* 1087:199-204.
Hirao I, Nishimura Y, Tagawa Y, Watanabe K and Miura K (1992) *Nucleic Acids Res* 20:3891-
    3896.
Hirao I, Kawai G, Yoshizawa S, NishimuraY, Ishido Y, Watanabe K and Miura K (1994) *Nucleic
    Acids Res* 22:576-582.
Jaccard P (1908) *Bull Soc Vaud Sci Nat* 44:223-270.
Jayarao BM, Bassam BJ, Caetano-Anollés G and Gresshoff PM (1992) *J Clin Microbiol* 30:1347-
    1350.
Lamboy WF (1994a) *PCR Methods Applic* 4:31-37.
Lamboy WF (1994b) *PCR Methods Applic* 4:38-43.
Lander ES, Green P, Abrahamson J, Barlow A, Daly A, Lincoln SE and Newburg L (1987)
    Genomics 1:174-181.
Liu BH and Knapp SJ (1990) *J Hered* 81:407.
Livak KJ, Rafalski JA, Tingey SV and Williams JG (1992) Process for detecting polymorphisms
    on the basis of nucleotide differences. US Patent 5,126,239.
Meyer W, Mitchel TG, Freedman EZ and Vilgalys R (1993) *J Clin Microbiol* 31:2274-2280.
McClelland M, Mathieu-Daude F and Welsh J (1995) *Trends Genet* 11:275-279.
Munthali M, Ford-Lloyd BV and Newbury HJ (1992) *PCR Methods Applic* 1:274-276.

Nadeau JH, Bedigian HG, Bouchard G, Denial T, Kosowsky M, Norberg R, Pugh S, Sargeant E, Turner R and Paigen B (1992) *Mammalian Genome* 3:55-64.

Nei M and Li WH (1979) *Proc Natl Acad Sci USA* 76:5269-5273.

Perring TM, Cooper AD, Rodriguez RJ, Farrar CA and Bellows TS (1993) *Science* 259:74-77.

Pluthero FG (1993) *Nucleic Acids Res* 21:4850-4851.

Prabhu R and Gresshoff PM (1994) *Plant Mol Biol* 26:105-116.

Rafalski JA and Tingey SV (1993) *Trends Genet* 9:275-279.

Salazar N and Caetano-Anollés G (1996) *Nucelic Acids Res* 24:5056-5057.

Scott K, Caetano-Anollés G and Trigiano RN (1996) *J Amer Soc Hort Sci* 121:1043-1048.

Stam P (1993) *Plant J* 3:739-744.

Suiter KA, Wendel JF and Case JS (1983) *J Hered* 74:203-204.

Taguchi G (1986) In: Asian Productivity Organization. UNIPUB, New York.

Vos P, Hogers R, Bleeker M, Reijans M, van de Lee T, Hornes M, Frijters A, Pot J, Peleman J, Kuiper M and Zabeau M (1995) *Nucleic Acids Res* 23:4407-4414.

Weaver KR, Callahan LM, Caetano-Anollés G and Gresshoff PM (1995) *Crop Sci* 35: 881-885.

Weising K, Atkinson RG and Gardner RC (1995) *PCR Methods Applic* 4:249-255.

Welsh J and McClelland M (1990) *Nucleic Acids Res* 19:861-866.

WilliamsJGK, Kubelik AR, Livak KJ, Rafalski JA and Tingey SV (1990) *Nucleic Acids Res* 18:6531-6535.

Wolff K, Schien ED and Peters-van Rijn J (1993) *Theor Appl Genet* 86:1033-1037.

Wu K, Jones R, Danneberger L and Scolnik PA (1994) *Nucleic Acids Res* 22:3257-3258.

Xodo LE, Manzini G, Quadrifoglio F, van der Marel GJ and van Boom JH (1991) *Nucleic Acids Res* 19:1505-1511.

Zabeau M and Vos P (1993) Selective restriction fragment amplification: a general method for DNA fingerprinting. EPO Patent No. 0534858A1.

Zietkiewicz E, Rafalski A, Labuda D (1994) *Genomics* 20:176-183.

AFLP analysis

Pieter Vos and Martin Kuiper
Keygene N. V., 6700 AE Wageningen, The Netherlands

Introduction

The AFLP™ technique is a powerful DNA marker technique. It was originally conceived to allow the construction of very high density DNA marker maps for application in genome research and positional cloning of genes. It is equally suitable for applications in genetic analysis which require more modest DNA marker densities. AFLP is based on the detection of DNA restriction fragments by PCR amplification (Vos et al. 1995, Zabeau and Vos 1993). Amplification of restriction fragments is accomplished by the ligation of double-stranded (ds) adapter sequences to the ends of the restriction sites which can subsequently serve as "universal" binding sites for primer annealing in PCR (Figure 1). In this way, restriction fragments of a particular DNA can be amplified with "universal" AFLP primers corresponding to the restriction site and adapter sequence. However, for most DNAs the number of fragments that will be simultaneously detected in this way will be too high to be resolved in any fragment analysis system, e.g. gels. Therefore, the AFLP primers have at their 3' end a number of selective bases that extend into the restriction fragments (Figure 2). This results in selective amplification of those fragments in which the primer extensions match the nucleotides flanking the restriction site. The number of fragments to be amplified can be "tuned" by the selection of the number of selective bases in the AFLP primers. In practice, the number of fragments simultaneously amplified will be limited to 50-100 to allow detection on denaturing polyacrylamide gels (sequence gels). In simple genomes the number of selective nucleotides required will be low because of the low number of restriction fragments available for amplification. Similarly, complex genomes will require more selective nucleotides to reduce the number of amplified fragments to a number suitable for resolution on sequence gels. Restriction fragment patterns generated by the AFLP technique are called AFLP fingerprints. These AFLP fingerprints are a rich source for restriction fragment polymorphisms, called AFLP markers. The frequency with which AFLP markers are found is dependent on the sequence polymorphisms between the tested DNA samples.

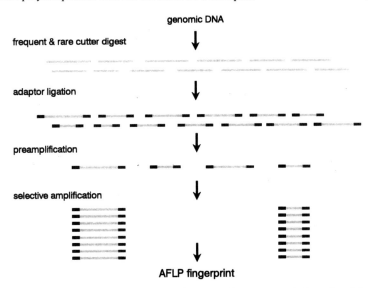

FIGURE 1. Flow chart of the AFLP protocol. Restriction fragments are indicated as light grey bars, the frequent cutter adapter is indicated as a dark grey box, the rare cutter adapter as a black box.

The first step of the AFLP procedure is restriction of the DNA with two different restriction enzymes, generally a rare-cutter and a frequent cutter, and the ligation of double-stranded adapters to the ends of the restriction fragments (Figures 1 and 2). The adapter and restriction site sequences serve as primer binding sites in the subsequent amplification steps. The frequent cutter is used to generate small fragments which amplify well and are in the optimal size range for separation on sequence gels. The rare cutter is used to limit the number of fragments to be amplified, because the AFLP amplification process results in preferential amplification of restriction fragments with two different ends (adapters) (Vos et al. 1995). Hence, the number of effective AFLP amplicons is determined by the number of restriction sites for the rare cutter. Furthermore, the use of two different restriction enzymes permits optimal flexibility in primer selection which determines the number of fragments that will be co-amplified. Alternatively, two rare cutter enzymes may also be used for AFLP fingerprinting. In this case, only a minor fraction of the fragments will be detected, i.e. small fragments, because the smaller fragments amplify with superior efficiency.

The second step of the AFLP procedure is the amplification of subsets of restriction fragments using selective AFLP primers (Figure 2). For this purpose, primers are used that correspond to the adapter and restriction site sequences and that have additional nucleotides at the 3' ends extending into the restriction fragments (Figure 2). These 3' extensions (called the selective nucleotides) assure that only a subset of the restriction fragments is amplified, i.e. those fragments of which the sequences adjacent to the restriction sites match the 3' primer extensions. With very small "genomes" (plasmids, cosmids, BACs) no selective nucleotides are required. AFLP fingerprinting of bacteria and fungi (or similar-sized genomes) is performed using up to 2 selective bases for each AFLP primer. Complex genomes require the use of more than 2 selective bases in one or both primers. In this case, the amplification is carried out in two consecutive steps, a first step called preamplification and a second step called selective AFLP amplification (Figure 1). The amplification in two steps warrants optimal primer selectivity, although the selectivity of primers having 3 selective bases is generally good (Vos et al. 1995). The preamplification may also be used to "increase" the amount of the template, i.e. by AFLP amplification with primers having no selective nucleotides. Detection of the AFLP fragments is achieved by labeling one of the two AFLP primers used in the selective amplification reaction. The experimental protocols in this chapter describe the use of radioactively labeled primers obtained by phosphorylating the 5'ends with γ-^{33}P-ATP and polynucleotide kinase. Other methods of labeling the AFLP primers may also be used. Only one of the two AFLP primers should be labeled, because this will result in only one of the two strands of the AFLP fragments being labeled and, hence, detected. Labeling both strands of the AFLP fragments will result in doublets on the sequence gels, because the fragments will generally have a slightly differently mobility (Vos et al. 1995). These doublets will also occur when an alternative method of labeling is used which detects both strands.

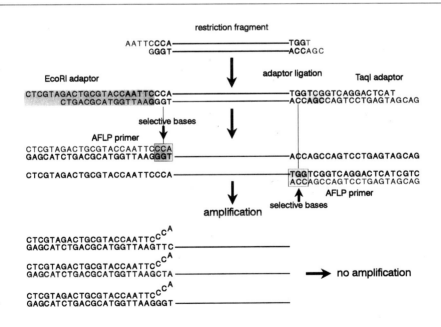

The final step of the AFLP technique is the analysis of the fingerprints. For this purpose the labelled reaction products are separated on denaturing polyacrylamide gels. These gels are very similar to those used for sequencing. After electrophoresis, gels are dried and exposed to X-ray films to visualize the AFLP fingerprints. Alternatively, AFLP images may be generated using phosphorimaging technology. Typically, fingerprints will contain 50-100 amplified restriction fragments.

The AFLP technique is a random fingerprinting technique that may be applied to DNA of any origin or complexity. The technique differs importantly from other random fingerprinting techniques (Williams et al. 1990, Caetano-Anollés et al. 1991, Welsh and McClelland 1990, 1991) by its robustness and reproducibility (Vos et al. 1995, Zabeau and Vos 1993, Riedy et al. 1992, Ellsworth et al. 1993, Muralidharan and Wakeland 1993, Micheli et al. 1994, Caetano-Anollés et al. 1992). The AFLP technique uses generic AFLP primers, which consist of two parts; one "common" part corresponding to the adapter and restriction site sequence and a unique part corresponding to the selective bases. AFLP primers are named "+0" when having no selective bases (only the common part), "+1" when having a single selective base, "+2" for having two selective bases etc. A limited set of AFLP primers will be sufficient to "create" a large set of different primer combinations, which in turn will each give a unique set of amplified fragments. This chapter describes a number of AFLP fingerprinting protocols for DNA varying widely in genome complexity. In the "overview" section the possibilities of the AFLP technique for application in genetic and genome analysis are addressed.

Materials

Laboratory equipment and materials

1. 0.2 mL PCR tubes (Perkin Elmer).
2. Base for 0.2 mL PCR tubes (Perkin Elmer).
3. Perkin Elmer 9600 thermal cycler (Perkin Elmer).
4. Gel unit for "sequencing" gels (e.g. Sequi-Gen 38 x 50 cm, BioRad).
5. High Voltage power supply (e.g. PowerPac 3000, BioRad).
6. X-ray film (Sakura) or phosphorimaging device (BAS 2000, Fujix; STORM 824, Molecular Dynamics).
7. Genomic DNA (*see* Notes 1 and 2).
8. *Eco*RI-adapter, 5 pmol/μL: 5′ -CTCGTAGACTGCGTACC
                                            CATCTGACGCATGGTTAA-5′
9. *Eco*RI-primer +0, 50 ng/μL: 5′ -GACTGCGTACCAATTC-3′
10. *Eco*RI-primer +1, 50 ng/μL: 5′ -GACTGCGTACCAATTCA-3′
11. *Eco*RI-primers +2, 50 ng/μL: 5′ -GACTGCGTACCAATTCAN-3′
12. *Eco*RI-primers +3, 50 ng/μL: 5′ -GACTGCGTACCAATTCANN-3′
13. *Pst*I-adapter, 5 pmol/μL: 5′ -CTCGTAGACTGCGTACATGCA
                                          CATCTGACGCATGT-5′
14. *Pst*I-primer +0, 50 ng/μL: 5′ -GACTGCGTACATGCAG-3′
15. *Pst*I-primer +1, 50 ng/μL: 5′ -GACTGCGTACATGCAGA-3′
16. *Pst*I-primers +3, 50 ng/μL: 5′ -GACTGCGTACATCGAGANN-3′
17. *Mse*I-adapter, 50 pmol/μL: 5′ -GACGATGAGTCCTGAG

TACTCAGGACTCAT-5'

18. *Mse*I-primer +1, 50 ng/μL: 5' -GATGAGTCCTGAGTAAC-3'
19. *Mse*I-primers +2, 50 ng/μL: 5' -GATGAGTCCTGAGTAACN-3'
20. *Mse*I-primers +3, 50 ng/μL: 5' -GATGAGTCCTGAGTAACNN-3'
21. *Taq*I-adapter, 50 pmol/μL: 5' -GACGATGAGTCCTGAG

TACTCAGGACTCGC-5'

22. *Taq*I-primer +A, 50 ng/μL: 5' -GATGAGTCCTGAGCGAA-3'
23. *Taq*I-primers +3, 50 ng/μL: 5' -GATGAGTCCTGAGCGAANN-3'

Solutions for AFLP reactions

1. 1 M Tris·HAc pH 7.5.
2. 1 M Tris·HCl pH 8.0.
3. 1 M Tris·HCl pH 8.3.
4. 0.1 mM $MgCl_2$.
5. 1 M KCl.
6. 0.5 M EDTA pH 8.0.
7. 1 M KAc.
8. 10× TE: 100 mM Tris·HCl, 10 mM EDTA pH 8.0.
9. $TE_{0.1}$: 10 mM Tris·HCl, 0.1 mM EDTA pH 8.0.
10. Milli-Q or double-distilled water (referred to as H_2O).
11. 100 mM DTT.
12. 10 mM ATP.
13. 5× RL (restriction-ligation) buffer: 50 mM Tris-HAc, 50 mM MgCl2, 250 mM KAc, 25 mM DTT, 250 ng/μL, pH 7.5.
14. 10× T4-buffer: 250 mM Tris·HCl pH 7.5, 100 mM $MgCl_2$, 50 mM DTT, 5 mM spermidine (3HCl-form).
15. 10× PCR buffer: 100 mM Tris·HCl pH 8.3, 15 mM $MgCl_2$, 500 mM KCl.
16. 5 mM of a mix of all 4 dNTPs (Pharmacia).
17. γ-^{32}P-ATP (~3000 Ci/mmol) or γ-^{33}P-ATP (~2000 Ci/mmol) (Amersham).
18. Restriction enzymes: *Eco*RI, *Pst*I, *Mse*I, *Taq*I (New England Biolabs).
19. T4 DNA ligase (Pharmacia).
20. T4 polynucleotide kinase (Pharmacia).
21. *Taq* DNA polymerase (Amplitaq; Perkin Elmer).
22. SequaMark™ molecular weight standard (10-base ladder; Research Genetics)

Solutions for gel electrophoresis

1. Formamide (deionized and filtered; Merck).
2. Urea (Gibco BRL).
3. 10× TBE (concentrated electrophoresis buffer): 1 M Tris base, 1 M boric acid, 20 mM EDTA.
4. 20% acrylamide, 1% methylene bisacrylamide in H_2O (Sequagel™, ready-to-use gelmix; National Diagnostics).
5. 4.5% acrylamide/methylene bisacrylamide (20:1) and 7.5 M urea in 0.5× TBE.
6. Ammonium persulphate (APS; Merck).

7. N,N,N',N'-tetramethylethylenediamine (TEMED; Pharmacia).
8. Bind silane (Pharmacia).
9. Repel silane (Pharmacia).
10. Bromophenol blue.
11. Xylene cyanol.
12. Loading dye: 98% formamide, 10 mM EDTA pH 8.0 and trace amounts of bromophenol blue and xylene cyanol.
13. 10% acetic acid solution.

Experimental protocol

Choice of methods

The method of choice depends largely on genomic size and structure. In the following table an overview is given of the correlation between genomic size and required template and amplification strategy.

TABLE 1. Examples of correlations between genome size, enzyme combination (EC), pre-amplification (PA) and amplification strategy (Amp) in various organisms. Genome size is indicated in megabases. Restriction enzymes include *Eco*RI (E), *Mse*I (M), *Pst*I (P) and *Taq*I (T). The number of selective bases for AFLP (pre)amplification is given in the last two columns.

Organisms	Genome size	EC	PA	AMP
1. Cosmids, BACs, PACs	0.01-0.1	E/M	–	0/0
2. YACs, microorganisms	0.1-1	E/M	–	0/1
3. Microorganisms	1-5	E/M	–	1/1
4. Microorganisms	5-20	E/M	–	1/2
5. Fungi	20-100	E/M	–	2/2
6. Plants, invertebrates	100-500	E/M	0/1	2/3
7. Plants	500-5000	E/M	1/1	3/3
8. Plants	›5000	P/M	1/1	3/3
9. Mammals, vertebrates	ca. 3000	E/T	1/1	3/3

Since the choice of DNA defines the required enzyme combination, preamplification strategy and amplification strategy, different protocols are given for five different applications (*see* Note 20):

    I.   Fingerprinting of microgenomes up to 100 megabases (Mb) (*see* Note 21).
    II. Fingerprinting of small genomes (100-500 Mb; *see* Note 22).
    III. Fingerprinting of medium-size genomes (500-5000 Mb; *see* Note 23).
    IV. Fingerprinting large genomes (over 5000 Mb; *see* Note 5).
    V. Fingerprinting mammalian genomes (*see* Note 24).

Modification of DNA and template preparation

A) Preparation of adapters

Site specific double-stranded (ds) adapters are made from the individual single-stranded (ss) oligonucleotides. The following procedures are used (*see* Note 3).

1. To prepare the *Eco*RI-adapter, mix 8.5 μg of the top strand and 9.0 μg of the bottom strand oligonucleotide (1500 pmoles of both oligonucleotides) in 300 μL of H_2O. This gives you 5 pmol/μL of *Eco*RI-adapter.

2. To prepare the *Pst*I-adapter, mix 10.5 μg of the top strand and 7.0 μg of the bottom strand oligonucleotide (1500 pmoles of both oligonucleotides) in 300 μL of H_2O. This gives you 5 pmol/μL of *Pst*I-adapter.

3. To prepare the MseI adapter, mix 8 μg of the top strand and 7 μg of the bottom strand oligonucleotide (1500 pMoles of both oligonucleotides) in 30 μL of H_2O. This gives you 50 pmol/μL of *Mse*I-adapter.

4. To prepare the *Taq*I-adapter, mix 8 μg of the top strand and 7 μg of the bottom strand oligonucleotide (1500 pmoles of both oligonucleotides) in 30 μL of H_2O. This gives you 50 pmol/μL of *Taq*I-adapter.

B) Digestion of DNA with restriction enzymes and adapter ligation

Procedure I, II and III: All three procedures use identical template preparation protocols and are optimized for the fingerprinting of: I. microgenomes up to 100 Mb, (e.g. cosmids, PACs, BACs, microorganisms, fungi); II. small genomes (100-500 Mb; e.g. arabidopsis, rice, cucumber, peach, nematodes); and III. medium-size genomes (500-5000 Mb; e.g. brassica, oilseed rape, tomato, lettuce, pepper, cotton, sunflower, potato, maize, rye, barley).

1. Digest genomic DNA with the corresponding restriction enzymes in a 40 μL reaction containing 8 μL 5× RL buffer, 5 units *Eco*RI, 2 units *Mse*I, H_2O and DNA to 40 μL (*see* Notes 1 and 2). Mix well and incubate for 2 h at 3°C.

2. To ligate the adaptors to the digested genomic DNA add the following components to the reaction mix and incubate another 2 h at 37°C: 1 μL *Eco*RI adapter (5 pmol), 1 μL *Mse*I adapter (50 pmol), 1 μL 10 mM ATP, 2 μL 5× RL buffer, 1 unit T4 DNA ligase, 5 μL H_2O (*see* Note 4).

3. After ligation the reaction mixture is diluted 10 times with $TE_{0.1}$. Generally 10 μL of the reaction mixture is diluted to 100 μL, and the remaining 40 μL is stored as backup. The diluted reaction mixture is used directly as template DNA for the AFLP reactions. Both the diluted DNA and undiluted DNA are stored at -20°C. This diluted reaction mixture will be referred to as 'template'.

Procedure IV: Fingerprinting large genomes (>5,000 Mb e.g. wheat, onion; *see* Note 5).

1. Digest genomic DNA with the corresponding restriction enzymes in a 40 μL reaction containing 8 μL 5× RL buffer, 5 units *Pst*I, 2 units *Mse*I, H_2O and DNA to 40 μL (*see* Note 2). Mix well and incubate for 2 h at 37°C.

2. To ligate the adaptors to the digested genomic DNA add the following components to the reaction mix and incubate another 2 h at 37°C: 1 μL *Pst*I adapter (5 pmol), 1 μL *Mse*I adapter (50 pmol), 1 μL 10 mM ATP, 2 μL 5× RL buffer, 1 unit T4 DNA ligase, 5 μL H_2O (*see* Note 3).

3. Dilution and storage are as described under procedure I.

Procedure V: Fingerprinting mammalian genomes and other vertebrates (e.g. humans, other primates, pig, cattle, horse, dog, mouse, rat, rabbit, chicken, duck; *see* Note 6).

1. Digest genomic DNA with the corresponding restriction enzymes in a 40 μL reaction containing 8 μL 5× RL buffer, and 5 units *Taq*I. Initially, the incubation is carried out at 65°C, for 90 min. Next, the reaction is transferred to 37°C, 5 units *Eco*RI are added and the reaction is continued for 1 h.
2. To ligate the adaptors to the digested genomic DNA add the following components to the reaction mix and incubate another 3 h at 37°C: 1 μL *Eco*RI adapter (5 pmol), 1 μL *Taq*I adapter (50 pmol), 1 μL 10 mM ATP, 2 μL 5× RL buffer, 1 unit T4 DNA ligase, 5 μL H_2O (*see* Notes 3 and 25).
3. Dilution and storage are as described under procedure I.

AFLP preamplification

Procedure I: No preamplification is required, the template can be used directly for AFLP. The genomic complexity is sufficiently low to get excellent quality fingerprints by amplifying directly from the template.

Procedure II:

1. The preamplification reaction is performed in 50 μL. A typical preamplification reaction contains the following components: 5 μL template-DNA, 1.5 μL *Eco*RI-primer +0 (75 ng), 1.5 μL *Mse*I-primer +C (75 ng), 2 μL 5 mM dNTPs, 0.2 μL Taq polymerase (1 unit), 5 μL 10× PCR-buffer, 34.8 μL H_2O (*see* Note 7). For preamplification 20 PCR cycles are performed using the following cycle profile: 30 s at 94°C; 60 s at 56°C; 60 s at 72°C (*see* Note 8).
2. After preamplification, 10 μL of the reaction is diluted with 190 μL of $TE_{0.1}$ to 100 μL which is sufficient for 40 AFLP-reactions +2/+3. The diluted reaction mix and the rest of the preamplification reaction is stored at -20°C. If necessary new dilutions of the preamplification reactions may be made to give additional template for the AFLP reactions.

Procedure III: The preamplification reaction is performed as described under II, with the exception that now the reaction mixture contains the *Eco*RI one-base extension primer - A; 1.5 μL *Eco*RI-primer +A (75 ng) is added together with 1.5 μL *Mse*I-primer +C (75 ng).

Procedure IV: The preamplification reaction is performed as described under II, with two exceptions: now the reaction mixture contains 1.5 μL of the *Pst*I one-base extension primer -A (75 ng) together with 1.5 μL *Mse*I-primer +C (75 ng), and the complete cycle profile is more extensive (*see* Note 9). The amplification program is: Cycle 1: 30 s 94°C, 60 s 65°C, 60 s 72°C; Cycle 2-13: Same PCR profile as cycle 1, except for a stepwise decrease of the annealing temperature in each subsequent cycle by 0.7°C during 12 cycles; and Cycle 14-36: 30 s 94°C, 30 s 56°C, 60 s 72°C.

Procedure V: The preamplification reaction is performed as described under II, with the exceptions that now the reaction mixture contains 1.5 μL of the *Eco*RI one-base extension primer -A (75 ng) together with 1.5 μL *Taq*I-primer +A (75 ng), and the number of preamplification cycles is 30.

AFLP amplification

A general AFLP amplification procedure is given. Two AFLP primers are used, one of which is labellled (*see* Note 10). These primers are the only variables in the different procedures. Their identity is given in Table 2 below.

1. Primers for selective AFLP amplification are labelled by phosphorylating the 5' end of the primers with γ-^{32}P-ATP or γ-^{33}P-ATP and polynucleotide kinase (*see* Note 10). Prepare the following primer labelling mixes using either γ-^{32}P-ATP or γ-^{33}P-ATP to generate sufficient labelled primer for 100 AFLP reactions. Labelling mix (40 μL) for ^{32}P: 20 μL γ-^{32}P-ATP (~3,000 Ci/mmol), 5 μL 10× T4-buffer, 2 μL T4-kinase (10 units/μL), 13 μL H$_2$O. Labelling mix (40 μL) for ^{33}P: 10 μl γ-^{33}P-ATP (~2,000 Ci/mmol); 5 μL 10× T4-buffer, 2 μL T4-kinase, 23 μL H$_2$O.

2. For primer labelling add the labelling mix to 10 μl of primer 1 (either *Eco*RI- or *Pst*I-primers at 50 ng/μL) to give 50 μL (*see* Note 11) and incubate 60 min at 37°C, followed by incubation at 70°C for 10 min (inactivation of the kinase). This gives a labelled primer with a concentration of 10 ng/μL.

3. Prepare the following AFLP reaction mixes for 10 reactions (*see* Note 12). Primers-dNTPs mix (50 μL): 5 μL labelled primer 1 (10 ng/μL), 6 μL unlabelled primer 2 (50 ng/μL), 8 μL 5 mM dNTPs, 31 μL H$_2$O. Taq polymerase mix (100 μL): 20 μL 10× PCR-buffer, 0.8 μL Taq polymerase (4 units), 79.2 μL H$_2$O.

4. The PCR reaction is assembled by adding 5 μL primers-dNTPs mix and 10 μL Taq-polymerase mix to 5 μL preamplified template DNA in the PCR tubes (*see* Notes 13, 14 and 15).

The AFLP amplification is carried out in a PE-9600 thermal cycler with the following cycle profile. Cycle 1: 30 s 94°C, 30 s 65°C, 60 s 72°C; Cycle 2-13: Same PCR profile as cycle 1, except for a stepwise decrease of the annealing temperature in each subsequent cycle by 0.7°C during 12 cycles, Cycle 14-36: 30 s 94°C, 30 s 56°C, 60 s 72°C (*see* Notes 16 and 17).

TABLE 2: Identity of primers used in the different procedures.

Procedure	Primer 1	Primer 2
A	E +0, +1, +2	M +0, +1, +2
B	E +2	M +3
C	E +3	M +3
D	P +3	M +3
E	E +3	T +3

E, *Eco*RI; M, *Mse*I; P, *Pst*I; T, *Taq*I

Overview of recommended primer extensions

In Table 3, a detailed diagram is provided that spells out the different primer combinations that we recommend for AFLP analysis of genomes of widely different complexity. These primers will ultimately all become available through proper distributors, in the form of AFLP kits (*see* Note 20).

TABLE 3: Overview of amplification strategies to be used for small genomes and plant genome (<5,000 Mb). Appropriate primer extensions for *EcoRI/MseI* templates are given for analysis of specific genomes. EcoRI primer extensions are given above the diagram, *MseI* primer extensions are given vertically, in front of the diagram.

MseI \ *Eco*RI	+0	+1 A, G, C, T	+2 AA, AC, AG, AT, TA, TC, TG, TT	+3 AAC, AAG, ACA, ACC, ACG, ACT, AGC, AGG
+0	cosmids PACs BACs	PACs BACs YACs		
+1 A C G T	PACs BACs YACs	YACs bacteria	yeast bacteria	
+2 AA AC AT AG CA CC CG CT		yeast bacteria	fungi bacteria microorganisms	
+3 CAA CAC CAG CAT CTA CTC CTG CTT			Arabidopsis rice cucumber peach	oilseed rape brassica tomato lettuce pepper maize barley sunflower potato

NOTE: YAC AFLP fragments may be detected in purified YACs or by fingerprinting of YAC clones and identification of YAC-specific AFLP fragments by comparison to yeast host strain fingerprints.

Gel analysis of AFLP reaction products

1. The AFLP reaction products are analyzed on 4.5% denaturing polyacrylamide gels (*see* Note 18). The gels should be cast at least 2 h before use and should be prerun for 0.5 h just before loading the samples, 1× TBE is used as running buffer (*see* Note 19). Prerunning and running of the gel is performed at 110 W with the BioRad system (no

limits for voltage and current). This warrants an even heat distribution during electrophoresis, which is crucial for good quality fingerprints.

2. Mix AFLP reaction products with an equal volume (20 μL) of loading dye after the PCR is finished.

3. Heat the resulting mixtures for 3 min at 90°C, and then quickly cool on ice. Beware that formamide is a harmful, teratogenic chemical substance so use a fume-hood.

4. Rinse the surface of the gel well with 1× TBE and push carefully two 24-well sharktooth combs about 0.5 mm into the gel surface to create the gel slots.

5. Rinse the gel slots formed in this way with 1× TBE and load 2 μL of each sample per well. Each gel allows for simultaneous electrophoresis of 49 samples.

6. After electrophoresis disassemble the gel cassette. The gel will stick to the front glass plate because of the silane treatment.

7. Fix the gel by soaking in 10% acetic acid for 30 min and dry it subsequently at room temperature in a fume hood for 10-20 h (Gel drying is faster at elevated temperatures).

8. Autoradiograph the gel. Generally, exposure of ^{32}P-gels to standard X-ray film overnight, without intensifying screens, gives good results. ^{33}P-gels take 2 to 3 d exposure to generate similar band intensities. Exposure times are reduced at least 2-fold using phosphorimaging technology.

Notes

1. DNA preparations need to be of sufficient quality to allow complete digestion by the restriction enzymes. Often, contaminating agents will be co-purified together with the DNA. When very little DNA is expected the ratio of contaminants over DNA may be such that it interferes with proper digestion. Especially DNA preparations of *Arabidopsis* (and possibly other plants with small genomes) are often of poor quality and have low yields. However, when only 50 ng of DNA is used for AFLP template preparation there usually is no problem. Complete restriction is crucial for good quality AFLP fingerprints. We have found that purification with glass beads (BIO101) generally provides a good way to remove impurities which cause incomplete restriction.

2. For all genomes larger than 500 Mb we use 100 ng DNA, for template preparations. DNA concentrations can be determined by measuring OD_{260}, but we advise estimating 'gellograms': the running of a small aliquot of DNA on a 1% agarose gel, next to a series of phage λ DNA dilutions ranging from 50 ng to 500 ng. The agarose image will also allow an inspection of the integrity of the DNA. Substantial smearing below the main band of high molecular weight DNA may be detrimental for AFLP fingerprinting quality.

3. The *Eco*RI, *Mse*I, *Pst*I, and *Taq*I adapters have double-stranded parts of 14, 12, 14 and 12 base pairs respectively. It appears unnecessary to perform a specific denaturation-renaturation procedure to anneal the two strands of the adapters.

4. DNA is incubated for a total of 4 h with restriction enzymes, the last 2 h in the presence of T4 DNA ligase and oligonucleotide adapters. The adapters do not restore the restriction sites and, hence, the presence of the restriction enzymes in the ligation

step results in almost complete adapter-to-fragment ligation because of the restriction of fragment concatemers formed by ligation. However, prolonged incubation with restriction enzyme is not recommended because of possible "star" activity of *Eco*RI giving reduced cleavage specificity and, ultimately, aberrant AFLP fingerprints.

5. It is difficult to generate good quality fingerprints from plants with very large genomes. Especially wheat is a difficult crop. Although procedure D may give satisfactory results with some varieties, it should not be considered a solid definite protocol.

6. In mammalian DNA the use of *Taq*I permits the specific targeting of CpG dinucleotide motifs. Since such motifs are prone to an increased level of mutation an increased polymorphism is observed. Although *Taq*I cuts the 4-base sequence TCGA, it is not a frequent cutter *per se*. *Taq*I fragments from most mammals are often in the same size range as *Eco*RI fragments. The combination of *Eco*RI and *Taq*I therefore does not produce fragments that can all be amplified well. A comparison of fingerprints based on the enzyme combinations *Eco*RI/*Mse*I and *Eco*RI/*Taq*I shows, that from the *Eco*RI/*Taq*I template about 5-fold fewer fragments can be visualized than from the *Eco*RI/*Mse*I template.

7. Working with mixes of reagents is preferred. The following procedure is suggested for a 50 μL reaction:

     5 μL template-DNA
    Mix A (25 μL): 1.5 μL *Eco*RI-primer +0 (75 ng)
                 1.5 μL *Mse*I-primer +C (75 ng)
                 2 μL 5 mM dNTPs
                 20 μL H_2O
    Mix B (20 μL): 0.2 μL *Taq* polymerase (1 unit)
                 5 μL 10× PCR-buffer
                 14.8 μL H_2O

8. Non-phosphorylated adapters have been used for adapter ligation. Therefore, only one strand of the adapter binds to the DNA, which means that the adapters are ligated to opposite strands. This would prevent amplification, if the template-DNA would be denatured prior to the start of PCR. This should never be done. The recessed 3' ends of the template DNA are filled-in by the Taq polymerase during heating to 94°C of the first PCR-cycle.

9. With highly complex genomes (>5000 Mb) 20 cycles of preamplification may be insufficient. To be sure that we get enough preamplified material an excess of cycles (13 touch-down cycles and 24 additional 56°C cycles) is used identical to the final AFLP cycle profile.

10. Only one of the two primers of the AFLP reaction should be labelled (there is no preference which one), generally the EcoRI-primer is selected. Mobility of the two strands of a DNA fragment on sequencing gels is generally slightly different. Therefore, care should be taken in comparing AFLP fingerprints obtained with

different labelled primers. ^{33}P-labelled primers are preferred because they give a better resolution of the PCR products on polyacrylamide gels. Also, the reaction products are less prone to degradation due to autoradiolysis. However, the use of ^{33}P-labelled primers is more expensive compared to ^{32}P-labelled primers.

11. It is also possible to divide the reaction mix over several primers if preferred, e.g. 5× 8 μL of the reaction mix may be added to 5× 2 μL (100 ng primer) of 5 different primers.

12. Working with AFLP reaction mixes is important for the reliability and reproducibility of AFLP reactions. It also facilitates the assembly of the AFLP reactions. Reaction mixes should be prepared for a minimum of 10 reactions to warrant accurate pipetting. Generally, several AFLP reactions are performed simultaneously and the quantities of the reaction mixes should be adjusted accordingly.

13. Two mixes are added to the preamplified template DNA: a mix with AFLP-primers and dNTPs and a mix with *Taq* polymerase (including the 2× concentrated PCR buffer).

14. Working with mixes is not only convenient but also essential to allow a rapid start of the AFLP reactions after all reagents have been pipetted together. The reaction is assembled at room temperature and although the activity of the Taq polymerase is low at ambient temperatures, the AFLP fingerprint quality suffers from a long lag time between assembly of the AFLP reaction mixtures and start of the PCR.

15. The PCR tubes are held by a "base plate" in a 8×12 format (96 tubes). The template DNA is pipetted first followed by the two mixes. The use of multi- channel pipettors is advised to allow rapid pipetting. The reagents are mixed by tapping the base plate with the tubes on the bench. After mixing of the reagents the AFLP reaction is started as soon as possible.

16. The AFLP-PCR is started at a rather high annealing temperature to obtain optimal primer selectivity. In the following steps the annealing temperature is lowered gradually to a temperature at which efficient primer binding occurs. This temperature is then maintained for the rest of the PCR cycles. The Perkin Elmer PE-9600 thermal cycler allows easy programming of the PCR profile described below.

17. The PE-9600 thermal cycler is favoured for performing AFLP reactions. Other thermal cyclers may be used but may not have the possibility to program the AFLP-PCR cycle profile. Alternatively, the annealing temperature may be decreased 1°C in each subsequent cycle during the first 10 cycles, gradually decreasing the annealing temperature from 65°C in the first cycle to 56°C from cycle 10 onwards.

18. These gels are essentially normal sequencing gels (Maxam and Gilbert 1980), with the exception that a lower percentage of polyacrylamide is used. Gels are cast using the instructions of the manufacturer of the gel system. We prefer a BioRad SequiGen sequencing gel system (38×50×0.04 cm), but there is no reason why other sequencing

gel systems should not work equally well. The back plate of the gels, the so called "integrated plate chamber" (IPC), is treated with 2 mL of repel silane. The front plate is treated with 10 mL of bind silane solution (30 μL acetic acid and 30 μL bind silane in 10 mL ethanol, freshly made immediately before use). The silane treatments cause the gels to stick to the front plate upon disassembly of the gel cassette after electrophoresis. The SequiGen sequence gels require about 100 mL of gel solution to which 500 μL of 10% APS (freshly made before use) and 100 μL of TEMED is added immediately before casting the gels.

19. The purpose of the prerun is to "warm up" the gel to about 55°C. This temperature is maintained throughout electrophoresis.

20. Both Applied Biosystems (ABI) and Life Technologies (LTI) have licenses for the production of specific AFLP kits. ABI focuses on the production of AFLP kits for fluorescent detection, LTI focuses on kits for radioactive detection. For a detailed protocol on how to prepare reactions for fluorescent detection ABI provides an excellent manual.

21. For the analysis of 'micro' genomes a special AFLP kit is under development (*see* Note 20). For such genomes a special set of primer combinations is needed, with shorter 3' extensions. This kit will contain adapters and a series of +0, +1 and +2 primers as described in Table 3.

22. For the analysis of plants with small genomes a kit is available both from ABI and LTI. These kits contain sets of 8 *Eco*RI +2 primers and 8 *Mse*I +3 primers, as described in Table 3. In addition, the kits contain adapters and preamplification primers. The small genome AFLP primer matrix can also be used on other crops with small genomes. We have excellent quality fingerprints with rice and cucumber. It is important to note that primer combinations described here should not be used on yeasts and microorganisms.

23. A kit for the analysis of plants with medium size genomes is available from both ABI and LTI. These kits contain sets of 8 *Eco*RI +3 primers and 8 *Mse*I +3 primers, as described in Table 3.

24. An "animal" kit is under development at ABI for the analysis of the mammalian genome. This kit will contain sets of 8 *Eco*RI +3 primers and 8 *Taq*I +3 primers, and the necessary adapters and pre-amplification primers.

25. Although *Taq*I is not a frequent cutter, for good quality fingerprints one needs to use a rather high concentration of adapters. Presumably this is because *Taq*I will not efficiently re-digest aberrant fragment-ligation products at 37°C.

26. We have developed an image analysis software package dedicated to the analysis of AFLP images. A full description of this software will be presented elsewhere (manuscript in preparation). Briefly, the software allows the identification and measurement of specific AFLP bands in a pixel image. The origin of the pixel image

can be either a phosphor-imager, a fluor-imager or a sequencer. All these scanner-types provide a dynamic range of sensitivity sufficient for accurate estimation of band intensities. Our software samples a pixel row over the peaks of bands and the total of the pixel information between lane boundaries is used for quantification of the individual bands. As a result, a band can be scored present or absent. This software is commercially available through proper distributors. With refined quantification procedures even heterozygosity (corresponding to band intensities 50% of homozygous bands) can be determined (manuscript in preparation).

27. AFLP is a trademark filed by Keygene N. V.

Overview

AFLP fingerprinting

In this chapter we have described a number of AFLP fingerprinting protocols for genomic DNA of varying origins and complexities. DNA fingerprints can be obtained from any DNA using a limited set of generic primers, clearly demonstrating the versatility of the AFLP technique. The AFLP technique has two key elements: (i) the template DNA are restriction fragments and the PCR products are amplified restriction fragments, and (ii) AFLP primers contain a variable 3' end which is employed to select a distinct set of fragments to be amplified. These two characteristics enable the AFLP reactions to be carried out at stringent PCR conditions which makes AFLP a robust and reliable technique. The drawback of using restriction fragments as the source for PCR amplification is that DNA of reasonable quality has to be used which has to be digested completely with the selected restriction enzymes.

One additional feature of AFLP is that the reaction appears to be quantitative. The quantitative nature of the AFLP reaction may be attributed to the observation that the labelled AFLP primer is completely consumed during the AFLP reaction at which point the amplification of labelled PCR products stops. Therefore, the reaction does not seem to reach a stage at which substantial differences in fragment amplification efficiencies occur. In addition, differences in DNA concentrations of individual samples at the start of AFLP reactions are generally not reflected in the final AFLP fingerprints because of the total "consumption" of the labelled primer. The most important application of this feature of AFLP is that it enables codominant scoring of AFLP markers (determine allele zygosity) based on band intensities.

AFLP markers

AFLP fingerprints will often be used as a source for DNA markers. The molecular basis for AFLP polymorphisms will most frequently be sequence polymorphisms at the nucleotide level. Single nucleotide changes will be detected by AFLP when either restriction sites themselves or nucleotides adjacent to the restriction sites are affected, causing the AFLP primers to mispair at the 3' end and preventing amplification (the selective nucleotides do not exactly match the sequence next to the restriction site). In addition, deletions, insertions and rearrangements affecting the presence or size of restriction fragments will lead to polymorphisms detected by AFLP. Most AFLP markers will be mono-allelic markers, the corresponding allele is not detected. With a low frequency bi-allelic markers will be identified due to small insertions or deletions in the restriction fragments.

The frequencies with which AFLP markers are identified are similar to what is found with RFLP and arbitrarily-primed PCR-based DNA marker techniques. The advantage of AFLP over these techniques lies in the high marker densities which can be obtained with AFLP with modest efforts. Most AFLP markers will inherit in a Mendelian way indicating that they are unique DNA fragments. This only means that the restriction fragments themselves are unique, the internal sequences will often contain repetitive elements as has been shown by hybridization studies (Vos and Frijters, unpublished results). The use of AFLP markers as hybridization probes, therefore, has certain limitations.

Applications of AFLP

AFLP markers can be used for any type of genetic study to which other DNA marker systems are employed. AFLP markers have a number of specific advantages:
1. No prior sequence knowledge is required. This makes the technique very suitable for applications like biodiversity studies, analysis of germplasm collections and genetic relationship studies.
2. AFLP markers can be scored codominantly. This makes the technique more suitable for genetic mapping studies than arbitrarily-primed PCR-based marker techniques.
3. High marker densities can be obtained with modest efforts. This makes the technique very useful in positional cloning and other types of research requiring marker saturation.
4. AFLP markers can be detected in almost any background or complexity. This enables the use of AFLP to detect DNA markers both in genomic DNA and in clones of genomic DNA or pools of these clones. This means that there is no need to convert the AFLP markers for identification of corresponding genomic clones.

The technique has also a number of limitations:
1. AFLP fingerprints will share very few common fragments when sequence homology is less than 90%. AFLP can, therefore, not be used for comparative genome analysis which requires a hybridization-based probe. Fingerprints of related bacteria may also show very few common bands, because these genomes evolve very quickly due to short generation times. It has indeed been shown in bacteria that relationships at the subspecies level may not be detected by AFLP (Janssen et al. 1996).
2. Detection of markers in genomic DNA with very little sequence variation may still be poor (like with most other marker systems) despite of the large amounts of fragments that can be tested for polymorphisms. In these cases, a marker system which targets hypervariable sequences (Jeffreys et al. 1985, 1991, Nakamura et al. 1987, Tautz 1989, Weber and May 1989, Edwards et al. 1991, Beyermann et al. 1992) may still be superior.

AFLP fragments may also be used as physical markers. AFLP fingerprints are generated from genomic clones and these fingerprints are subsequently used to determine the overlaps and positions of the clones. AFLP is presently one of the most suitable techniques for YAC fingerprinting and YAC mapping. For this purpose also complete YAC clones may be used, each YAC clone fingerprint displaying a few YAC-specific AFLP fragments in the background of yeast genomic AFLP fragments.

The AFLP technique has the unique property that the AFLP markers can be used as both genetic and physical DNA markers, which allows the integration of genetic and physical maps. This makes AFLP a powerful new enabling technology in genome research.

References

Beyermann B, Nurnberg P, Weihe A, Meixner M, Epplen JT and Bîrner T (1992) *Theor Appl Genet* 83:691-694.
Caetano-Anollés G, Brassam BJ and Gresshoff PM (1991) *Bio/Technology* 9:553-557.
Caetano-Anollés G, Brassam BJ and Gresshoff PM (1992) *Mol Gen Genet* 235:157-165.
Edwards A, Civitello A, Hammond HA and Caskey CT (1991) *Am J Hum Genet* 49:746-756.
Ellsworth DL, Rittenhouse KD and Honeycutt RL (1993) *BioTechniques* 14:214-217.
Janssen P, Coopman R, Huys G, Swings J, Bleeker M, Vos P, Zabeau M and Kersters K (1996) *Microbiology* 142:1881-1893
Jeffreys AJ, McLeod A, Tamaki K, Neil DL and Monckton DG (1991) *Nature* 354:204-209.
Jeffreys AJ, Wilson V and Thein SL (1985) *Nature* 314:67-73.
Maxam AM and Gilbert W (1980) *Methods Enzymol* 65:499-560.
Micheli MR, Bova R, Pascale E and D'Ambrosio E (1994) *Nucleic Acids Res* 22:1921-1922.
Muralidharan K and Wakeland EK (1993) *BioTechniques* 14:362-364.
Nakamura Y, Leppert M, O'Connell P, Wolff R, Holm T, Culver M, Martin C, Fujimoto E, Hoff M, Kumlin E and White R (1987) *Science* 235:1616-1622.
Riedy MF, Hamilton WJ III and Aquadro CF (1992) *Nucleic Acids Res* 20:918.
Tautz D (1989) *Nucleic Acids Res* 17:6463-6472.
Vos P, Hogers R, Bleeker M, Reijans M, van de Lee T, Hornes M,Frijters A, Pot J, Peleman J, Kuiper M and Zabeau M (1995) *Nucleic Acids Res* 23:4407-4414.
Weber JL and May PE (1989) *Am J Hum Genet* 44:388-396.
Welsh J and McClelland M (1990) *Nucleic Acids Res* 18:7213-7218.
Welsh J and McClelland M (1991) *Nucleic Acids Res* 19:861-866.
Williams JGK, Kubelik AR, Livak KJ, Rafalski JA and Tingey SV (1990) *Nucleic Acids Res* 18:6531-6535.
Zabeau M and Vos P (1993) Selective Restriction Fragment Amplification: A general method for DNA fingerprinting. European Patent Application EP 0534858.

Direct amplification from microsatellites: detection of simple sequence repeat-based polymorphisms without cloning

Julie M. Vogel[1] and Pablo A. Scolnik[2]
Agricultural Products[1], and Central Research and Development[2], E. I. du Pont de Nemours and Co. Inc., Wilmington, DE 19880-0402, USA

Introduction

Reliable methods for estimating genetic diversity within and between species are crucial for studying populations, classifying germplasm, breeding and identifying genes conferring particular traits. A comparison of the common DNA marker systems has been presented elsewhere (Rafalski et al. 1996; *see* Chapter 1). Of the available methods to identify alleles simultaneously at multiple, interspersed loci in a genome, perhaps the most widely applied involves polymerase chain reaction (PCR) with single primers of arbitrary sequence. This general method, variously termed random amplified polymorphic DNA (RAPD; Williams et al. 1990), arbitrarily primed PCR (AP-PCR; Welsh and McClelland 1990), or DNA amplification fingerprinting (DAF; Caetano-Anollés et al. 1991), provides good discrimination within and between species and allows analysis of whole genomes without prior DNA sequence information. Despite the multiplexed DNA fingerprints produced, these methods generate dominant markers, reveal few alleles and usually require careful optimization to assure reproducibility (Penner et al. 1993, Williams et al. 1993).

Methods that detect codominant polymorphisms at simple sequence repeat (SSR) microsatellite loci offer good reproducibility and generate large numbers of detectable alleles (Litt and Luty 1989, Tautz 1989, Weber and May 1989, Weber 1990). SSR repeats are arrays of mono-, di-, tri-, tetra- or penta-nucleotide units widely dispersed throughout animal (Beckmann and Weber 1992, Hudson et al. 1992, Stallings et al. 1992, Buchanan et al. 1993, Ostrander et al. 1993, Dietrich et al. 1994) and plant (Morgante and Olivieri 1993, Wu and Tanksley 1993, Bell and Ecker 1994, Cregan et al. 1994, Wang et al. 1994) genomes that display high levels of genetic variation based on differences in the number of tandemly repeating units at a locus. Oligonucleotides composed wholly of these defined short tandem repeat sequences [for example $(GATA)_4$] have been used extensively as hybridization probes on Southern blots to generate highly discriminating fingerprints (Ali and Epplen 1991, Vosman et al. 1992, Ramakishana et al. 1994). Alternatively, SSR length polymorphisms at individual loci are detected by PCR using locus-specific flanking region primers (Tautz 1989, Weber and May 1989). These assays typically carry a high information content (Weber and May 1989, Weber 1990, Cregan et al. 1994, Morgante et al. 1994) and have been extremely useful for mapping and gene discovery efforts in humans (Beckmann and Weber 1992) and rodents (Love et al. 1990, Cornall et al. 1991)[reviewed in Hearne et al. 1992, Todd 1992]. However, these markers require precise DNA sequence information for each marker locus, from which a pair of identifying flanking primers are designed. This is impractical for many plant and animal species that are not already well characterized genetic systems and for which the resources needed to carry out a labor-intensive SSR cloning and sequencing effort are not available. Nevertheless, the high levels of polymorphism inherent to SSRs has prompted an alternative approach: the detection of SSR-derived polymorphisms without the need for up-front cloning. These methods constitute variations on a common theme: PCR-based multiplexed detection of SSR-based genetic polymorphisms without the need to discover the unique genomic sequences flanking each repeat. This review will focus on these microsatellite-based DNA fingerprinting methods.

The approaches described below all involve PCR amplification fingerprinting with oligonucleotide primers corresponding directly to simple sequence repeat target sites. Some of these SSR-based methods, here collectively termed microsatellite-primed PCR (MP-PCR) are reminiscent of similar approaches using PCR with single primers targeted to defined interspersed repetitive elements, such as *Alu* (Nelson et al. 1989, Sinnett et al. 1990, de Souza et al. 1994) or SINE repeats (Ellegren and Basu 1995). Collectively, these methods are based on three principles: (i) that any particular repeat is widely represented and interspersed in a genome; (ii) that at some measurable frequency, adjacent copies of a particular repeat will be in inverse orientation within a PCR-amplifiable distance from one another in the genome; and (iii) that a detectable polymorphism will result from insertions-deletions between conserved repeats or sequence variation at a priming site. Some methods involve inter-repeat amplification using a single primer, whereas others involve PCR between a repeat and a non-repeat priming site. In any case, repeat-directed amplification reactions have proved extremely useful for DNA fingerprinting, for new polymorphic marker identification and for germplasm discriminations in plant, animal and fungal genomes.

In this review, we will provide an overview and comparision of the different ways in which microsatellite-primed PCR has been applied, and will attempt to offer practical suggestions on the best uses, and limitations, of each. A schematic representation of these related methods is shown in Figure 1.

Linear primer extension

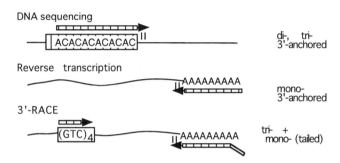

DNA sequencing

ACACACACACAC

di-, tri-
3'-anchored

Reverse transcription

AAAAAAAAA

mono-
3'-anchored

3'-RACE

(GTC)₄   AAAAAAAAA

tri- +
mono- (tailed)

PCR amplification (MP-PCR)

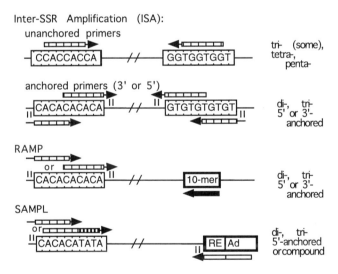

Inter-SSR Amplification (ISA):

unanchored primers

CCACCACCA   //   GGTGGTGGT

tri- (some),
tetra-,
penta-

anchored primers (3' or 5')

CACACACACA   //   GTGTGTGTGT

di-, tri-
5' or 3'-
anchored

RAMP

or

CACACACACA   //   10-mer

di-, tri-
5' or 3'-
anchored

SAMPL

or

CACACATATA   //   RE Ad

di-, tri-
5'-anchored
or compound

FIGURE 1. Schematic representation of applications using direct priming with microsatellite based primers. Types of primers reported as successful are listed at right. See text for references. RE, restriction enzyme; Ad, synthetic adaptor, 10-mer, arbitrary sequence decamer.

The simplest case: SSR-directed primer extension for DNA sequencing and cDNA cloning

During the course of conventional SSR marker discovery, it is necessary to determine the sequence of cloned DNA fragments containing microsatellite repeats. Although vector-directed primers are used most often, an alternative approach has been to use primers that correspond directly to either the 'forward' or 'complementary' strand of the repeat itself to primer extend the polymerization into the nearby flanking DNA (*see* Figure 1). This approach has been applied to DNA sequencing both with T7 polymerase (Sequenase) in a conventional sequencing reaction (Yuile et al. 1991, Browne and Litt 1992) and with *Taq* polymerase during cycle sequencing (Hawkins et al. 1994). In all cases, it was necessary to anchor the repeat primer to a fixed position at the target site through the addition of nonrepeat 3'-base(s) on the primer. This ensures that all the extension events initiate at a common nucleotide, rather than in 2 bp intervals, and thus gives uniform sets of dideoxy-terminated fragments that resolve as single bands on the gel. First reported was the use of dinucleotide repeat primers carrying a single, nondegenerate 3'-anchor nucleotide (Yuile et al. 1991). With this primer set, $(CA)_7D$, $(AC)_7B$, $(GT)_7H$ and $(TG)_7V$ (where B=any base except A, D=any base except C, H=any base except G, V=any base except T), it was necessary to sequence each unknown template individually with a set of 12 primers (6 for each strand), even though only 2 primers would successfully prime (one per strand) from any given template. Later approaches used primer mixtures carrying degeneracy at the penultimate 3'-positions with nondegeneracy at the 3'-ultimate position (Browne and Litt 1992), or approaches carrying only a single, nondegenerate 3'-base (Hawkins et al. 1994). For the latter, only 6 primers are required to ensure complete sequence from every possible template. These methods were reported to generate flanking sequence information successfully, although all leave the repeat itself not sequenced. Such approaches may be most useful when the flanking sequences of a repeat are needed from a long fragment, such that sequencing in from the vector ends of a large insert clone may not provide sufficient information close to an SSR several hundred nucleotides away. This general approach is useful for directly sequencing uncloned DNA fragments containing SSR repeats, such as are generated from many of the amplification-based methods described below.

Primer extensions directly from repeat sequence priming sites also have been routinely performed during cDNA cloning and mRNA analysis. The use of oligo-d$(T)_n$ to prime from the poly$(A)_n$ tails of mRNA is widely used for conventional first-strand synthesis of cDNA for library construction, differential display (Liang and Pardee 1992) and 3'-RACE PCR (Frohman et al. 1988; *see* Figure 1). In many cases, the d$(T)_n$ primer used is 'un-anchored', although many more recent methods involve the use of d$(T)_n$ primers carrying non-T bases of either degenerate or nondegenerate composition at their 3'-ends (Liang and Pardee 1995). At least one variation of this standard cDNA cloning protocol is a 3'-RACE PCR method that used a $(CAG)_4$ primer to prime second strand synthesis and isolate from human cells those cDNA carrying specifically this repeat (Carney et al. 1995). Although not reported, second strand synthesis should also be possible using additional repeat types as forward primers.

Microsatellite-directed PCR: unanchored primers

Oligonucleotides composed wholly of defined, short tandem repeat sequences and representing a variety of different microsatellite types, have been used as generic primers in PCR amplification-based applications for DNA fingerprinting in plants and animals. Most often, an SSR oligonucleotide is used in a single-primer amplification reaction (SPAR; Gupta et al. 1994), an approach reminiscent of both RAPD and interspersed repeat-based PCR (REP-PCR; Versalovic and Lupski 1995). Microsatellite repeat-primed PCR (MP-PCR; Sharma et al. 1995) reactions are similar to a RAPD reaction in that exponential amplification will occur from these single primer reactions only when the particular repeat used as a primer is represented in multiple copies, which are closely spaced (<2-3kb) and inversely oriented in the template DNA (Figure 1). Typically, these reactions are multiplexed; multiple products are simultaneously co-amplified, and therefore multiple loci can be detected from a genome using a single PCR reaction. SSR-based primers representing tri- tetra- and penta-nucleotide repeats have been used successfully to generate distinct banding patterns that are resolvable on low-resolution agarose gels using ethidium bromide staining (Meyer et al. 1993, Perring et al. 1993, Gupta et al. 1994, Sharma et al. 1995, Weising et al. 1995), or on high-resolution polyacrylamide gels by silver staining (Buscot et al. 1996) or through primer radiolabeling followed by autoradiography (Gupta et al. 1994). As would be predicted, the best product size discrimination is obtained with polyacrylamide-based gel analysis, although agarose gel resolution is sufficient for many applications. Apart from the SSR sequence used as a primer, there is no other DNA sequence information required to design and perform these amplifications. The same set of SSR primers can be used across species and even kingdoms, and therefore constitutes a general-purpose primer set.

The number of amplification products generated by MP-PCR depends on the frequency of occurrence in the target genome of the particular repeat used as the primer. In humans, it has been estimated that a dinucleotide repeat occurs on average once every 30-50 kb, and trinucleotide repeats every 300-500 kb (reviewed in Hearne et al. 1992). In plant genomes, the overall frequency of microsatellite repeats appears to be generally lower (Morgante and Olivieri 1993, Wu and Tanksley 1993), although the incidence of closely spaced repeats has been bourne out experimentally (Gupta et al. 1994, Zietkiewicz et al. 1994). The level of polymorphism observed in the amplification products can be related to the genomic diversity within a species, and the amplification products reflect both dominant and codominant polymorphisms. As for RAPD, dominant polymorphisms result primarily from mismatch differences between primer and target site between genomes, and codominant polymorphisms from insertions/deletions between conserved priming sites. Unlike RAPD, additional codominant polymorphisms can result from SSR length differences at a target locus that become incorporated into the products. Furthermore, it has been argued that MP-PCR reactions can be more specific than RAPD reactions since the longer SSR-based primers (15-mers to 20-mers) enable higher-stringency amplifications (Gupta et al. 1994, Weising et al. 1995). This may result in fewer problems with reproducibility, a complaint frequently leveled against the low-stringency RAPD assay with shorter primers (Penner et al. 1993; see below). Amplification involving single, unanchored SSR primers has been used successfully for a variety of applications, including species identification in whitefly (Perring et al. 1993), discrimination of human fungal pathogen strains (Meyer et al. 1993), chickpea and tomato germplasm analysis (Sharma et al. 1995), analysis of morel mushroom populations (Buscot et al. 1996), and a polymorphism survey of diverse plant and animal genomes (Gupta et al. 1994).

The specific repeats that will be most informative for a particular genome are not likely to be absolutely predictable. Rather, the best primers, the ones that produce the most informative fingerprints, must be determined empirically for every new species investigated. It has been found, for example, that in the morel genome, $(GACA)_4$ did not amplify well whereas $(GTG)_5$ gave good fingerprints (Buscot et al. 1996). In chickpea and related species, distinct reproducible patterns were obtained with primers representing both tetranucleotides and GC-rich trinucleotides (Sharma et al. 1995). A third study revealed that no single primer gives polymorphism in every plant and animal species tested, although tetranucleotide repeat primers were the most effective at amplifying polymorphic patterns, and the best fingerprints for plants were produced with $(GACA)_4$ or $(ACTG)_4$ (Gupta et al. 1994). This study also demonstrated the Mendelian inheritance of polymorphic bands generated in maize.

Although discriminating fingerprints are generated, MP-PCR reactions must be optimized carefully to assure reproducibility and prevent artifactual results. Several studies have shown that, despite the higher annealing conditions used, these amplifications behave much like RAPD reactions in that the products obtained often vary with the precise conditions used (reviewed by Weising et al. 1995, Buscot et al. 1996). Annealing temperature, primer concentration and magnesium concentration, for example, all affect the quality of the resulting banding patterns (Weising et al. 1995). In addition, there is a demonstrated problem with mismatch priming even at high annealing temperatures, which normally are sufficient for good primer specificity. Finally, amplification products are produced from *E. coli* template DNA, even though this genome likely does not contain any SSR repeats matching the primers tested (Weising et al. 1995). Two alternative experimental steps have been offered as solutions to these primer specificity problems. In one case, empirically adjusting the annealing temperature led to more reproducible patterns (Gupta et al. 1994), and in another case the application of touchdown PCR thermocycling (Don et al. 1991) increased the specificity (Sharma et al., 1995). It is interesting to note that the banding patterns produced using touchdown PCR, while more reproducible, were also less polymorphic than those generated with a conventional, constant annealing temperature profile, indicating that much of the apparent polymorphism observed in many of these studies may not have resulted from true SSR-based genetic variability, but rather from local differences in primer-template affinity between genomes. Using such information in a DNA fingerprinting study can be risky for making definitive conclusions about genome relatedness, particularly if the patterns are not entirely reproducible, or if SSR-based products are desired.

A second general problem associated with MP-PCR is that unanchored dinucleotide and AT-rich trinucleotide primers do not produce distinct, resolvable products (Sharma et al. 1995, Weising et al. 1995). Both cases are probably the result of poor primer anchoring to the numerous target sites in a genomic template. The unanchored primer can anneal in many possible registers at each target site, resulting in multiple heterogeneous products from each locus, with the net result visualized as a smear on the gel. Successive cycles of annealing in multiple registers to the target SSR also results in the generation of smaller and smaller products in successive cycles, such that the smallest possible SSR-to-SSR bands predominate in the final product mixture. In some cases, even AT-rich tetranucleotide repeats produced smears, similarly indicating primer annealing in multiple registers (Gupta et al. 1994). In all reported cases, $(AT)_n$ dinucleotide repeat primers produced no detectable products, a consequence of secondary structure problems resulting from primer self-annealing. Although not reported, some of these problems might be alleviated with the inclusion of tetramethylammonium chloride (TMAC) in the amplification reactions; this compound serves to negate the weak hydrogen bonding effects of A:T base pairing, and has been shown to promote hybridization specificity of excessively AT-rich and GC-rich oligonucleotides (Ossewaarde et al. 1992). The smear generated by dinucleotide repeat oligonucleotides priming at multiple staggered sites at a locus, however, is an inherent effect of repeat unit length and therefore probably will not be remedied with TMAC.

It was initially predicted that amplification products resulting from single-primer PCR with SSR-directed primers would result in an enrichment for genomic fragments containing SSR repeats, and thus would be a way to selectively amplify and isolate SSR sequences of desired types (Weising et al. 1995). However, hybridization studies showed that the mixture of amplified bands from MP-PCR generally appear to be no more enriched for tri, tetra and pentanucleotides, and barely so for dinucleotide repeats, than are random genomic fragments amplified with arbitrary 10-mer primers (Sharma et al. 1995, Weising et al. 1995). Although some MP-PCR studies found good polymorphism between different genotypes of some plants and animals (Gupta et al. 1994), the results of at least two studies, with chickpea (Sharma et al. 1995) and morels (Buscot et al. 1996), revealed problems with obtaining sufficient intraspecific polymorphism. In both cases, interspecific polymorphism was well represented. In contrast, direct Southern blot hybridization with SSR probes (Ramakishana et al. 1994) and rDNA internal transcribed spacer fingerprinting (Buscot et al. 1996), revealed good intraspecific discrimination on the same genomic templates. Some intraspecific variation was detected using MP-PCR in *Actinidia* species (Weising et al. 1995). It may well be that, since the unanchored SSR primers frequently anneal nonspecifically to non-SSR target sites and since genetic SSR length variation between genomes is not captured in the MP-PCR products, this approach is probably not suitable for identifying differences within species that inherently carry a low level of genetic variation. Methods using SSR primers that can incorporate SSR length variation into the PCR products are more suitable for intraspecific analyses.

Overall, MP-PCR assays with unanchored SSR-based primers are simple to perform and require little up-front locus-specific or species-specific information, yet still provide a large amount of information if certain cautionary measures are followed. MP-PCR appears to be subject to irreproducibility effects, and requires careful optimization of conditions for specific primer-genome combinations to ensure that the heterogeneity of banding patterns observed is the result of true genetic polymorphism at fixed loci. Much of the polymorphism observed may not be SSR-based. The unanchored SSR primers appear to behave during PCR much like a RAPD primer, with considerable dependence on specific amplification conditions. Possible ways to increase primer specificity would be to perform these reactions using hot-start initiation or pre-amplification heating prior to the thermocycling, to add spermidine or TMAC to the reaction (Ossewaarde et al. 1992), or to use a touchdown thermocycling program (Don et al. 1991). With any or all of these modifications, successful and reproducible fingerprinting with tri- tetra- and pentanucleotide based primers should be possible. A nondenaturing polyacrylamide (6-8%) gel should be used to visualize the products if extremely high discrimination between individuals is required; otherwise, agarose gel (1.4-2%) resolution is generally suitable. Even with these modifications, the use of dinucleotide-based primers, and even those that represent AT-rich trinucleotides, should be avoided since their anchoring ability is extremely poor and distinct bands will not be seen.

Microsatellite-directed PCR: anchored primers

As a way to improve the specificity of SSR primers in MP-PCR reactions, and to provide a means to employ dinucleotide-based primers for DNA fingerprinting, methods were developed that use SSR primers modified by the addition of either 3'- or 5'-anchor sequences composed of nonrepeat bases (Zietkiewicz et al. 1994; see Figure 1) One of the first examples of the use of anchored simple sequence repeat primers was for first-strand cDNA synthesis with oligo-d(T) primers carrying one or two non-T bases at the 3' end (Liang and Pardee 1992). In some cases these 3' positions are degenerate and in other cases nondegenerate. Whether the primer is used for linear primer extension or for PCR amplification of the intervening region between neighboring SSRs, the anchor serves to fix the annealing of the primer to a single position at each target site on the template, such that every new polymerization event initiates at the same target position. Thus, there is little or no chance for primer slippage on the template, and problems with priming out of register are drastically minimized. This modification was critical for enabling dinucleotide repeat primers to perform well in MP-PCR reactions, and reduced product heterogeneity resulting from primer slippage.

This method, known variously as anchored MP-PCR, inter-microsatellite PCR and inter-SSR amplification (ISA) was first reported by Zietkiewicz et al. (1994). For consistency in this review, we will use the ISA term to refer to the DNA fingerprinting application of 5'- or 3'-anchored SSR primers in inter-repeat amplifications (Figure 1). These investigators reasoned that since dinucleotide repeats are far more abundant in most genomes than other types of microsatellites, and at least $(CA)_n:(GT)_n$ repeats appear to be ubiquitously represented in eukaryotic genomes, a small defined set of primers representing these repeats can be developed that would produce complex fingerprints from distantly related plant and animal species. They also predicted that such dinucleotide-based primers, if anchored, would generate distinct bands from individual loci. This survey of plant and animal genomes, and a later report demonstrating *Chysanthemum* cultivar identification (Wolff et al. 1995), were performed using the 3'-anchored primers, $(CA)_8RG$, $(CA)_8RY$ and $(CA)_7RTCY$, and the 5'-anchored primers, $BDB(CA)_7C$, $DBDA(CA)_7$, $VHVG(TG)_7$ and $HVH(TG)_7T$ (where Y=pyrimidine and

R=purine). In both reports, the primer was end-labeled with ^{32}P using polynucleotide kinase, and the co-amplified products separated on 6% nondenaturing polyacrylamide gels and visualized by autoradiography. A later study demonstrating popcorn cultivar identification also used the 3'-anchored trinucleotide primer, $(AGC)_4GY$ (Kantety et al. 1995), and a third report used primers carrying three-fold degeneracy at the penultimate 3' position and nondegeneracy at the 3'-terminal base, with the unlabeled products detected by silver staining (Ishibashi et al. 1996). A study of Douglas fir and sugi tree diversity resolved ISA polymorphisms on agarose gels (Tsumura et al. 1996), and a report on pine tree diversity used polyacrylamide gels and autoradiography (Kostia et al. 1995).

Typically, dozens of bands are co-amplified from the genomes of mammals and other vertebrates, as illustrated in Figure 2. Mendelian segregation of $(CA)_n$-based ISA polymorphisms was demonstrated using a three-generation human pedigree (Zietkiewicz et al. 1994). In contrast, fewer total bands are generated from most plant genomes tested (Figure 2), although this result is consistent with the predicted difference in abundance of $(CA)_n:(GT)_n$ repeats in mammalian compared to plant genomes. Nevertheless, the amplification products generated from both animal and plant genomes are reproducible and a large percentage of them are polymorphic between even closely related genotypes. Successful use of the ISA approach for DNA fingerprinting and cultivar discrimination in soybean (Zietkiewicz et al. 1994), maize (closely related popcorn lines; Kantety et al. 1995), *Chrysanthemum* (Wolff et al. 1995) and conifer trees (Kostia et al. 1995, Tsumura et al. 1996) is good evidence for its overall usefulness for genome analysis in plants.

The types of products obtained with 3' versus 5'-anchored primers differ somewhat. Both types generate products resulting from amplification between inversely oriented, nearby repeat target sites, and these products can reflect the same types of polymorphism as seen in unanchored MP-PCR and even RAPD bands. However, only the 5'-anchored primers generate products that capture within the amplified product any allelic length variation that may exist within an SSR target site. Unlike 3'-anchored primers that begin priming from the downstream flank of the SSR, the 5'-anchored primers position themselves at the upstream boundary of the target SSR, and primer extend across the remaining part of the target repeat into the downstream flanking region (Figure 1). Thus, in principle, the products generated from 5'-anchored primers should exhibit more codominant polymorphisms. There have not been enough segregation analyses reported with 5'-anchored primers to either substantiate or contradict this prediction.

Perhaps the major limitations to the ISA approach are the extreme care with which the amplifications need to be performed to prevent stuttering and primer nonspecificty, and the degree to which one is limited to visualizing polymorphism only at SSR loci that are in inverse orientation and within PCR-amplifiable distances from one another in the genome. Any isolated SSR in the genome will not be detected. The use of 3'-anchored primers also prevents the visualization in the products of any allelic SSR length variation that may exist at the SSR target sites; however, the use of 5'-anchored primers usually presents more problems with primer specificity, since the 3'-end of the primer still can anneal out of register if conditions are not optimized to prevent nonspecific annealing of the 5'-anchored primer. Nevertheless, this approach has been shown to be useful for both inter- and intraspecific germplasm discrimination in a variety of applications.

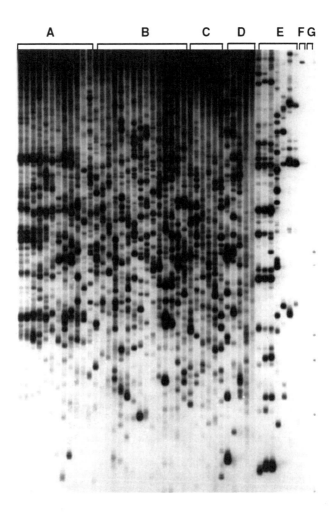

FIGURE 2. Inter-SSR amplificationfingerprints of diverse animal and plant species using (CA)$_8$RY primer (reprinted with permission from Zietkiewicz et al. 1994). Lanes correspond to primates (A), other mammals and marsupials (B), rodents (C), reptiles and birds (D), plants (E), *E. coli* (F), and no DNA (G).

Microsatellite-priming combined with arbitrary sequence primed PCR: random amplified microsatellite polymorphisms (RAMP)

Independently, a method was developed that would combine the best features of the RAPD assay and SSR detection. This approach, termed random amplified microsatellite polymorphisms (RAMP; Wu et al. 1994), is based on the random distribution of nucleotide sequences immediately flanking a simple sequence repeat. Amplification is performed between a 5'-anchored mono-, di- or trinucleotide repeat primer and a random sequence decamer to generate co-amplified products capable of carrying codominant length polymorphism that may be present at the SSR target site (*see* Figure 1). The RAPD primer binding site serves as an arbitrary endpoint for the SSR-based amplification product, and therefore the products obtained are not as restricted by the relative genomic position of SSRs in the genome as they are with ISA. Individual combinations of different SSR-based primers with different decamers allows for the generation of nearly limitless numbers of unique amplification products, all of which are bordered at one end by a microsatellite repeat.

The SSR primer is ^{32}P- or ^{33}P-labeled and the amplification products resolved on polyacrylamide gels. Denaturing gel conditions can be used since only one strand of every amplified fragment is labeled. To facilitate specific annealing of the two primers having widely differing T_m and to reduce the amount of amplification resulting from the 10-mer alone, a modified asymmetric thermocycling profile was adopted (Wu et al. 1994). This program toggles between high- and low-temperature annealing, allowing the SSR primer at high annealing temperature to generate a set of linear fragments that then serve as preferential templates for the 10-mer when the program switches to the low annealing temperature. This scheme results in the preferential amplification from SSR loci, apparently with minimal competition from single RAPD primer amplification at non-SSR sites in the genome.

The RAMP method was shown to produce polymorphic co-amplification products in *Arabidopsis*. Both dominant and codominant polymorphisms were detectable, many were shown to exhibit Mendelian segregation and were mapped as distinct, apparently random loci in the *Arabidopsis* genome (Wu et al. 1994). The 5'-anchored SSR primers reported all carried nondegeneracy within the 3-4 base anchor sequence, and many hundreds of different primers are possible. Successful primers included the mononucleotide-based primers, $CCGG(T)_{10}$, $GCGC(A)_{10}$ and $CACGA(T)_{15}$, as well as dinucleotide-based primers such as $CAA(CT)_6$, $GCCA(GA)_7$ and $CCAG(GT)_7$. Depending on the primer combinations chosen, between 5 and 50 bands were resolvable from a single reaction, and at least one band from each reaction appeared to be codominantly polymorphic (Figure 3A). With one SSR primer studied extensively [$CAA(CT)_6$], 104 polymorphic bands were identified, of which 67 appeared to be polymorphic. Cloning and sequencing of a representative polymorphic band confirmed that the amplified RAMP fragment was bordered at one end by the predicted SSR and at the other end by the arbitrary 10-mer sequence. Although these results with *Arabidopsis* generate predictions for other genomes, application of the RAMP method to other species will require an empirical test of different anchored SSR and decamer primers to first identify the most informative primer combinations.

Two other recent reports demonstrated the applicability of RAMP analysis for barley cultivar identification and mapping (Becker and Heun 1995, Sanchez de la Hoz et al. 1996). In one study (Sanchez de la Hoz et al. 1996), random sequence 10-mers were used singly for RAPD analysis, 5'-anchored $(CA)_n$ repeat 10-mers were used individually for inter-SSR repeat analysis, and the two primer types were also combined for RAMP analysis. All thermocycling with these short primers was performed using a low (36°C) annealing temperature, and the products for all three primer combination types were resolved on polyacrylamide minigels and visualized by silver staining. This study showed that RAMP analysis with these short SSR primers could generate large numbers of polymorphisms, of which a subset segregated codominantly. Interestingly, genetic similarities calculated using RAMP faithfully approximated the known geneologies of the different barley cultivars, whereas RAPD analysis did not lead to the predicted dendogram. In a second independent study of barley germplasm, RAMP products generated with primers $CAA(GA)_5$ and $GGGC(GA)_8$ were examined for polymorphism by *Mse*I restriction enzyme digestion (Becker et al. 1995). The RAMP and digested RAMP (dRAMP) reactions generated up to 11 polymorphisms per primer combination, and 40 new RAMP/dRAMP markers were mapped to the barley RFLP map. Thus study revealed that restriction digestion of RAMP amplification products can generate additional polymorphisms. A third variation of this general SSR-to-random sequence amplification method, termed double-stringency PCR, used $(GACA)_4$ primers in single primer reactions at high stringency for 15 cycles, then a random 10-mer was added and thermocycling continued for 25 additional cycles at a low annealing temperature (Matioli and deBrito 1995). This two-step cycling method was shown to produce genetically informative markers that were more reproducible than RAPD alone. However, this analysis was not extended to other types of SSR sequences as primers, and it was not demonstrated that the amplified products were SSR-based.

Overall, the RAMP method and its variations has demonstrated utility for generating codominant markers that are based on polymorphic simple sequence repeats. With 5'-anchored primers, allelic variation at the target SSR locus can be captured within the amplification products, and fewer restrictions are imposed on the genomic sequences that are amplified (i.e. "stand-alone" SSRs in the genome can be detected), compared to inter-SSR amplification which can only detect the subset of genomic SSRs within amplifiable distances from one another. Despite these advantages, problems can result from the low annealing temperatures used in the RAMP amplifications. The asymmetric thermocycling profile uses low temperatures early in the cycling, which can allow for primer nonspecificity; the 5'-anchored SSR primers can exhibit nonspecificity at low temperatures, leading to unreproducible banding patterns.

Microsatellite-primed PCR: selective amplification of microsatellite polymorphic loci (SAMPL)

A new genetic fingerprinting method, amplified fragment length polymorphism (AFLP, Vos et al. 1995) is based on the detection of genomic restriction fragments by PCR amplification with primers corresponding to synthetic adaptors that are attached to the ends of the fragments. This method uses a set of generic adaptor-directed primers, requires no prior DNA sequence information for a target genome, results in the simultaneous co-amplification of multiple arbitrary regions of the genome, and reveals mainly dominant polymorphisms that result from either restriction site presence/absence or single nucleotide variaton between genomes. An SSR-based modification of the AFLP approach, termed selective amplification of microsatellite polymorphic loci (SAMPL, Vogel and Morgante 1995, 1997, Morgante and Vogel 1996) was developed as a way to build into the method a greater level of codominance and to take advantage of the length polymorphism of SSR sites in the genome. The SAMPL method uses the same adaptor-modified restriction fragment templates as are used for AFLP, but the amplification is performed using a ^{32}P or ^{33}P-labeled SSR primer combined with an unlabeled adaptor primer (Figure 1). Like AFLP, a touchdown thermocycling program (Don et al. 1991) is used to promote maximal primer specificty. Only one strand of each amplification product becomes labeled, and the products are resolved on denaturing polyacrylamide gels. Also like AFLP, the adaptor primer can carry 1-3 nondegenerate, arbitrary nucleotides at its 3' end, thus restricting any individual primer to only a subset of genomic target fragments and allowing different subsets of amplification products to be amplified by altering not only the SSR primer but also the adpator primer 3' extension. Many different combinations of restriction endonucleases, SSR primers and selective adaptor primers are possible with SAMPL, allowing for the detection of nearly limitless numbers SSR-based polymorphisms in the genome. Consistent with the prediction that the SSRs used as primers are ubiquitously represented in eukaryotes, complex fingerprints were demonstrated from plant as well as animal genomes (Morgante and Vogel 1996). Both codominant and dominant polymorphisms were detected among diverse soybean genotypes, as shown in Figure 3B. Most of these polymorphisms showed Mendelian segregation and mapped to apparently random sites on the soybean RFLP map.

For detection of maximal amounts of codominant polymorphism, it was recognized that the SSR primers should be made to anneal somewhere inside the target SSR, and primers extend through it to incorporate SSR length variation in the amplified products. 5'-anchored dinucleotide repeat primers similar to those used for ISA and RAMP are suitable only if hot-start thermocycling is performed (J. Vogel, unpublished data), as the 5'-anchor is not completely specific at ambient temperatures prior to the first denaturation. A low level of primer specificity and the resulting high level of stuttering make cold-start amplification completely unacceptable. Primers corresponding to compound repeats (two repeats directly abutted to one another, with no intervening nonrepeat bases) were recognized to naturally carry a repeat-based anchor segment; these compound repeat primers were shown to be perfectly self-anchoring and their use, even in cold-start amplification, results in a very low level of stuttering (illustrated in Figure

3). A defined subset of all possible permutations of compound repeats were found to be effective as primers for both SAMPL and ISA. These include $A(CA)_7(TA)_2T$, $T(GT)_7(AT)_2T$, $A(GA)_7(TA)_2$, $A(CA)_4(GA)_4G$, $C(CT)_4(GT)_4G$, $A(GA)_4(CA)_4C$, $T(GT)_4(CT)_4C$, $G(TG)_4(AG)_4A$, $A(GA)_4(GT)_4G$, $A(CA)_4(CT)_4C$, $C(TC)_4(AC)$ and $(CT)_8ATA$ (Morgante and Vogel 1996).

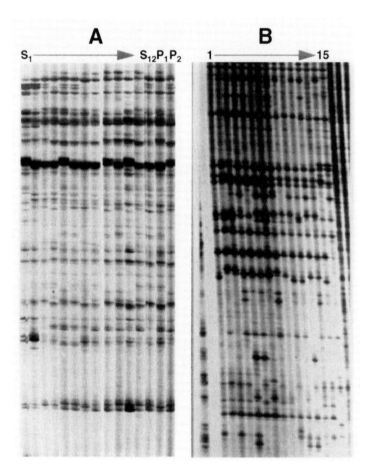

FIGURE 3. RAMP and SAMPL analysis of plant genomes. Panel A, RAMP analysis of an Arabidopsis mapping population using $CAA(CT)_6$. P_1 and P_2 are the parents, S_1 through S_{12} are 12 segregant lines. Panel B, SAMPL analysis of diverse *Glycine* species using $A(CA)_7(TA)_2T$ primer. Lanes 1-13 are *Glycine max* lines, lanes 12-15 are *Glycine soja* accessions.

These 12 primers represent compound SSR permutations that are present at the highest abundance among cloned plant and animal genomic sequences and that have the greatest likelihood for producing complex and polymorphic fingerprints from diverse genomes (Vogel and Morgante 1997). Typically, from 10 to more than 50 bands are simultaneously co-amplified, of which many are polymorphic. Both intra- and interspecific distinctions among soybean cultivars were detected, and intraspecific differences among maize genotypes were easily found. In an interspecific soybean cross, at least 10 polymorphic bands resulting from each of several specific primer combinations [$A(CA)_7(TA)_2T$ and individual adaptor primers] could be scored, and mapped to apparently random genomic sites. For every new genome under investigation, pairwise combinations of this or other SSR primers, with individual adaptor primers, should be tested empirically for those producing the most informative patterns. As for MP-PCR, ISA and RAMP, the Mendelian nature of the observed SAMPL polymorphisms should be confirmed using a segregating population or a well characterized pedigree.

With different restriction enzyme, SSR primer and specific adaptor-primer combinations, the SAMPL method has the capability of detecting very large numbers of SSR-based, codominant polymorphisms. Individual amplified bands have been isolated and sequenced, and shown to carry the predicted SSR at one end and the restriction site/adaptor at the other end (Vogel and Morgante 1997). In most cases, the codominant polymorphism observed between genotypes was confirmed to correspond to SSR length variation. These amplified fragments can be isolated and sequenced, from which flanking primers specific to an amplified SSR can be designed following a two-step PCR and sequencing procedure (Vogel and Morgante 1997). Thus, a desired polymorphic band on a SAMPL fingerprint gel can be converted to a conventional, single-locus SSR marker with little effort.

This SSR-to-restriction site amplification method is a simple extension of the already established AFLP method. It has been shown to identifiy new polymorphic markers in genomic regions previously devoid of known markers. However in at least one case, the SAMPL bands generated appeared to be no more polymorphic than AFLP products derived from the same germplasm (Matteo et al. 1996). Of all the MP-PCR methods outlined above, SAMPL is probably the most technically complex. If the user is unfamiliar with AFLP, then the multiple steps involved in the SAMPL method may take some time to master.

Conclusions

The choice of an appropriate DNA marker system for a particular application will depend on a combination of factors, including technical ease, informativity, and level of discrimination possible. Whether the final goal is map construction, trait marking or germplasm discrimination, methods involving the simultaneous detection of multiple polymorphic loci are extremely useful, especially when little or no DNA sequence or defined marker information exists or can be generated easily for a particular genome. Since simple sequence repeats are nearly ubiquitous in plant and animal genomes, and since they constitute highly polymorphic repeating arrays with high allelic variation, the SSR-based direct amplification methods discussed here represent simple and inexpensive ways to visualize these highly polymorphic sequences without the need for up-front cloning and sequencing SSRs and their flanking regions. Instead, a wide and sweeping screen can be performed with these direct amplification methods to identify the best or most polymorphic SSR-based genomic fragments on a target set of germplasm, and then only those "best" fragments need be cloned, sequenced and converted into more convenient, single locus PCR-based markers.

It is not possible to state definitively which of the methods outlined above would be the most suitable for every application. Using unanchored SSR primers for any method requires a heavy note of caution, since even tri, teta- and pentanucleotide-based oligonucleotides have been shown to exhibit a great deal of nonspecificity, and hence the bands generated may not be SSR-based and may not be reproducible. Inter-repeat amplifications with anchored primers appear to be well suited for fingerprinting and germplasm discrimination, but the bands produced are always limited to SSRs within amplifiable distances of one another in the genome. Nevertheless, ISA was shown to be suitable for differentiation of closely related genotypes. If greater numbers of markers are desired, such as for map construction or for trait marking, then either of the two methods, RAMP and SAMPL, involving amplification between a single SSR and an arbitrary fixed site will allow for the detection of nearly limitless numbers of SSR-based markers; higher levels of polymorphism may be obtained by digesting the amplification products with restriction enzymes prior to gel fractionation. If RAMP is used, caution should be exercised so as to minimize the amount of nonspecific priming by the anchored SSR primer at low temperatures. The difficulty with primer specificity for both RAMP and other MP-PCR methods may be alleviated or eliminated by using compound SSR primers, already found to be most suitable for good specificty in the SAMPL assay. SAMPL may prove to be well-suited for detecting large amounts of diverse SSR-based codominant polymorphism, although the assay itself is somewhat more technically demanding than any of the others.

Acknowledgments

We wish to thank Cathy Kershaw for assistance with the figures, Carol Starkey for assistance with the manuscript, and Antoni Rafalski and Kun-sheng Wu for helpful comments.

References

Ali S and Epplen JT (1991) *Indian J Biochem Biophys* 28:1-9.
Becker J and Heun M (1995) *Genome* 38:991-998.
Becker J, Vos P, Kuiper M, Salamini F and Heun M (1995) *Mol Gen Genet* 249:65-73.
Beckmann JS and Weber JL (1992) *Genomics* 12:627-631.
Bell CJ and Ecker JR (1994) *Genomics* 19:137-144.
Browne DL and Litt M (1992) *Nucleic Acids Res* 20:141.
Buchanan FC, Littlejohn RP, Galloway SM and Crawford AM (1993) *Mammalian Genome* 4:258-264.
Buscot F, Wipf D, di Battista C, Munch J-D, Botton B and Martin F (1996) *Mycol Res* 100:63-71.
Caetano-Anollés G, Bassam BJ and Gresshoff PM (1991) *Plant Mol Biol Rep* 4:294-307.

Carney JP, McKnight C, VanEpps S and Kelley MR (1995) *Gene* 155:289-292.

Cornall RJ, Aitman TJ, Hearne CM and Todd JA (1991) *Genomics* 10:874-881.

Cregan PB, Bhagwat AA, Akkaya MS and Rongwen J (1994) *Meth Mol Cell Biol* 5:49-61.

de Souza AP, Allamand V, Richard I, Brenguier L, Chumakov I, Cohen D and Beckmann JS (1994) *Genomics* 19:391-393.

Dietrich WF, Miller JC, Steen RG, Merchant M, Damron D, Nahf R, Gross A, Joyce DC, et al. (1994) *Nature Genetics* 7:220-225.

Don RH, COx PT, Wainwright BJ, Baker K and Mattick JS (1991) *Nucleic Acids Res.* 19:4008.

Ellegren H and Basu T (1995) *Cytogenet Cell Genet.* 71:370-373.

Frohman MA, Dush MK and Martin GR (1988) *Proc Natl Acad Sci USA* 85:8998-9002.

Gupta M, Chyi Y-S, Romero-Severson J and Owen JL (1994) *Theor Appl Genet* 89:998-1006.

Hawkins GA, Bishop M, Kappes S and Beattie CW (1994) *BioTechniques* 16:418-420.

Hearne CM, Ghosh S and Todd JA (1992) *Trend Genet* 8:288-294.

Hudson TJ, Engelstein M, Lee MK, Ho EC, Rubenfield MJ, Adams CP, Housman DE and Dracopoli NC (1992) *Genomics* 13:622-629.

Ishibashi Y, Abe S and Yashida MC (1996) *Jap J Genet* 70:75-78.

Kantety RV, Zeng Z, Bennetzen JL and Zehr BE (1995) *Mol Breeding* 1:365-373.

Kostia S, Sirkka-Lilsa V, Vakkari P and Pulkkinen P (1995) *Genome* 38:1244-1248.

Liang P and Pardee AB (1992) *Science* 257:967-971.

Liang P and Pardee AB (1995) *Current Opinion Immunol* 7:274-280.

Litt M and Luty JA (1989) *Am.J Hum.Genet* 44:397-401.

Love JM, Knight AM, M.A. M and Todd JA (1990) *Nucleic Acids Res* 18:4123-4130.

Matioli SR and deBrito RA (1995) *Biotechniques* 19:752-758.

Matteo D, Marocco A, Lucchin M, Parrini P and Morgante M (1996) *Plant Genome IV. Abstract* p. 91.

Meyer W, Mitchell TG, Freedman EZ and Vilgalys R (1993) *J Clin Microbiol* 31:2274-2280.

Morgante M and Olivieri AM (1993) *Plant J* 3:175-182.

Morgante M, Rafalski JA, Biddle P, Tingey S and Olivieri AM (1994) *Genome* 37:763-769.

Morgante M and Vogel JM (1996) *International Patent Publicication* No. WO96/17082.

Nelson DL, Ledbetter SA, Corbo L, Victoria MF, Ramirez-Solis R, Webster TD, Ledbetter DH and Caskey CT (1989) *Proc Natl Acad Sci USA* 86:6686-6690.

Ossewaarde JM, Rieffe M, Rozenbergarska M, Osenkoppele PM, Nawrocki RP and Vanloon AM (1992) *J Clin Microbio.* 30:2122-2128.

Ostrander EA, Sprague GF and Rine J (1993) *Genomics* 16:207-213.

Peneer GA, Bush A, Wise R, Kim W, Domier L, Kasha K, Laroche A, Scoles G, et al. (1993) *PCR Methods Applic* 2:341-345.

Perring TM, Cooper AD, Rodriguez RJ, Farrar CA and Bellows TS (1993) *Science* 259:74-77.

Rafalski JA, Vogel JM, Morgante M, Powell W and Tingey SV (1996) In: Birren B and Lai E (eds), *Analysis of Non-Mammalian Genomes: A Practical Approach.* Academic Press Inc., Orlando, FL. In press.

Ramakishana W, Lagu MD, Gupta VS and Ranjekar PK (1994) *TheorAppl Genet* 88:402-406.

Sanchez de la Hoz MP, Davila JA, Loarce Y and Ferrer E (1996) *Genome* 39:112-117.

Sharma P, Huttel B, Winter P, Kahl G, Gardner RC and Weising K (1995) *Electrophoresis* 16:1755-1761.

Sinnett D, Deragon J-M, Simard LR and Labuda D (1990) *Genomics* 7:331-334.

Stallings RL, Ford AF, Nelson D, Torney DC, Hildebrand CE and Moyzis RK (1992) *Genomics* 10:807-815.

Tautz D (1989) *Nucleic Acids Res* 17:6463-6471.

Todd JA (1992) *Hum Mol Genet* 1:663-666.

Tsumura Y, Ohba K and Strauss SH (1996) *Theor Appl Genet* 92:40-45.

Versalovic J and Lupski JR (1995) *Meth Mol Cell Biol* 5:96-104.

Vogel JM and Morgante M (1995) *Plant Genome III. Abstract* p. 84.

Vogel JM and Morgante M (1997) Mns. in preparation.

Vos P, Hogers R, Bleeker M, Reijans M, van de Lee T, Hornes M, Firjters A, Pot J, et al. (1995) *Nucleic Acids Res* 23:4407-4414.

Vosman B, Arens P, Rus-Kortekaas W and Smulders MJM (1992) *Theor Appl Genet* 85:239-244.

Wang Z, Weber JL, Zhong G and Tanksley SD (1994) *Theor Appl Genet* 88:1-6.

Weber J and May PE (1989) *Am J Hum.Genet.* 44:388-396.

Weber JL (1990) *Genomics* 7:524-530.

Weising K, Atkinson RG and Gardner RC (1995) *PCR MethodsApplic* 4:249-255.

Welsh J and McClelland M (1990) *Nucleic Acids Res* 18:7213-7218.

Williams JGK, Kubelik AR, Livak KJ, Rafalski JA and Tingey SV (1990) *Nucleic Acids Res* 18:6531-6535.

Williams JGK, Rafalski JA and Tingey SV (1993) In: Wu R (Eds), *Genetic analysis using RAPD markers.* Academic Press, Orlando, FL., pp. 704-740.

Wolff K, Zietkeiwicz E and Hofstra H (1995) *Theor Appl Genet* 91:439-447.

Wu K-S, Jones R, Dannenberg L and Scolnik PA (1994) *Nucleic Acids Res* 22:3257-3258.

Wu K-S and Tanksley SD (1993) *Mol Gen Genet* 241:225-235.

Yuile MAR, Goudie DR, Affara NA and Ferguson-Smith MA (1991) *Nucleic Acids Res* 19:1950.

Zietkiewicz E, Rafalski A and Labuda D (1994) *Genomics* 20:176-183.

Characterization and classification of microbes by rep-PCR genomic fingerprinting and computer assisted pattern analysis

Jan L. W. Rademaker[1] and Frans J. de Bruijn[1,2]
MSU-DOE Plant Research Laboratory,[1] Department of Microbiology, NSF Center for Microbial Ecology,[2] Michigan State University, East Lansing, MI 48824, USA

Introduction

The identification and classification of bacteria are of crucial importance in environmental, industrial, medical and agricultural microbiology and microbial ecology. A number of different phenotypic and genotypic methods are presently being employed for microbial identification and classification (*see* Figure 1, Louws et al. 1996). Each of these methods permits a certain level of phylogenetic classification, from the genus, species, subspecies, biovar to the strain specific level (Figure 1). Moreover, each method has its advantages and disadvantages, with regard to ease of application, reproducibility, requirement for equipment and level of resolution (Akkermans et al. 1995). Generally, DNA-based methods are emerging as the more reliable, simple and inexpensive ways to identify and classify microbes. In fact, the assignment of genera/species has traditionally been based on DNA-DNA hybridization methods (Wayne et al. 1987) and modern phylogeny is increasingly based on 16S rRNA sequence analysis (Woese 1987, Stackebrandt and Goebel 1994). Here, we describe a method referred to as rep-PCR genomic fingerprinting, a DNA amplification based technique, which has been found to be extremely reliable, reproducible, rapid and highly discriminatory (Versalovic et al. 1994, Louws et al. 1996).

Rep-PCR genomic fingerprinting makes use of DNA primers complementary to naturally occurring, highly conserved, repetitive DNA sequences, present in multiple copies in the genomes of most Gram-negative and several Gram-positive bacteria (Lupski and Weinstock 1992). Three families of repetitive sequences have been identified, including the 35-40 bp repetitive extragenic palindromic (REP) sequence, the 124-127 bp enterobacterial repetitive intergenic consensus (ERIC) sequence, and the 154 bp BOX element (Versalovic et al. 1994). These sequences appear to be located in distinct, intergenic positions around the genome. The repetitive elements may be present in both orientations, and oligonucleotide primers have been designed to prime DNA synthesis outward from the inverted repeats in REP and ERIC, and from the boxA subunit of BOX, in the polymerase chain reaction (PCR) (Versalovic et al. 1994). The use of these primer(s) and PCR leads to the selective amplification of distinct genomic regions located

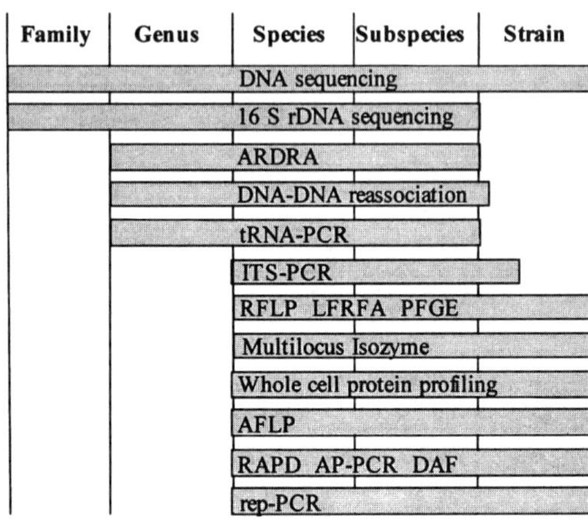

Family	Genus	Species	Subspecies	Strain
DNA sequencing				
16 S rDNA sequencing				
ARDRA				
DNA-DNA reassociation				
tRNA-PCR				
ITS-PCR				
RFLP LFRFA PFGE				
Multilocus Isozyme				
Whole cell protein profiling				
AFLP				
RAPD AP-PCR DAF				
rep-PCR				

FIGURE 1. Relative resolution of various fingerprinting and DNA techniques.

between REP, ERIC or BOX elements. The corresponding protocols are referred to as REP-PCR, ERIC-PCR and BOX-PCR genomic fingerprinting respectively, and rep-PCR genomic fingerprinting collectively (Versalovic et al. 1991, 1994). The amplified fragments can be resolved in a gel matrix, yielding a profile referred to as a rep-PCR genomic fingerprint (Versalovic et al. 1994; see Figure 2). These fingerprints resemble "bar code" patterns analogous to UPC codes used in grocery stores (Lupski 1993). The rep-PCR genomic fingerprints generated from bacterial isolates permit differentiation to the species, subspecies and strain level.

Rep-PCR genomic fingerprinting protocols have been developed in collaboration with the group led by Dr. J.R. Lupski at Baylor College of Medicine (Houston, Texas) and have been applied successfully in many medical, agricultural, industrial and environmental studies of microbial diversity (Versalovic et al. 1994). In addition to studying diversity, rep-PCR genomic fingerprinting has become a valuable tool for the identification and classification of bacteria, and for molecular epidemiological studies of human and plant pathogens (van Belkum et al. 1994, Louws et al. 1996 and references therein, Versalovic et al. 1997).

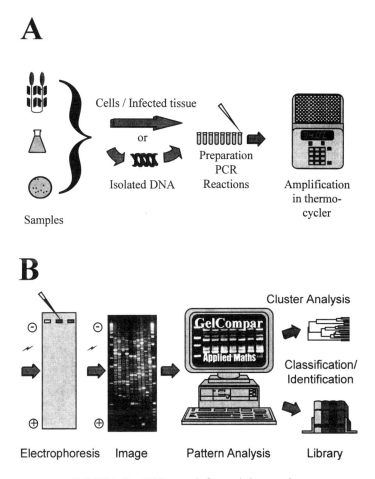

A

Cells / Infected tissue

or

Isolated DNA

Preparation
PCR
Reactions

Amplification
in thermo-
cycler

Samples

B

Cluster Analysis

GelCompar

Applied Maths

Classification/
Identification

Electrophoresis   Image   Pattern Analysis   Library

FIGURE 2. Rep-PCR genomic fingerprinting overview.

This chapter also describes the application of computer assisted analysis of rep-PCR generated genomic fingerprints for the identification and classification of microbes using cluster analysis algorithms. Cluster analysis is the art of finding groups in data, and bacterial classification and taxonomy are principal applications of this methodology (Figure 2B). We will describe the generation of raw data, the comparison of fingerprints, and the different algorithms used to find groupings in the data, and to identify specific strains in a database using their genomic fingerprints.

Experimental protocols

In this second section we will provide an overview of the different methodologies and protocols used to implement rep-PCR genomic fingerprinting. One distinct advantage of the rep-PCR genomic fingerprinting method is that the primers used work in a variety of Gram-negative and Gram-positive bacteria (*see* Versalovic et al. 1991, 1994, Louws et al. 1996). This means that no previous knowledge of the genomic structure or nature of indigenous repeated sequences is necessary. It also bypasses the need to identify suitable arbitrary primers by trial and error, that is inherent in the RAPD protocol (Welsh and McClelland 1990).

Sample preparation is simple and rapid and genomic fingerprints can be obtained from a variety of different templates (see next section). Many samples can be prepared in a short time. PCR amplification requires 5-7 h. Electrophoresis on agarose gels can be performed in 8 h, but 18 h is preferred for better resolution of the complex fingerprints on long (24 cm) gels. Therefore rep-PCR fingerprinting, including pattern analysis by eye or using a computer, can be performed in two days. In this section, we will primarily focus on examples involving the analysis of plant-associated and other soil bacteria. For a discussion of medical applications, consult reviews by Versalovic et al. (1994, 1997).

Correct pattern imaging, visual interpretation or conversion to computer processable data, will be described in the third section. Important parameters that will be discussed include choice of size marker standards for multiple gel comparison and database construction, determination of proximity coefficients, and use of appropriate clustering methods for phylogenetic analysis.

Template preparation for rep-PCR genomic fingerprinting.

Several methods of template preparation can be used for rep-PCR mediated genomic fingerprinting. The method used depends on the nature of the microbes to be analyzed, their receptiveness to lysing (releasing DNA), size of pools to be analyzed, level of resolution desired and time available. Rep-PCR genomic fingerprints have been obtained from purified DNA, whole cells from pure liquid cultures or cultures from plates, as well as directly from extracts of plant lesions or nodules (Versalovic et al. 1994, 1997, Louws *et al.* 1994, 1995, 1996, Nick and Lindstrom 1994, Schneider and de Bruijn 1996, Vera Cruz et al. 1996). Here we will focus on whole cell and purified DNA based methods, and will not discuss the plant tissue related approaches. For a detailed description of the latter, see Schneider and de Bruijn (1996) and Louws et al. (1996).

(i) *Whole cells from pure liquid cultures:* Whole cells obtained from a liquid culture can be directly used in rep-PCR amplification reactions. Generally, washing the cells improves the quality of the rep-PCR reactions. Using this method rep-PCR genomic fingerprinting have been generated from for example *Azorhizobium caulinodans* and *Rhizobium meliloti* (Schneider and de Bruijn 1996).
1. Take 3 mL of a liquid culture (OD$_{600}$ 0.65-0.95).
2. Spin down the cells.

3. Wash the cell pellet with 1 M NaCl. Repeat the washing several times for cultures producing a lot of polysaccharides.
4. Resuspend the cell pellet in 100 µL double distilled water. Store aliquots -20°C (optional).
5. Use 1-2 µL of template per PCR reaction (see below).

(ii) *Whole cells from single colonies on plates:* Whole cells obtained from single colonies on plates can also be used directly in the rep-PCR reaction. Using this method rep-PCR genomic fingerprints have been obtained from *Rhizobium* sp., *Clavibacter michiganensis* subsp., *Escherichia coli*, various Xanthomonads and Pseudomonads, as well as from a large collection of unidentified 3 CBA degraders and subsurface microbes (de Bruijn 1992, Louws et al. 1994, 1995, Judd et al. 1993, Zlatkin et al. 1996, Schneider and de Bruijn 1996; M.H. Schultz, J.L.W. Rademaker and F.J. de Bruijn, unpublished results).

1. Remove a small portion of a well-defined single colony directly off a fresh plate using a 1 µL disposable inoculation loop (Simport L200-1).
2. Insert loop into 25 µL PCR mix and whisk to resuspend the cells.
NOTE: Up to 4 and even 12 weeks old plates have been used successfully (Schneider and de Bruijn 1996; M.H. Schultz, J.L.W. Rademaker and F.J. de Bruijn unpublished results). Relatively few cells yield enough DNA for a rep-PCR reaction; in fact, using too many cells results in the generation of a background smear.

(iii) *Whole cell after alkaline lysis:* Whole cells of microbial species that are difficult to lyse and do not release DNA early during the PCR cycles may be pretreated by the alkaline lysis method.
1. Take 10 µL from a cell suspension (10^2-10^7 bacteria) or a portion of one colony.
2. Add 100 µL of 0.05 M NaOH and incubate at 95°C for 15 min.
3. Centrifuge for 2 min at 14,000 rpm.
4. Use 1 µL of the supernatant per rep-PCR reaction.

(iv) *Purified genomic DNA:* The procedure described is based on Pitcher et al. (1989), and modified by Dr. Luc Vauterin (personal communication). This method describes the extraction of DNA from solid media (agar plates), instead of liquid media.

A) Materials
1. TSA (Tryptone 15 g/L, Soy peptone 5 g/L, NaCl 5 g/L, agar 15 g/L, pH 7.3) or LB (*see* Sambrook et al. 1989) plates.
2. 10 mL disposable inoculation loops (Simport, L200-2)
3. 0.5 M EDTA pH 8.0.
4. 100× TE buffer: 1 M Tris, 0.1 M EDTA, pH 8.
5. TE buffer: 10 mM Tris, 1 mM EDTA, pH 8.
6. Guanidine thiocyanate-EDTA-Sarkosyl (GES) solution: 60 g guanidine thiocyanate (Sigma, cat. no. G-9277), 20 mL of 0.5 M EDTA pH 8, in 100 mL sterile water. Facilitate dissolving of all components by heating to 65°C; cool down the solution and add 1 g N-lauroyl sarkosine (Sigma, cat. no. L-5125). Adjust to 100 mL with water, filter sterilize using a 0.45 µm filter and store at room temperature.
CAUTION: Guanidine thiocyanate is harmful, wear suitable protective clothing.

7. Resuspension buffer (RB): 0.15 M NaCl, 0.01 M EDTA pH 8.
8. 7.5 M ammonium acetate: 578,1 g/L NH$_4$Ac.
9. Chloroform-isoamyl alcohol 24:1 (v/v)
   CAUTION: Chloroform-isoamyl alcohol is poisonous; wear suitable protective clothing and use in a well ventilated area, such as a fumehood.
10. RNase solution (2.5 mg/mL): 50 mg RNase in 20 mL water.
    Incubate 10 min at 100°C, aliquot and store at -20°C. Dilute 10× before use.
    NOTE: All solutions and media except for those described in 6, 9, and 10, need to be sterilized by autoclaving.

B) Experimental protocol
1. Use as starting material a well-grown 24-48 h old pure culture from a TSA or LB agar plate.
2. Remove cells with a sterile loop, taking one loop full, spherical on one side and flat on the other side, and resuspend in 0.5-1 mL RB.
3. Pellet cells by centrifugation and remove the supernatant using a 1000 μL tip.
4. Pellet cells again and remove the supernatant using a 200 μL tip.
5. Add 100 μL TE and mix using a pipette.
   NOTE: For isolates that produce excess polysaccharides such as several Xanthomonads, use 1 mL RB or repeat wash with RB and centrifuge for 10 min or even longer. An enzyme treatment can be added for microbes that are difficult to lyse; 10 mg/mL lysozyme dissolved in TE can be added in step 5, and incubated for 30 min at 37°C. Alternatively a chromopeptidase, mutanolisine or pronase may be used (Versalovic et al. 1994).
6. Add 500 μL GES and mix the reaction vials gently.
7. Incubate the mixture 5 min on ice.
8. Add 250 μL cold (-20°C) ammonium acetate (7.5 M), mix the tubes by shaking the tubes gently, and incubate them 5 min on ice.
9. Add 500 μL chloroform-isoamyl alcohol (24:1), shake vigorously until the solution is homogeneously milky, and centrifuge the mixture 10 min or until the upper phase is clear.
10. Prepare numbered tubes with 378 μL isopropanol at -20°C.
11. Carefully remove 700 μL DNA-solution of the upper phase using a 1000 μL tip and add the DNA solution to the tubes containing isopropanol.
    NOTE: To facilitate removing the upper phase without including some of the inter-face material, and to prevent shearing of the DNA, the end of the tips used should be cut off.
12. Shake *gently* until a white cloud of precipitated DNA becomes visible.
    NOTE: The reaction vials can now be stored at -20°C.
13. Centrifuge the DNA and remove the supernatant as described in step 3 and 4.
14. Add 150 μL 70 % ethanol but do not mix, centrifuge briefly and remove ethanol with 200 μL pipette, repeat centrifugation and remove residual ethanol.
15. Air dry the DNA pellet and redissolve in 200 μL TE, pH 8.
16. Incubate at room temperature or at 4°C until the DNA is dissolved, add 25 µL RNAse (250 μg/mL) and mix gently.
17. Incubate 1 h at 37°C and store the DNA at 4°C or at -20°C.
18. Determine the DNA concentration using a spectrophotometer at 260 nm (1 OD$_{260}$ = 50 μg/mL) and adjust it to 50 ng/μL.

Rep-PCR fingerprinting protocol

Often a detailed characterization of strains can be obtained by applying one primer (set). The BOX primer is generally recommended since it generates robust fingerprints and yields a highly complex fragment pattern. The REP primer set generates a lower level of complexity, but still yields reproducible and differentiating fingerprints (*see* Figure 7). The ERIC primer set is more sensitive to sub-optimal PCR conditions, such as the presence of contaminants in the DNA preparations but also generates highly discriminatory patterns (Figure 3A). A small pilot experiment is usually carried out to find the optimum primer set for a given application.

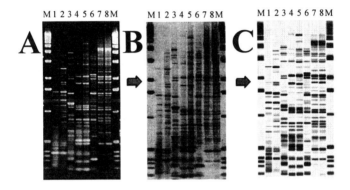

FIGURE 3. A. Scanned photograph of ERIC genomic fingerprints resolved on an ethidium bromide stained gel. B. Black and white inverted scan of A, converted in tracks and normalized. C. Image B after background substraction in GelCompar. *X. vesicatoria* LMG 911 and LMG 920 (lanes 1 and 2), *X. c. aberrans* LMG 9037 (lane 3), *X. c. barbareae* LMG 547 and LMG 7385 (lanes 4 and 5), *X. c. campestris* LMG 568 (lane 6), *X. c. incanae* LMG 7421 and LMG 7490 (lanes 7 and 8), and 1 kb size ladder (M).

PCR reactions and stock solutions are prepared on the bench. Filtertips are used to prepare stock solutions only. Wearing gloves is not essential. Care is taken that everything is efficiently organized to ensure a swift preparation of the reactions, and to avoid DNA contamination.

A) Materials
1. 5× Gitschier Buffer (Kogan et al. 1987): Prepare the following 4 stock solutions and autoclave them separately: 1 M $(NH_4)_2SO_4$, 1 M Tris-HCl pH 8.8, 1 M $MgCl_2$, 0.5 M EDTA pH 8.8. To prepare 200 mL of 5× Gitschier, combine 16.6 mL of 1 M $(NH_4)_2SO_4$, 67 mL of 1 M Tris-HCl (pH 8.8), 6.7 mL of 1 M $MgCl_2$, 1.3 mL of a 1:100 dilution of 0.5 M EDTA pH 8.8, and 2.08 mL of a 14.4 M commercial stock of β-mercapto-ethanol, stored at -4°C. Adjust finally to 200 mL with approximately 106 mL water and mix. Store at -20°C in 1 mL aliquots.
   NOTE: BSA is added while preparing a master mix.

2. Ultra pure 100 mM dNTP set (Pharmacia; cat. no. 272035-1): 100 mM stocks are mixed at equimolar ratios to obtain a solution with 25 mM of each nucleotide. The solution is divided in 100 μL aliquots and stored at -20°C.
3. Nuclease-free BSA (Boehringer; cat. no. 711454): A 850 μg/mL solution is divided in 20 μL aliquots and stored at -20°C.
4. 100% DMSO (Fluka, cat. no. 41640): The solution is divided in 0.5 mL aliquots and stored at -20°C. One working solution is kept at room temperature.
5. Autoclaved ddH$_2$O: 2.5 mL aliquots in 5 mL screw cap vials are stored at room temperature.
6. Primer 1 (BOX A1R, ERIC 1R or REP 1R) and primer 2 (ERIC 2 or REP 2I): 0.3 μg/μL primer solutions are divided in 200 μL aliquots and stored at -20°C. Primers have the following sequences:
   BOX A1R (Versalovic et al. 1994): 5'-CTACGGCAAGGCGACGCTGACG-3'.
   ERIC 1R (Versalovic et al. 1991): 5'-TGTAAGCTCCTGGGGATTCAC-3'.
   ERIC 2 (Versalovic et al. 1991): 5'-AAGTAAGTGACTGGGGTGAGCG-3'.
   REP 1R (Versalovic et al. 1991): 5'-IIIICGICGICATCIGGC-3'.
   REP 2I (Versalovic et al. 1991): 5'-ICGICTTATCIGGCCTAC-3'.
7. *Taq* DNA polymerase (5 U/μL; e.g. Perkin-Elmer, cat. no. N8080070) stored at -20°C.
8. Mineral oil (Sigma, cat. no. M-3516) stored at room temperature.
9. Thermal cycler (e.g. Perkin Elmer 480 or MJ Research PTC 100 or 200).

B) Experimental protocol
   1. Number the reaction tubes.
   2. Prepare the master mix.

Stock solution	μL per *one* 25 μL reaction	For ... 25 μL reaction
5× Gitschier Buffer	5	... μL
20 mg/mL BSA	0.2	... μL
100% DMSO	2.5	... μL
ddH$_2$O	12.65	... μL
	13.65 (for BOX)	
dNTP mix	1.25	... μL
primer 1	1	... μL
primer 2 (not for BOX)	1	... μL
Taq polymerase (5 U/μL)	0.4	... μL

   3. Mix gently.
   4. Collect by centrifugation.
   5. Aliquot 24 μL master mix to each tube.
   6. Overlay the mix with a drop of mineral oil to reduce evaporation.
      NOTE: Mineral oil can be applied in sample tubes anywhere between step 4 and 8. To prevent cross contamination of the samples it should be applied as early as possible.
   7. Add 1 μL of the samples to be fingerprinted to the reaction mix.
   8. Collect by centrifugation.
   9. Insert samples and control reactions in thermal cycler.

10. Start the PCR with an incubation of 7 min at 95°C [time delay file #2 for Perkin Elemer 480 thermocycler]. Continue using 30 cycles or 35 cycles for whole cell rep-PCR, using a step cycle file (#4) consisting of: 1 min at 94°C, 1 min at 53°C, 52°C or 40°C for BOX, ERIC or REP primers, and 8 min at 65°C. Terminate the PCR reactions with an extension of 16 min at 6 5°C (time delay file #2), and store at 4°C (soak file #1).

NOTE: Instead of the Perkin Elmer 480, the Peltier element based thermocyclers such as the MJ Research PTC 100 and PTC 200 with a 96 well format and a heated lid can be used. These machines are faster, use smaller tubes with thinner walls, do not "over-shoot" or "under-shoot" target temperatures and do not need mineral oil. Therefore the following cycles in the "block" mode are used to obtain similar profiles in a shorter time. The PCR is started with an incubation of 2 min at 95°C, and continued using 30 cycles or 35 cycles, for whole cell rep-PCR, consisting of: 3 s at 94°C and 30 s at 92°C, 1 min at 50°C (BOX and ERIC) or 40°C (REP), and 8 min at 65°C. Terminate the PCR reaction with an extension of 8 min at 65°C and store at 4°C.

11. Store the completed PCR reactions at 4°C and whole cell rep-PCR products preferably at -20°C, to prevent breakdown of the amplified products, or use directly for gel electrophoresis, as described below.

Separation of rep-PCR amplified genomic fragments by gel electrophoresis

Normally, an agarose gel is sufficient to separate the rep-PCR generated fragments (*see* Figure 3A), although polyacrylamide gel electrophoresis may also be used.

A) Materials and reagents
1. Agarose LE (Seakem, cat. no. 50004).
2. 50× TAE electrophoresis buffer: 242 g Tris-HCl, 57.1 mL glacial acetic acid, 18.61 g Na$_2$EDTA and ddH$_2$O up to 1 L; store at room temperature.
3. Ethidium bromide (10 mg/mL) (Sigma); store at 4°C.
4. 6× loading buffer: 0.25% bromophenol blue, 0.25% [w/v] xylene cyanol FF and 15% [w/v] Ficoll 400 in ddH$_2$O; store at -2 0°C.
5. DNA 1 kb size ladder (Gibco BRL; cat. no. 15615-016); store at -20°C.
6. Horizontal gel electrophoresis apparatus H4 (Gibco BRL; cat. no.11025012).
7. Gel tray 20×25 cm (Gibco BRL).
8. Comb: 30 tooth, 1 mm thick (Delrin; cat. no. 1951CO).
9. Electrophoresis power supply (Gibco BRL, model 250).
10. UV transilluminator (wavelength 312 nm) (Fotodyne; 3-3000).
11. Face protecting shield (Fotodyne; model K).
12. Camera and suitable filters (Kodak, MP4).
13. UV-sensitive film (Polaroid 55 or 57 film).
14. Oven or water bath at 50-70°C.

B) Experimental protocol
1. In a 500 mL bottle add 3.75 g agarose to 250 mL 0.5× TAE and dissolve the agarose by microwaving.
2. Pour the agarose (50-70°C) into the gel tray and insert the comb.
3. After the gel has completely hardened, carefully remove the comb and fill the wells with 0.5× TAE.

4. Mix a 6 μL PCR sample with 1.2 μL loading buffer on a piece of Parafilm and load the gel; load 2 μL 1 kb DNA size ladder, mixed with 4 μL ddH$_2$O and 1.2 μL loading buffer into the terminal wells and in the middle.

5. Run the gel in the cold room for 18-19 h at a constant 70 V, 25-30 mA. This corresponds to 2 V/cm, measured as the distance between the electrodes.

6. Stain the gel for 30 min in an ethidium bromide solution of 0.6 μg/mL in 0.5× TAE (60 μL of a 10 mg/mL stock solution in 1 L 0.5× TAE), and destain for 30 min in 0.5× TAE.
   CAUTION: Be careful and wear gloves at all times to avoid touching the agarose gel. Ethidium bromide is a very powerful mutagen.

7. Visualize the bands on the gel under ultraviolet light and take a photograph (*see* Figure 3A) or capture the image by a video camera and print (or store as a TIFF file).
   CAUTION: Ultraviolet light is dangerous, particularly to the eyes. To minimize exposure wear a safety mask that efficiently blocks ultraviolet light.

When more than one gel needs to be analyzed and compared, using for example the GelCompar computer program, the details of this protocol should be followed carefully or standardized in other ways. It is important to use at least 3 marker lanes in a 30 lane gel, one at both ends and one in the middle. This allows for correction of possible "smiling" effects.

Computer-assisted rep-PCR genomic fingerprint analysis

When a high number or diverse rep-PCR fingerprints need to be compared, computer assistance becomes essential. When the quality of the raw data is high, computer assisted analysis generally generates high quality data as well. Therefore care should be taken to produce quality primary data. Many hard and software combinations are available for fragment (pattern) analysis. In our laboratory, two commercial software packages have been extensively applied to the analysis of rep-PCR fingerprints, namely the AMBIS system (Scanalytics, Waltham, MA, USA) and GelCompar (Applied Maths, Kortrijk, Belgium; Vauterin and Vauterin 1992). Computer programs, such as AMBIS and GelCompar normalize fingerprints according to intra-gel size standards (Versalovic et al. 1994, Rossbach et al. 1995, de Bruijn et al. 1996a, 1996b, Schneider and de Bruijn 1996). In this section, we will first discuss some of the general parameters of computer assisted (phylogenetic) analysis and then focus on the use of the GelCompar system for the analysis of rep-PCR generated genomic fingerprints (Louws et al. 1996, Schneider and de Bruijn 1996, de Bruijn et al. 1996a, 1996b).

General parameters: band or curve based characterization of fingerprints

Cluster analysis of a collection of genomic fingerprints obtained by rep-PCR can be carried out in different ways. The input of a clustering method is a proximity or resemblance matrix, the output a dendrogram, (Jardine and Sibson 1971), or 3D presentation (PCA; Hope 1968, Cooley and Lohnes 1971). Proximities can be described by a broad array of coefficients comparing one or more of the features of the fingerprints and resulting in similarity or dissimilarity units. Fingerprints in general can be analyzed on a band-based or curve-based pattern. Bands can be used to characterize a well defined fingerprint of low complexity as an array of peak positions alone, or combined with the height or area of the peak. Using a band-based method a collection of these fingerprints can be described as a matrix of binary variables, band present 1, band absent 0. Bands can be assigned by hand (Woods et al. 1992, Judd et al. 1993, Reboli et al. 1994, Versalovic and Lupski 1995, Koeuth et al. 1995) or by a computer program according to preset band searching settings (*see* Fig. 4; Versalovic et al. 1994, de Bruijn et al. 1996a; Schneider and de Bruijn 1998).

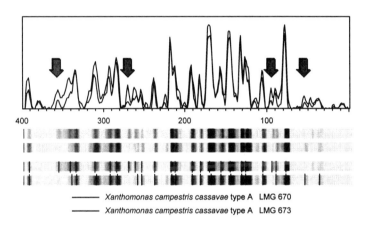

———— *Xanthomonas campestris cassavae* type A LMG 670
———— *Xanthomonas campestris cassavae* type A LMG 673

FIGURE 4. Densitometric curves, geltracks and scored bands of rep-PCR genomic fingerprint pattern. Arrows indicate positions of differentially scored bands. This illustrates why we prefer the characterization of the fingerprint patterns by whole densitometric analysis.

Manual as well as band-based computer-assisted analysis methods often require tedious and laborious band assignment or checking steps and can be subjective in nature (Figure 4). Information contained in fingerprints of high complexity, such as those generated in rep-PCR, is captured by the number and position of peaks and by different ratios in peak heights and areas (Figures 3-6). Therefore, a binary system is not sufficient to describe these highly complex fingerprint patterns. Preferably these fingerprint patterns are analysed using a curve-based protocol. The full complexity of rep-PCR genomic fingerprints can only be characterized by the densitometric curves, described as an array of densitometric values (J.L.W. Rademaker, F.J. Louws, U. Rossbach and F.J. de Bruijn, unpublished results). The product moment, see below, allows for the direct comparison of these whole densitometric curves.

Proximity coefficients.

The analysis of rep-PCR genomic fingerprints generally requires a simplification of the original data and can be used to calculate a proximity matrix. This calculation can either be based on dissimilarity, or similarity criteria (Figure 5). These (dis)similarities can be established using a wide array of coefficients.

A   Product moment correlation
     Original Ordering

1 *X. vesicatoria*	100							
2 *X. vesicatoria*	69.3	100						
3 *X.c. aberrans*	32.2	17.6	100					
4 *X.c. barbareae*	11.1	2.3	32.5	100				
5 *X.c. barbareae*	3.9	-0.3	26.6	91.0	100			
6 *X.c. campestris*	26.7	16.5	51.3	26.7	33.0	100		
7 *X.c. incanae*	0.9	7.5	14.2	28.5	31.5	23.9	100	
8 *X.c. incanae*	0.8	8.7	15.7	29.6	31.5	27.9	96.1	100

B   Product moment correlation
     Clustering: UPGMA

4 *X.c. barbareae*	100							
5 *X.c. barbareae*	91.0	100						
7 *X.c. incanae*	28.5	31.5	100					
8 *X.c. incanae*	29.6	31.5	96.1	100				
3 *X.c. aberrans*	32.5	26.6	14.2	15.7	100			
6 *X.c. campestris*	26.7	33.0	23.9	27.9	51.3	100		
1 *X. vesicatoria*	11.1	3.9	0.9	0.8	32.2	26.7	100	
2 *X. vesicatoria*	2.3	-0.3	7.5	8.7	17.6	16.5	69.3	100

C   Dendrogram correlation
     Clustering: UPGMA

4 *X.c. barbareae*	100							
5 *X.c. barbareae*	91.0	100						
7 *X.c. incanae*	30.3	30.3	100					
8 *X.c. incanae*	30.3	30.3	96.1	100				
3 *X.c. aberrans*	25.1	25.1	25.1	25.1	100			
6 *X.c. campestris*	25.1	25.1	25.1	25.1	51.3	100		
1 *X. vesicatoria*	10.7	10.7	10.7	10.7	10.7	10.7	100	
2 *X. vesicatoria*	10.7	10.7	10.7	10.7	10.7	10.7	69.3	100

FIGURE 5. Similarity matrices of ERIC-PCR genomic fingerprints of 8 *Xanthomonas* strains belonging to DNA homology groups 14 and 15 (Vauterin et al. 1995) using the product moment and the UPGMA method. Lines start with lane numbers (*see* Figure 3).

The band-based similarity coefficient defined by Jaccard (1908) is solely based on the presence of a band and its position as a binary variable. The coefficient derived by Dice (1945) also uses the band position, but adds more weight to matching bands. A more sophisticated "area-sensitive" similarity coefficient (GelCompar 4.0), takes into account the correspondence of bands expressed as in the coefficient of Jaccard, as well as the differences of the relative areas under each of the corresponding bands.

The product moment or Pearson correlation is a curve-based coefficient applied to the array of densitometric values formed by the fingerprint (Figures 4 and 5A). The product moment is a more robust and objective coefficient since whole curves are compared and subjective band-scoring is omitted. The product moment is independent of the relative concentrations of fingerprints and fairly insensitive to differences in background. Patterns such as the more complex rep-PCR genomic fingerprints benefit especially from these characteristics.

Clustering methods.

Figure 3 is an example of a gel with ERIC-PCR genomic fingerprints of 8 bacterial strains. In this small data set one can clearly discriminate the groups of fingerprints {1,2}, {4,5} and {7,8} and two more individual fingerprints in lane 3 and 6 which share some bands. These groups are called clusters (Figure 6). Finding these groups is the aim of cluster analysis. The basic assumption is that subsets can be characterized by possession of properties of coherence and isolation (Jardine and Sibson 1971). The goal is to form groups with highly similar fingerprints in such a way that the fingerprints in different groups are as dissimilar as possible. The example of the fingerprints in Figures 3 and 6 is simple. There are few fingerprints and the differences are clear. When the number of fingerprints is higher, the complexity is higher, and the fingerprints are more similar it is more difficult and tedious to assign groups and mathematical algorithms become necessary to perform cluster analyses (Figures 7 and 8).

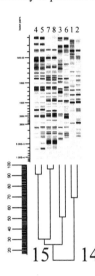

FIGURE 6. Cluster analysis of ERIC-PCR genomic fingerprints of 8 *Xanthomonas* strains belonging to DNA homology groups 14 and 15 (Vauterin et al. 1995) using the product moment and UPGMA method. See Figure 3 for lane numbers.

The choice of a clustering method is not always obvious and depends on the nature of the original data and the purpose of the analysis. Cluster analysis, including PCA, is mostly used to describe, present and explain data. It is not a statistical test to prove or disprove a preconceived hypothesis (Jardine and Sibson 1971, Kendal 1975, Kaufman and Rousseeuw 1990). The application of more than one clustering method, and comparison of the resulting classifications, can aid in the process of choosing the most appropriate representation of the data, and be of confirmatory importance.

Several algorithms for hierarchical or divisive clustering analyses leading to dendrograms are available. The unweighted pair-group method, using arithmetic averages (UPGMA) described by Sneath and Sokal (1973) is frequently used. This method has also been applied to the analysis of rep-PCR genomic fingerprints (Versalovic et al. 1994, Versalovic and Lupski 1995, Koeuth et al. 1995, Louws et al. 1995, van Belkum et al. 1996). Examples of this type of analysis are shown in Figures 5 (B and C), 6 and 7, where the rep-PCR genomic fingerprint patterns of a variety of *Xanthomonas* strains are investigated (F.J. Louws, J.L.W. Rademaker, L. Vauterin, J. Swings and F.J. de Bruijn, unpublished results). The strains are clustered according to DNA-homology groups, as previously assigned by Vauterin et al. (1995). The fingerprint patterns shown in Figure 7 (as well as in Figure 6), represent computer generated fingerprints, rearranged according to the "phylogenetic" tree.

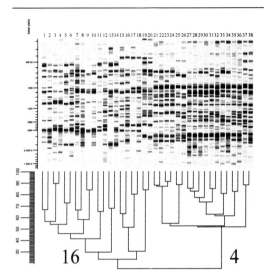

FIGURE 7. Cluster analysis of REP-PCR genomic fingerprints of 38 non-specified *Xanthomonas* strains belonging to DNA homology groups 4 and 16 (Vauterin et al. 1995) using product moment and UPGMA method.

The Neighbour Joining approach (Saitou and Nei 1987), is a method which attempts to reflect evolutionary distances that can be used for reconstructing phylogenetic trees. Alternatively, the method of Ward (1963) can be used, which is intended for interval scaled measurements and makes use of Euclidean distances (Kaufman and Rousseeuw 1990).

The agglomerative or partitioning principal component analysis (PCA) can also be applied (Versalovic et al. 1994, Vera Cruz et al. 1995, 1996). As a non-hierarchic clustering method, the PCA method is an interesting alternative to the hierarchical methods described above (UPGMA, Ward's and Neighbour Joining). For example, using GelCompar, PCA can be started directly from the densitometric curves, without applying any similarity coefficient, but using the original arrays of densitometric values instead. A set of Eigenvalues, derived from these curves, is the basis for calculating the three principal discriminating axes in a multi-dimensional space. Groups of entities (e.g. taxonomical units, strains) can be represented as clouds of dots in a spatial conformation (Figure 8). From a mathematical point of view, PCA is the most genuine grouping method and is an excellent method to discriminate between two to five groups. However, it is less suited for the discrimination of more than five groups, since the first three dimensions of a multidimensional system do not allow satisfactory representation of such complex structures.

Phylogenetic studies

Phylogeny is the evolutionary history of lineages (Hillis 1993). Phylogenetic trees are a specific type of relational graphic representations, reflecting these genealogical or evolutionary connections in a group of organisms. Phylogenetic trees are a specific kind of dendrograms, not only because of the cluster analysis methods employed, but especially due to the nature of the raw data and the collection of strains studied. The anomalous phylogenies based on bacterial catalase gene sequences which do not demonstrate relationship to phylogenies based on rDNA sequences (Mayfield and Duvall 1996) form an interesting illustration of this phenomenon.

Information of phylogenetic relations in a collection of closely related (sub) species, pathovars or strains can be obtained by cluster analysis of rep-PCR genomic fingerprints (Figures 6 and 7). Using rep-PCR, it is virtually impossible to obtain information on phylogenetic relationships between genera. However, rep-PCR is an excellent tool to asses diversity in a collection containing many genera. An interesting place to obtain a broad overview about phylogeny, including references, is the "Tree of life" website of the university of Arizona (Maddison and Maddison, http://phylogeny.arizona.edu/tree/phylogeny.htmL).

Database generation, microbial identification methods and available software

Specific data sets in microbial classification or taxonomy can be used to order the units of these groups into reference collections. These organized reference collections of fingerprints, or libraries, can be extremely useful for taxonomic and identification purposes. Currently several computer programs are available to facilitate the storage and analysis of rep-PCR generated fingerprints. GelCompar (Applied Maths, Kortrijk, Belgium; Vauterin and Vauterin 1992) is a state-of-the-art package for comparative analysis of electrophoresis patterns. It consists of extensive basic software, which allows for normalization of fingerprints according to intra-gel size standards. Additional comprehensive modules are available, comprising hierarchical cluster analysis, principal component analysis, library management and identification, and comparative quantification and polymorphic analysis. Five band-based similarity coefficients (Jaccard, Dice, an area sensitive, fuzzy logic and Jeffreys x coefficient) are provided together with the curve-based product moment correlation coefficient, which is based on the whole densitometric curves. The unweighted pair group method (UPGMA), Ward's and Neighbour Joining algorithms can be used to perform the cluster analyses. The GelCompar program has been succesfully applied to rep-PCR genomic fingerprinting analysis (de Bruijn et al. 1996a, 1996b, Louws et al. 1996, Schneider and de Bruijn 1996, Zlatkin et al. 1996).

Other available programs include the Whole Band Analyser (BioImage, Ann Arbor, MI, USA) for a Sun station and other image analysis software for Windows and Macintosh. Moreover, Dendron (Solltech, Oakdale, IA, USA) is a computer-assisted gel analysis system operating on Macintosh. This package allows the generation of data bases and hierarchical cluster analysis. Four similarities based upon band position alone or band position and intensities are present, and subsequently either the weighted pair group method (WPGM) or a proprietary method can be used to generate dendrograms (Snelling et al. 1996). RFLPscan (Scanalytics, Waltham, MA, USA) is the new generation of the AMBIS system, it allows creation of databases and calculation of one band-based similarity coefficient between lanes. As pointed out in the introduction to this section, we will focus on the GelCompar system as applied to rep-PCR genomic fingerprinting pattern analysis.

Image capture and standardization of rep-PCR genomic fingerprint patterns for computer-assisted analysis.

When one wants to compare many different gels with each other, the best results are obtained when all experimental parameters are standardized as much as possible. This is especially important when large databases are to be generated or when data generated by different laboratories need to be compared. The standardized conditions used should include sample preparation and processing, use of similar growth conditions, the same DNA isolation methods and use of the same rep-PCR conditions. Moreover the use of standardized electrophoresis conditions and size markers is essentials. Lastly, image capturing should be standardized. A video camera based on charged coupled device (CCD) photography, can be used and the digitized images of ethidium bromide stained gels can be saved directly as TIFF files. Alternatively photographs can be taken using conventional photographic cameras, such as the KODAK MP4 landcamera, and the resulting photographs can be scanned by normal flatbed scanners or laser scanners, and saved as TIFF files. We will describe our method below.

Experimental protocol
1. Place the photograph (*see* Figure 3A) on the top right hand corner of the HP Scanjet 3c, with the gel lanes parallel to the edge of the scanner and start Deskscan II version 2.2 software.
2. Use the image type defined as "Sharp B. and W. Photo", with "256 grays" and a "normal" sharpening. The path type has a resolution of 75 halftones and 100 dpi.
3. Activate "Preview" and frame the image within the borders of the picture, using only informative bands.
4. Activate "Zoom" and correct the frame when necessary, activate the ❻ sign for automatic contrast and brightness correction.
5. Activate the "Final" function and save the image as a TIFF file in the subdirectory of GelCompar called "images" or "gels.ima".

Analysis of rep-PCR generated fingerprint patterns using GelCompar.

The gel images are further processed by defining and naming patterns (lanes) in the Conversion (Figure 3B), followed by normalization and subtracting the background of the fingerprints (Figure 3C). Subsequently, databases can be generated, which allow editing, and searching and selecting lists of fingerprints. Fingerprints can be grouped by numerical comparison or identified by comparison to lists, or to more sophisticated specific libraries.

A) Materials
1. GelCompar version 4.0 program (Applied Maths, Kortrijk, Belgium).
2. PC-compatible computer with a 80486 DX cpu or better and minimal 8 MB RAM, running Windows 3.1 or 95.
   NOTE: GelCompar requires a basic understanding of Windows 3.1. or 95 and some understanding of the MS-DOS directory system, file naming and management. We run GelCompar on a Pentium PC (100 MHz, 16 MB RAM) which allows rapid processing of hundreds of fingerprints.

B) Analytical protocol
(I) *Conversion of images into a GelCompar format:* In this part the whole gel image is converted to separate lanes or gel tracks (*see* Figure 3B) and specific information is added to these lanes.
   1. Click on "Convert" in the start up menu and then choose "File" and "Load Image"
   2. Choose a TIFF-file from the subdirectory "images" or "gels.ima" and check "Negative" for light bands on a dark background, such as ethidium bromide stained gels.
   3. Click on "Edit" and "Settings" and customize the 'Track scanning settings'. We typically use for rep-PCR a 'Track resolution' (the number of points each densitometric track will consist of) of 400, 'Curve smoothing' of 0, 'Spline thickness' (indicates the number of points, to be averaged for a more stable profile, at either side horizontally of the center of the spline) of 2, and 'Gelstrip thickness' (indicates the number of points horizontally at either side of the center of the spline to be taken for the image of a gelstrip) of 5. We also activate the 'Rescaling' 'Whole gel' and the 'Number of tracks' dependent on the samples on the gel and 'Number of nodes' depending on the curvature of the lanes. We normally apply three nodes. Either of the 'Track search algorithm' I or II can be applied. 'Track resolution', 'Curve smoothing' and 'Spline thickness' and preferably also 'Gelstrip thickness' should be fixed per database.
   4. Fit the limits of the green frame snugly around the gel.
   5. Assign tracks with the help of either one of the automated track search routines defined in the settings, use "Add group" or assign tracks manually.
   6. Position the splines on the tracks using the left mouse button and the "Shift" key. Apply more nodes using "Page up", when necessary, to follow the lanes distortion.
   7. Apply the "Edit" and "Zoom box" commands to check or edit results.
   8. When satisfied with the results activate "Scan" and *do not* close the window, leave the window by minimizing it and enter descriptive information for each lane.
   9. For Marker lanes check "Reference" to ensure recognition of the internal standards by GelCompar and finish by saving the gel file in the appropriate "gels.raw" subdirectory.
(II) *Normalization of fingerprints:* In this part the lanes are normalized to an absolute database standard (Figures 3A-3B), using the reference tracks, and the background of the geltracks is subtracted (*see* Figures 3C, 6 and 7).
   1. Click on "Normalize" in the start up menu and subsequently activate "Edit" and "Settings" to customize Normalization and Background Subtraction. For rep-PCR fingerprint analysis, we typically use a 'Resolution' of 400 pt, a 'Smoothing' of 3 pt and a 'Rolling Disk' background subtraction with an 'Intensity Setting' of 12.
   2. Choose "File" and "Load" and a file from the "gels.raw" directory.
   3. For the first gel in this database, choose "Edit" and "Reference positions", move cursor to a well defined, single and reproducible band in a typical reference track and choose "Peak" and "Add". Assign all well defined and reproducible

marker bands in the standard which are useful, and finish by choosing "OK". When the standard reference is already chosen, proceed to step 5.

5. To apply this standard reference to the database choose "Edit", "Settings" "Use current standard reference" and "OK".

6. Choose "Associate" and "By pattern recognition". Check all associated bands, and make necessary improvements by selecting a standard reference band with the left mouse button, and assigning the corresponding new marker band with the right hand mouse button. When the marker lanes are very different from the absolute database standard, visible in the left part of the window, it is necessary to press "Control" and/or "Shift" to associate corresponding bands.

7. When satisfied with the association choose "Alignment" and "Align associated peaks".

8. Check the normalized pattern again.

9. When satisfied with the normalization choose "File" and "Save", the background is now subtracted and the normalized profiles are saved in the appropriate "gels.int" subdirectory.

(III) *Analysis:* The analysis starts with loading or creating a list. A list is a series of lanes from one or more gels from the same database to be compared with each other. Gels can be reconstructed (Figures 6 and 7) or composite gels can be assembled using a list. Dendrograms can be generated using several methods and libraries can be built.

(i) Creating a list:

1. Click on "Analyze" in the start up menu and double click a gel name in the analysis window.

2. Select all wanted lanes with the right hand mouse button or by applying the "Search" and "Topic" functions. Lists can be saved by choosing "List" and "Save".

(ii) Assigning bands: The program features a helpful automated gel searching option to assign bands in a lane.

1. Double click the gel name of your choice; when the window has opened choose "Gel", "Bands" and "Auto Search".

2. Check assigned bands, and select more or less by moving the cursor to the band in question. The form of the densitometric peak can be adjusted by replacing the pink squares.

3. Select band with left mouse button and click on right hand mouse button to assign it.
   NOTE: Clicking on the lane number above a lane reveals the number of bands in this lane in the bottom of the window.

4. Alternatively changes can be made by selecting a lane and choosing "Bands" and "Edit".

5. When content with the assigned bands choose "OK" or "exit" and "OK" and the adjusted peak assignment is saved.

(iii) Combining or superimposing gels: Rep-PCR genomic fingerprint patterns of different gels can be combined linearly. Gels can be linked to allow the analysis of associated genotypic fingerprints. This may lead to a higher contrast in the cluster analysis of the strains, as described by de Bruijn et al. (1996).

1. Click on "Analyze" in the start up menu then "Database" and "New combined gels", "File" and "Add new component".
2. Choose gel with correct descriptive information and double click on the name.
3. When necessary, exchange unwanted gel tracks by wanted lanes using "File", "Change link", type number of wanted lane in "Entry number" and double click the name of the gel of choice.
4. Click "Add new component" to add more fingerprints.
5. Repeat 3 and 4, until all your fingerprints are combined.
6. Choose "File" "Save combined gel" or "Save superimposed gel" to save the gel of choice.
    NOTE: Check the combined gel using cluster analysis or other analysis tools before you close the 'Combined Gel' window. This way you are able to save an improved version of the combined gel without going through the whole routine again. When satisfied, close the 'Combined Gel' window.

(iv) Hierarchical cluster analysis using GelCompar: Fingerprints can be grouped by similarity based on their whole densitometric curve.

1. When necessary assign bands as described above.
2. Create a list as described above.
3. Choose band based (comparison bands) or curve based (comparison correlation) comparison. For rep-PCR genomic fingerprints we prefer the product moment (comparison correlation) (*see* Fig. 5, 6, 7).
4. Choose similarity coefficient.
5. Choose clustering algorithm, UPGMA (Figures 5-7), Ward's or Neighbour Joining.
6. Evaluate dendrograms.
    NOTE: Similarity coefficients are relative values and not absolute percentages, but for convenience they are given as a value between 0-100 (Figures 5C, 6 and 7).

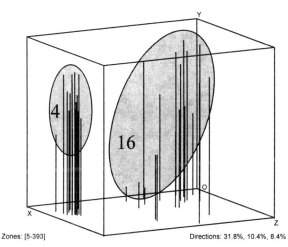

Zones: [5-393]   Directions: 31.8%, 10.4%, 8.4%

FIGURE 8. PCA of REP-PCR genomic fingerprints of 38 *Xanthomonas* strains belonging to DNA homology groups 4 and 16 (Vauterin et al. 1995).

(v) Principal Component Analysis: As pointed out above, non-hierarchic clustering methods such as Principal Component Analysis (PCA) are useful alternatives to the hierarchical methods. For an example of this analysis see Figure 8.
1. Create list as above described.
2. Choose "Comparison", "PCA" or click on the icon.
3. A three-dimensional representation of the taxonomical units as clouds of dots in spatial conformation is now seen. Samples in PCA of GelCompar are labeled by the names assigned to the tracks in "Conversion"; in this way, an unique label is automaticly assigned to each group.
4. It is often helpful to choose "Show", "construction lines" to get a better impression of the spatial distribution of the presented entities (see Fig. 8).
5. Use Page Up or Page Down to zoom in and out respectively and the arrows on the keyboard to turn the spatial array.
6. Evaluate spatial distribution of entities (see Fig. 8).
(vi) Microbial identification: GelCompar allows for the creation of lists (described above) or more sophisticated, specific libraries of rep-PCR patterns for identification. Either of the described correlation coefficients can be used for comparison and a detailed identification report can be obtained.

Acknowledgments

The development of the rep-PCR genomic fingerprinting method for the analysis of plant associated and soil microbes has been supported by the DOE (DE FG 0290ER20021), the NSF Center for Microbial Ecology (DIR 8809640), Heinz Inc., Roger Seeds Co. and the Consortium for Plant Biotechnology Research (DE-FC05-92OR22072). We also gratefully acknowledge Frank Louws, Uwe Rossbach, Maria Schneider and Jim Lupski (Baylor College of Medicine, Houston) for help and many useful discussions on rep-PCR. We would also like to thank Luc Vauterin and Jon Stoltzfus for critical reading of the manuscript.

References

Akkermans ADL, van Elsas JD and de Bruijn FJ (1995) *Molecular Microbial Ecology Manual* Kluwer Academic Publishers, Dordrecht, The Netherlands, 1–488.
Cooley WW, and Lohnes PR (1971) *Multivariate data analysis.* Wiley, New York.
de Bruijn FJ (1992) *Appl Environ Microbiol* 58:2180-2187.
de Bruijn FJ, Schneider M, Rossbach U and Louws FJ (1996a) In: Proceedings of the 7th International Symposium on Microbial Ecology. Brazil. In press.
de Bruijn FJ, Rademaker JLW, Schneider M, Rossbach U and Louws FJ (1996b) In: Stacey G Mullin B and Gresshoff P (eds) *The Biology of Plant-Microbe Interactions.* APS Press. In press.
Dice LR (1945) *J Ecology* 26:297-302.
Hillis DM, Allard MW, Miyamoto MM (1993) *Meth Enzymology* 224:456-487.
Hope K (1968) *Methods of Multivariate Analysis.* University of London Press, London.
Jaccard P (1908) *Bull Soc Vaud Sci Nat* 44:223-270.
Jardine and Sibson (1971) *Mathematical Taxonomy*, John Wiley & Sons, New York.
Judd AK, Schneider M, Sadowsky MJ, de Bruijn FJ (1993) *Appl Environ Microbiol* 59:1702-1708.
Kaufman L, and Rousseeuw PJ (1990) *Finding groups in data.* John Wiley & Sons, New York.
Kendal M (1975) *Multivariate analysis.* Griffin, London.
Koeuth T, Versalovic J and Lupski JR (1995) *Genome Res* 5:408-418.

Kogan SK, Doherty MD, Gitschier JG, (1987) *New England J Medicine* 317:16:985-990.
Louws FJ, Fulbright DW, Stephens CT and de Bruijn FJ (1994) *Appl Environ Microbiol* 60:2286-2295.
Louws FJ, Fulbright DW, Stephens CT, and de Bruijn FJ (1995) *Phytopathology* 85:528-836.
Louws FJ, Schneider M, and de Bruijn FJ (1996) In: Toranzos G (ed), *Nucleic Acid Amplification Methods for the Analysis of Environmental Samples*. Technomic Publishing Co. pp. 63-94.
Lupski, JR (1993) *JAMA* 270:363-1364.
Lupski JR and Weinstock GM (1992) *J Bacteriol* 174:4525-4529.
Mayfield JE and Duvall MR (1996) *J Mol Evolution* 42:469-471.
Nick G and Lindstrom K (1994) *Syst Appl Microbiol* 17: 265-273.
Pitcher DG, Nauders, NA and Owen RJ (1989) *Lett Appl Microbiol* 8:151-156.
Reboli AC, Houston ED, Monteforte JS, Wood CA and Hamill RJ (1994) *J Clin Microbiol* 32:2635-2640.
Rossbach SR, Rasul G, Schneider M, Eardley B, and de Bruijn FJ (1995) *Mol Plant–Microbe Interact* 8:549–559.
Saitou N and Nei M, (1987) *Mol Biol Evol* 4:406-425.
Sambrook J, Fritsch EF, Maniatis T (1989) *Molecular Cloning,* Cold Spring Harbor Laboratory Press, New York.
Schneider M and de Bruijn, FJ (1996) *World J Microbiol Biotechnol.* 12:163- 174.
Sneath PHA and Sokal RR (1973) *Numerical Taxonomy*, Freeman, San Francisco.
Snelling AM, Gerner-Smidt P, Hawkey PM, Heritage J, Parnell P, Porter C, Bodenham AR and Inglis T (1996) *J Clin Microbiol* 34:1193-1202.
Stackebrandt E, and Goebel BM (1994) *Int J Syst Bacteriol* 44:4:846-849.
van Belkum A (1994) *J Clin Microbiol Rev* 7:174-184.
van Belkum A, Kluytmans J, Van Leeuwen W, Bax R, Quint W, Peters E, Fluit A, Vandenbroucke-Grauls C, van den Brule A, Koeleman H, Melchers W, Meis J, Elaichouni A, Vaneechoutte M, Moonens F, Maes N, Struelens M, Tenover F and Verbrugh H (1995) *J Clin Microbiol* 33-6:1537-1547.
van Belkum A, Sluijter M, de Groot R, Verbrugh H and Hermans PWM (1996) *J Clin Microbiol* 34, 1176-1179.
Vauterin L and Vauterin P (1992) *European Microbiol* 1:37-41.
Vauterin L, Hoste B, Kersters K, and Swings J (1995) *Int J Syst Bact* 45: 472–489
Vera Cruz CM, Halda-Alija L, Louws F, Skinner DZ, George ML, Nelson RJ, de Bruijn FJ, Rice C and Leach JE (1995) *Intl Rice Res Notes* 20:4:23-24.
Vera Cruz CM, Ardales EY, Skinner DZ, Talag J, Nelson RJ, Louws FJ, Leung H, Mew TW, Leach JE (1996) *Phytopathology*, in press.
Versalovic J, Koeuth T, and Lupski JR (1991) *Nucleic Acids Res.* 19: 6823-6831
Versalovic J, Schneider M, de Bruijn FJ, and Lupski, JR (1994) *Meth Cell Mol Biol* 5: 25-40.
Versalovic J and Lupski JR (1995) *Meth Mol Cell Biol* 5:96-104.
Versalovic J, Kapur V, Koeuth T, Mazurek GH, Whittam TS, Musser JM, and Lupski JR (1995) *Arch Pathol Lab Med* 119:23-29.
Versalovic J de Bruijn FJ and Lupski JR (1997) In: de Bruijn FJ, Lupski JR and Weinstock GM (eds), *Bacterial Genomes: Physical Structure and Analysis*. Chapman and Hall, New York. In press.
Wayne LG, Brenner DJ, Colwell RR, Grimont PAD, Kandler P, Krichevsky MI, Moore LH, Murray RGE, Stackebrandt E, Starr MP and Trüper HG (1987) *Int J Syst Bacteriol* 37:463-464.
Ward JH (1963) *J Am Statist Assoc* 58:236-244.
Welsh J and McClelland M (1990) *Nucleic Acids Res* 18:7213-7218.
Woese CR (1987) *Microbiol Rev* 51:2:221-271.
Woods C, Versalovic J, Koeuth T and Lupski JR (1993) *J Clin Microbiol* 31:1927-1931.
Zlatkin IV, Schneider M, de Bruijn FJ and Forney LJ (1996) *J Ind Microbiol* 16: in press.

Simple sequence repeat DNA marker analysis

Perry B. Cregan and Charles V. Quigley
Soybean and Alfalfa Research Laboratory, Beltsville Agricultural Research Center, USDA-ARS, Beltsville, MD 20705, USA

Introduction

Simple sequence repeat (SSR) or microsatellites are DNA sequences that consist of two to five nucleotide core units such as (AT), (CTT) and (ATGT) which are tandemly repeated. These small repetitive DNA sequences which are spread throughout the genomes of eukaryotes provide the basis of a polymerase chain reaction (PCR) based, multi-allelic, co-dominant genetic marker system. The regions flanking the microsatellite are generally conserved among genotypes of the same species. PCR primers to the flanking regions are used to amplify the SSR-containing DNA fragment. Length polymorphism is created when PCR products from different individuals vary in length as a result of variation in the number of repeat units in the SSR. Microsatellite markers are rapidly becoming the predominant type of DNA marker used by human geneticists for linkage map development (Hudson et al. 1995) and for identification of individuals (Hammond et al. 1994, Kimpton et al. 1993). While plant geneticists still rely on restriction fragment length polymorphism (RFLP) and random amplified polymorphic (RAPD) markers, investigators in many plant species have begun to develop and use SSR markers in a wide range of plant species.

Microsatellite markers are being used for the construction of linkage maps in *Arabidopsis thaliana* (Bell and Ecker 1994), soybean [*Glycine max* (L.) Merr.; Akkaya et al. 1995], barley (*Hordeum vulgare* L.; Liu et al. 1996) and maize (*Zea mays* L.; Senior et al. 1996). SSRs also serve as a tool in genotype identification and for other purposes including population genetic studies in plant species such as soybean (Akkaya et al. 1992, Maughan et al. 1995, Morgante and Olivieri 1993), maize (*Zea mays* L.; Senior and Heun 1993), *Brassica* (Lagercrantz et al. 1993), wheat (*Triticum aestivum* L.; Devos et al.,1994; Ma et al. 1996), grape (*Vitis vinifera L.* ; Thomas and Scott 1993), barley (Saghai-Maroof et al., 1994), rice (*Oryza sativa* L.; Wu and Tanksley 1993, Zhao and Kochert 1993), sunflower

(*Helianthus annuus* L.; Brunel 1994), avocado (*Presea americana*; Lavi et al. 1995), banana (*Musa* spp.; Jarret et al. 1994), rapeseed (*Brassica napus* L.; Kresovich et al. 1995), *Pinus radiata* (Smith and Devey 1994), and wild yam (*Dioscorea tokoro*; Terauchi and Konuma 1994). SSRs are reported to exhibit high levels of length polymorphism with as many as 37 alleles at individual loci in barley (Saghai-Maroof et al. 1994) and 26 alleles in soybean (Rongwen et al. 1995). In most plant species the level of polymorphism is considerably higher than that found with RFLP markers (Diwan and Cregan 1997, Plaschke et al. 1995, Saghai-Maroof et al. 1994). Given this high level of informativeness, abundance and apparent random distribution in plant genomes combined with reliable amplification via PCR, SSR markers are likely to become an important and widely used DNA marker system in plants.

It is the intention of this chapter to outline some of the basic methodologies that are available for the detection of SSR allele length variation and to briefly discuss the research applications in which these methodologies may be useful and those situations in which their application is likely to be less successful. The techniques described include those which require a minimum of basic equipment to the somewhat more sophisticated requiring equipment that is likely to only be available in core facilities.

Applications and analysis requirements

There are numerous applications in plant genetic research for which microsatellite markers are an appropriate tool. These include exact allele sizing in variety identification and population genetic studies, genetic mapping, and marker assisted selection. The degree of precision in allele sizing and the rate of analysis and required sample throughput may vary greatly with the application and the individual researcher. The likely SSR analysis requirements for the different applications are described below.

Exact allele sizing

Genotype identification using an allelic profile at a number of microsatellite loci will soon be an accepted means of cultivar identification for some crop plants (Diwan and Cregan 1997). In order to protect the rights of the developer of a plant cultivar under the Plant Variety Protection Act, accurate and reproducible allele size determination is required. Likewise, in population genetic studies where allele frequencies are being estimated, exact allele sizing is a necessity.

Genetic mapping

Genetic mapping using F_2 populations requires distinguishing between two homozygotes and the heterozygote, in backcross populations between the homozygote and the heterozygote, and in recombinant inbred lines between two homozygotes. Obviously, when the sizes of segregating alleles differ by only a few base pairs nearly exact allele sizing is required. However, when the SSR alleles being assayed vary in size by more than a few bases genetic mapping is a relatively simple task. In this instance, the only requirement is to determine if a particular segregant carries the smaller allele, the larger allele, or both. Because genetic mapping populations may contain as many as 300-500 individuals high throughput is often a requirement in mapping studies. One way to increase throughput is via multiple gel loadings. For example, each gel lane can be reloaded periodically in order to allow determination of the genotype of a number of different members of the mapping population simultaneously.

Marker assisted selection

The use of microsatellite markers in marker assisted selection has analysis requirements similar to those of genetic mapping. If alternative allele sizes vary by more than a few base pairs selection is simply a manner of identifying the genotype with the larger or smaller allele with no need to determine exact allele size. The need for high throughput genotyping is likely to be great in this application as many plant breeding programs evaluate large segregating populations.

SSR analysis in high resolution agarose

For those accustomed to RFLP and RAPD markers, agarose gels provide a familiar method to visualize SSR length polymorphisms which does not require the use of radiolabeled compounds or sophisticated or expensive equipment.

Materials and Reagents

1. MetaPhor (FMC Bioproducts) or Synergel (Diversified Biotech) agarose.
2. Standard PCR reagents including: 25 mM MgCl$_2$, PCR primers, NTPs, 10× PCR buffer (500 mM KCl, 100 mM Tris-HCl pH 8.3), and *Taq* DNA polymerase.

Experimental protocol

1. *PCR reaction mix and conditions for the amplification of SSR alleles:* Standard PCR reactants are used including 30 ng of genomic DNA template, 1.5 mM Mg^{2+}, 0.15 μM of 3' and 5' end primers, 100 μM of each nucleotide, 1× PCR buffer, and 2 units of *Taq* DNA polymerase in a total volume of 20 μL.
2. *PCR cycling:* In the case of soybean with a 35% GC content we have designed primers to all microsatellite loci with identical optimal annealing temperatures of 47°C, thus cycling conditions are as follows: 25 s denaturation at 94°C, 25 s annealing at 47°C, and 25 s extension at 68°C for 32 cycles on a model PTC-100 thermocycler (MJ Research). Add 4 μL of Type II dye (Sambrook et al. 1989) to each reaction before loading.
3. *Agarose gel electrophoresis:* MetaPhor agarose is usually cast in 3 or 4% gels. In the case of Synergel, gels consist of 1.25% Synergel and 1.25% agarose in TBE buffer. Gels are cast in a standard horizontal gel frame and products are visualized by incorporating 5 μg ethidium bromide per 10 mL of gel and viewing on a UV transilluminator.

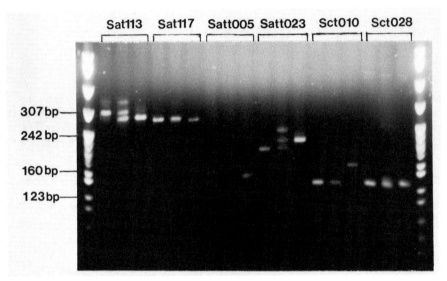

FIGURE 1. 4% MetaPhor agarose with ethidium bromide staining. PCR products were produced using Jackson (J), the F₁ of Jackson x Williams (F₁), and Williams (W) soybean genomic DNA as template with primers to soybean SSR loci Sat113, Sat117, Satt005, Satt123, Sct010, and Sct028, respectively. *Mbo*I digested DNA of pBR322 was included as a size standard in the first and last lanes of the gel.

Limitations

There are difficulties associated with the use of high resolution agarose for the analysis of SSR markers that may or may not be of significance in particular applications. Firstly, exact sizing of microsatellite alleles cannot be accomplished on agarose. Secondly, it is difficult to distinguish two, three, or four base pair differences in DNA fragment length on agarose. Thus, for successful genetic mapping of microsatellite loci, alleles must vary by more than 6 or 8 bp in length. This is illustrated in Figure 1. At soybean SSR locus Sct028 the alleles from Jackson and Williams soybean vary by 4 bp, 127 and 131 bp, respectively. It is quite difficult to detect this difference in 4% MetaPhor agarose and it is at least as difficult to distinguish the F₁ from either homozygous parent. A third difficulty with high resolution agarose is that amplification products from some SSR loci appear to contain "extra" bands when resolved on agarose. This is most common with $(AT)_n / (TA)_n$ SSRs (Figure 1, locus Sat113). Also, extra bands occur more frequently with heterozygotes (Figure 1, the F₁ of Jackson x Williams at the Satt023 and Sct010 loci). Despite these difficulties, high resolution agarose can in some cases serve well for microsatellite mapping and marker assisted selection.

SSR analysis in DNA sequencing gels

Because microsatellite alleles may vary in length by only one or two base pairs, DNA sequencing gels have been the most commonly used method to separate SSR-containing amplification products.

Materials and Reagents

1. With the exception of the sequencing gel formulation, standard materials and reagents are used.
2. Modified sequencing gel formulation: Rather than the standard sequencing gel mix (6% polyacrylamide, 8 M urea) a modified gel mixture is used to separate SSR-containing PCR products. The gel mixture (60 mL) consists of 6% acrylamide:bis-crylamide (19:1), 5.6 M ultra-pure urea, and 30% formamide in TBE buffer to which TEMED (65 μL) and 10% APS (750 μL) are added. This gel formulation eliminates extra, poorly defined, and slightly faster migrating "smears" of signal that are associated with the clearly defined microsatellite-containing amplification product(s). Any standard sequencing gel frame can be employed.

Experimental protocol

1. *^{32}P-labeling of microsatellite-containing PCR product:* Labeling of SSR-containing PCR products is done either by 5'-end labeling of one or both PCR primers or by incorporation of ^{32}P-labeled nucleotides in the PCR product. The 5'-end labeling is a standard procedure using γ-^{32}P-dATP (Sambrook et al. 1989) followed by removal of unincorporated ^{32}P using a BioSpin 6 column (BioRad). The PCR reactants and conditions are as described for high resolution agarose except for the inclusion of one or both ^{32}P end labeled primers and the use of a 10 μL reaction volume. In the case of internal labeling of SSR-containing fragments, the PCR reactants and cycling conditions are also identical except that 0.1 μL of 3,000 Ci/mmol α-^{32}P-dATP is added to the reaction mix and a 10 μL reaction volume is used.
2. *Separation of SSR-containing PCR product:* After PCR amplification, 4 μL of stop solution (90% formamide, 20 mM EDTA, 0.1% bromophenol blue, 0.1% xylene cyanol FF) is added to each reaction followed by denaturation at 95°C for 2 min. Products (3 μL) are loaded per lane. To estimate the exact length of the denatured PCR product and/or to allow a comparison with predicted length, sequencing reactions of M13 ssDNA can be used as molecular weight standards. The A sequencing reaction is run in one lane and the G, C, and T reactions combined and run in a second lane (Figure 2, lanes 1 and 2). Gels are run at 60 W constant power for 2-3 h. Bromophenol blue migrates with ssDNA fragments of approximately 80 nucleotides and xylene cyanol FF with 120-130 nucleotide fragments. Gels are dried and exposed to X-ray film.

Comments

The use of DNA sequencing gels for the separation of SSR-containing DNA fragments that are either internally labeled or end-labeled allows the exact sizing of alleles after careful comparison with a size standard (Figure 2). It is also possible to use multiple loadings with DNA sequencing gels (Figure 3) thereby allowing the genotyping of many individuals. In Figure 3, the parents and F₁ hybrid, as well as 45 F₂ individuals from a mapping population were evaluated in each loading. In this manner the genotype of 240 individuals was determined using five gel loadings at 40 min intervals.

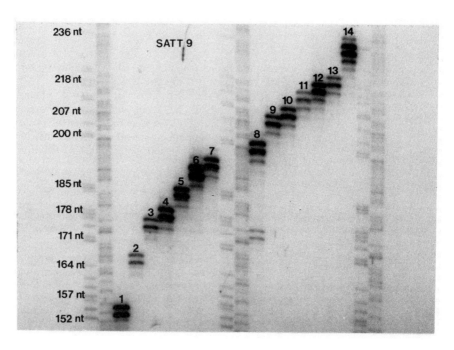

FIGURE 2. The Satt009 microsatellite locus showing 14 of the 19 alleles detected among 96 soybean genotypes (*from* Cregan et al. 1995). The denatured DNA fragments in lanes 1 through 14 are SSR alleles from the soybean genotypes indicated with the estimated allele size (based upon the upper band of the two distinct bands): 1, Williams (154 bp); 2, PI 101404B (167 bp); 3, Lee (175 bp); 4, PI 180501 (178 bp); 5, PI 86063 (184 bp); 6, PI 326582A (190 bp); 7, PI 200492 (193 bp); 8, PI 240664 (200 and 175 bp); 9, Aoda (209 bp); 10, S-100 (212 bp); 11, Wuchang (218 bp); 12, Centennial (221 bp); 13, Haberlandt (224 bp); and 14, Tracy (236 bp). The two lanes containing fragment size standards at the left, center, and right are derived from sequencing reactions of M13 ssDNA. The reaction using ddATP is run in the first of each pair of lanes and the reactions using ddGTP, ddCTP, and ddTTP as terminators were combined and run in the second lane of each pair. Sizes of the standard fragments in nucleotides (nt) are indicated at the left.

There are a number of techniques available to visualize microsatellite alleles on sequencing gels that do not require the use of radio labeled compounds. Hazan et al. (1992) developed a protocol that was refined and described in detail by Vignal et al. (1993). PCR products from as many as 16 SSR loci from the same individual were combined and separated in a single lane on a sequencing gel. The gel was blotted onto a

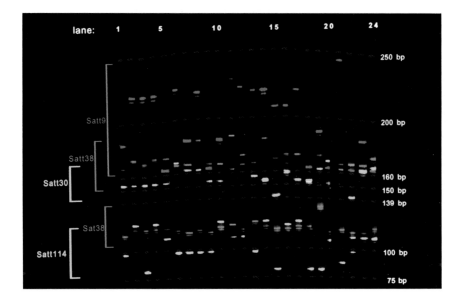

CHAPTER 11, FIGURE 6. A gel image of 24 soybean genotypes with 5 fluorescent labeled SSR loci taken from the ABI Prism 373A DNA sequencer using GeneScan analysis software. Fragment size standards (in red) are indicated on the gel image.

Nylon membrane and sequentially probed with one of the primers of each pair that had been used in the PCR amplifications. While the primers could be radiolabeled, a nonradioactive probing technique was suggested and described by Vignal et al. (1993). This technique was successfully used for the large-scale human mapping project at Genethon in Evry, France (Weissenbach et al. 1992).

Silver staining is another alternative that avoids the use of radiolabeled compounds. Various silver staining protocols have been outlined including those by Bassam et al. (1991), Klinkkicht and Tautz (1992), and Von-Deimling et al. (1993).

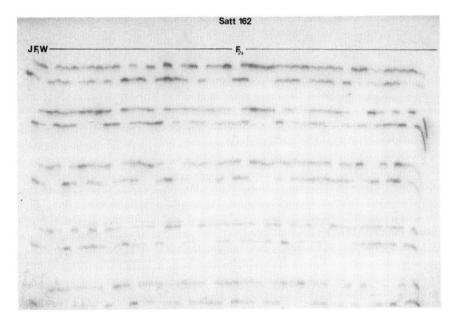

FIGURE 3. A sequencing gel loaded at 40 min intervals with internally labeled products from soybean microsatellite locus Satt162. The first three lanes of each loading were Jackson (J) soybean, the F_1 of Jackson x Williams (F_1), and Williams (W) soybean, respectively. The remaining 45 lanes were F_2 plants of Jackson x Williams mapping population. A total of 225 F_2 individuals were genotyped.

Limitations

When using sequencing gels for exact allele sizing, single primer end labeling gives better results than internal labeling because signal is produced by only one strand of the denatured PCR product. However, as compared to internal labeling with α-^{32}P-dNTP, longer exposure to X-ray film is required due to the weaker signal. In the case of high throughput genotyping with SSR-containing fragments labeled internally or with labeled primers, multiple loadings can be problematic because of the presence of unincorporated nucleotides or primers. This is because of the potential of co-migration of nucleotides or primers from later loadings with products from earlier loadings. Another obvious limitation of ^{32}P is radiation safety and the associated costs of procuring and disposing of ^{32}P-labeled compounds. The use of ^{33}P would have the advantage of somewhat reduced safety concerns.

SSR analysis in denaturing polyacrylamide gel electrophoresis (1.5 mm-thick) and SYBR-green I staining

A system using the modified sequencing gel formulation described above with SYBR-green I staining meets a number of needs for high throughput microsatellite genotyping without the use of radio labeled compounds. SYBR-green I is DNA specific, but unlike ethidium bromide functions well with single stranded DNA. This gel system accommodates loading directly from a microtiter plate with a multi-channel pipette.

Materials and Reagents

1. The gel apparatus (Owl Scientific) accommodates a 1.5 mm thick, 35cm x 30cm, vertical gel. A custom 64-tooth square toothed comb (0.45 cm well center to well center) permits the loading of alternate lanes directly from a microtiter plate.
2. SYBR-green I stain (FMC Bioproducts).

Experimental protocol

1. *PCR cycling:* The PCR reaction mix and cycling conditions are the same as that described above for high resolution agarose. Reactions are run in a 96-well thin walled polycarbonate V-bottom microtiter plate (Costar) on a MJ Research model PTC-100 thermocycler (MJ Research).
2. *Gel preparation:* The modified sequencing gel mixture described above (150 mL in TAE, rather than TBE buffer) is poured up to the bottom of the wells and is followed by a second gel intended to provide good well integrity. This top gel (20 mL) consists of 20% polyacrylamide gel (9:1 polyacrylamide:bisacrylamide), 5.6 M urea, 30% formamide, 170 μL 10% APS, and 14 μL TEMED in 1 x TAE buffer. TAE, rather than TBE buffer is used because SYBR-green I is bound by borate.
3. *Gel loading and electrophoresis:* A total of 30 μL (20 μL PCR product(s) plus 10 μL loading buffer [90% formamide, 10% ficoll, 20 mM EDTA, 0.1% bromophenol blue, 0.1% xylene cyanol FF]) is heat denatured at 95°C for 2 min and loaded per lane. Electrophoresis is at 60 W constant power.
4. *SYBR-green I staining:* After electrophoresis, remove the back plate and spacers and place a piece of UV transparent plexiglass on the gel. Turn the plexiglass, gel, and glass plate sandwich over so that the plexiglass is on the bottom and remove the glass plate from the gel. Pour a thin film of 1x SYBR-green I dye on the entire surface of the gel. Protect from light and allow staining to proceed for 30 min. View on UV transilluminator and photograph using a SYBR-green filter (FMC Bioproducts).

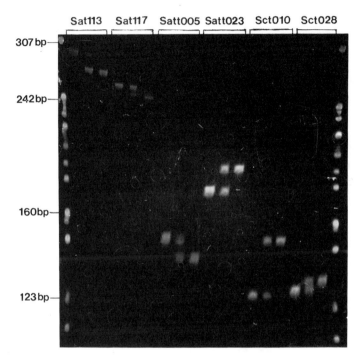

FIGURE 4. A 1.5 mm thick denaturing polyacrylamide gel with SYBR-green I staining. PCR products and the size standard are identical to those shown in Figure 1.

Comments

Separation of microsatellite alleles on the 1.5 mm thick TAE gel followed by SYBR-green I staining allows the resolution of SSR alleles varying in length by only 4 bp (Figure 4, locus Sct028). The two homozygotes and the heterozygote can all be distinguished from each other. Another important feature of this separation and staining technique is the absence of higher molecular weight bands as occurs with high resolution agarose. The absence of these extra bands makes sequential multiple loadings of each lane possible.

The 64 lane version of the same gel system can be loaded with 30 μL of PCR mix and stop solution in about 2 min. Results obtained with this gel configuration are shown in Figure 5. The microsatellite allelic constitution of two parents, their F_1, and 61 F_2 plants from a soybean mapping population was determined at three microsatellite loci. Because the alleles at these loci vary considerably in size, the PCR product(s) from the 3 loci could be loaded at 5 min intervals and the genotype of each individual at each locus could be readily determined.   When used with SYBR-green I staining, the 64 lane denaturing gel apparatus provides a high throughput system for SSR genotyping that can be used for genetic mapping and marker assisted selection.

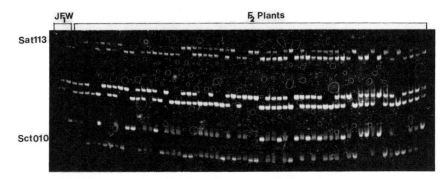

FIGURE 5. A 64-well denaturing polyacrylamide gel stained with SYBR-green I. Products amplified from SSR loci Sat113 (highest molecular wt.), Satt123 (medium molecular wt.), and Sct010 (lowest molecular wt.) were loaded 10 minutes apart. Lanes 1, 2, and 3, contain products amplified from Jackson, F_1 Jackson x Williams, and Williams DNA and lanes 4-64 contain products amplified from DNA of F_2 plants from Jackson x Williams.

Limitations

This system will not allow exact allele sizing and does not have sufficient resolution to permit mapping or marker assisted selection with microsatellite alleles that vary by only 2 bp in length.

Automated fluorescent labeled SSR allele sizing

The Perkin-Elmer Prism 373A DNA Sequencer with GeneScan 672™ software [Applied Biosystems (ABI)], is widely employed by human geneticists using microsatellite markers. As currently configured, the system permits accurate sizing of SSR alleles in 36 gel lanes with a number of loci per lane. DNA fragments of standard size are run in each lane along with the microsatellite alleles whose size is being determined. The size standards are labeled with one fluorescent dye while the unknowns are labeled with one of three other dyes. Amplification products from different SSR loci carrying the same fluorescent label can be simultaneously analyzed in the same gel lane if allele size ranges do not overlap. Thus, alleles from many loci can be sized simultaneously.

Materials and Reagents

1. Perkin-Elmer Prism 373A of 377 DNA Sequencer with GeneScan 672™ software (ABI).
2. PCR primer pairs of which the forward primers are labeled with either blue (FAM), green (TET), or yellow (HEX) fluorescent tags (ABI). Fluorescent labeled primers can be synthesized by a number of commercial sources.

3. Prepared sequencing gel formula such as 6% SeqQuate DNA Sequencing Solution (Sooner Scientific). Alternatively, individual gel mix reagents per the recommendation of ABI can be purchased.
4. Red (TAMRA) labeled size standards (ABI).
5. Standard PCR reagents.

Experimental protocol

PCR reaction mixes are constituted as described above except that the forward primer is fluorescently labeled and 0.1% Triton X-100 is included in a 10 μL reaction. Cycling conditions are the same as those described above. Primer pairs that function well in allele sizing using ^{32}P generally function well on the ABI system although small alterations in annealing temperature or Mg^{+2} concentration are sometimes needed to optimize PCR amplification.

After cycling, the fluorescently labeled PCR products are separated on the Perkin-Elmer Prism 373A or 377 DNA Sequencer. The fluorescent red (TAMRA) labeled internal size standard is included in each lane. GeneScan 672™ software (ABI) is used for gel analysis. GeneScan 672™ software is used for computation of allele size, and Genotyper™ software (ABI) is applied for accurate visualization of the alleles, and for automated data output.

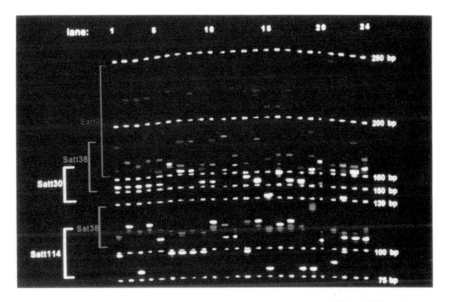

FIGURE 6. A gel image of 24 soybean genotypes with 5 fluorescent labeled SSR loci taken from the ABI Prism 373A DNA sequencer using GeneScan analysis software. Fragment size standards (in red) are indicated on the right of the gel image. See color plate.

Comments

The automated sizing of microsatellite alleles with the Perkin-Elmer DNA Sequencer is now commonly used for human identification (Kimpton et al. 1993). A similar system is now available for the genetic profiling of soybean populations and the identification of individual soybean cultivars. Diwan and Cregan (1997) demonstrated the use of allelic profiles derived from 20 SSR loci to distinguish soybean cultivars which were indistinguishable based upon standard morphological and pigmentation traits. The ability of alleles from a number of loci to be sized in one lane with an internal size standard offers rapid data acquisition (Figure 6) with accuracy of about 0.5 bp from gel to gel and 0.2 bp within a gel. Techniques are available to further improve gel to gel precision (Smith 1995).

One disadvantage of the ABI system is the cost associated with the synthesis of fluorescently labeled PCR primers. This cost can be avoided by the use fluorescently labeled nucleotides. dGTP is available with three different fluorescent labels. These are used in the PCR reaction mix along with unlabeled dGTP. Both strands of the amplification product produced in this manner will fluoresce, thereby reducing the resolution of the ABI system. Nonetheless, fairly accurate allele size data can be obtained.

Other techniques for SSR allele detection: capillary electrophoresis

Capillary electrophoresis is a promising method for performing high-speed separation and sizing of DNA fragments. A number of different separation matrices including acrylamide, liquefied agarose, and polyvinyl alcohol have been used. Marino et al. (1994) separated SSR alleles from the soybean genotypes Williams and Jackson that differed by only 9 base pairs. The two alleles present in the F_1 of Jackson × Williams were easily distinguished. Capillary electrophoresis used a 100 μL, 3% polyacrylamide gel capillary with an effective length of 40 cm. Detection was by absorbance at 260 nm. ABI has a commercially available capillary electrophoresis system that uses fluorescent chemistry and detection similar to the DNA sequencer described above. This system can be used for microsatellite allele sizing.

The possibility of high speed parallel capillary electrophoresis separation and sizing of DNA fragments was suggested by Clark and Mathies (1993). Such a system would allow simultaneous electrophoretic separation and sizing of SSR alleles in a number of capillaries using a sieving medium such as hydroxyethyl-cellulose. The combining of this type of parallel separation technology with highly sensitive fluorescence detection has promise to dramatically increase the speed and sensitivity SSR allele size determination. These and other technologies that will result from the major investment in the Human Genome Project will have a major impact on plant genome research. In the near future rapid, accurate, and efficient molecular genotyping of large numbers of plant genotypes at multiple loci will be easily accomplished.

References

Akkaya MS, Shoemaker RC, Specht JE, Bhagwat AA, and Cregan PB (1995) *Crop Sci* 35:1439-1445.

Akkaya MS, Bhagwat AA and Cregan PB (1992) *Genetics* 132:1131-1139.

Bassam BJ, Caetano-Anollés G and Gresshoff PM (1991) *Anal Biochem* 196:80-83.

Bell CJ and Ecker JR (1994) *Genomics* 19:137-144.

Brunel D (1994) *Plant Mol Biol* 24:397-400.

Clark SM and Mathies RA (1993) *Anal Biochem* 215:163-170.

Cregan PB, Bhagwat AA, Akkaya MS and Rongwen J (1994) *Mol Cell Biol* 5:49-61.

Devos KM, Bryan GJ, Collins AJ, Stephenson P and Gale MD (1995) *Theor Appl Genet* 90:247-252.

Diwan N and Cregan PB (1997) *Theor Appl Genet,* mns submitted.

Hammond HA, Jin L, Zhong Y, Caskey CT and Chakraborty R (1994) *Am J Hum Genet* 55:175-189.

Hazan J, Dubay C, Pankowiak M-P, Becuwe N and Weissenbach J (1992) *Genomics* 12:183-189.

Hudson TJ, Stein LD, Gerety SS, Ma J, Castle AB, et al. (1995) *Science* 270:1945-1954.

Jarret RL, Bhat KV, Cregan P, Ortiz R and Vuylsteke D (1994) *Info Musa* 3:3-4.

Kimpton CP, Gill P, Walton A, Urquhart A, Millican ES and Adams M (1993) *PCR Methods Applic* 3:13-22.

Klinkicht M and Tautz D (1992) *Mol Ecol* 1:133-134.

Kresovich S, Szewc-McFadden AK and Bliek SM (1995) *Theor Appl Genet* 91:206-211.

Lavi U, Akkaya MS, Bhagwat AA, Lahav E and Cregan PB (1995) *Euphytica* 80:171-177.

Lagercrantz U, Ellegren H and Andersson L (1993) *Nucleic Acids Res* 21:1111-1115.

Liu Z-W, Biyashev RB and Saghai Maroof MA (1996) *Theor Appl Genet,* in press.

Ma ZQ, Roder M and Sorrells ME (1996) *Genome* 39:123-130.

Marino MA, Turni LA, DelRio SA, Williams PE and Cregan PB (1995) *Appl Theor Electrophoresis* 5:1-5.

Maughan PJ, Saghai Maroof MA and Buss GR (1995) *Genome* 38:715-723.

Morgante M and Olivieri AM (1993) *Plant J* 3:175-182.

Plaschke J, Ganal MW and Roder MS (1995) *Theor Appl Genet* 91:1001-1007.

Rongwen J, Akkaya MS, Lavi U, and Cregan PB (1995) Theor Appl Genet 90:43-48.

Saghai Maroof MA, Biyashev RB, Yang GP, Zhang Q and Allard RW (1994) *Proc Natl Acad Sci USA* 91:5466-5470.

Sambrook J, Fritsch EF and Maniatis T (1989) *Molecular Cloning: a laboratory manual* Second Edition. Cold Spring Harbor Laboratory, Cold Spring Harbor, NY.

Senior ML, Chin ECL, Lee M, Stuber CW and Smith JSC (1996) *Crop Sci*, in press.

Senior ML and Heun M (1993) *Genome* 36:884-889.

Smith DN and Devey ME (1994) *Genome* 37:977-983.

Smith RN (1995) *Biotechniques* 18:122-128.

Terauchi R and Konuma A (1994) *Genome* 37:794-801.

Thomas MR and Scott NS (1993) *Theor Appl Genet* 86:985-990.

Vignal A, Gyapay G, Hazan J, Nguyen S, Dupraz C, Cheron N, Becuwe N, Tranchant M, Weissenbach J (1993) In Adolph KW (ed), *Methods in Molecular Genetics—Gene and Chromosome Analysis*, Part A. Academic Press, New York. pp. 211-221.

Von-Deimling A, Bender B, Louis DN and Wiestler OD (1993) *Neuropathol Appl Neurobiol* 19:524-529.

Weissenbach, J. G. Gyapay, C. Dib, A. Vignal and J. Morissette (1992) *Nature* 359:794-801.

Wu K-S and Tanksley, SD (1993) *Mol Gen Genet* 241:225-235.

Zhao X and Kochert G (1993) *Plant Mol Biol* 21:607-614.

Use of cleaved amplified polymorphic sequences (CAPS) for genetic mapping and typing

Eliana Drenkard[1], Jane Glazebrook[2], Daphne Preuss[3], and Frederick M. Ausubel[1]
Department of Genetics, Harvard Medical School, and Department of Molecular Biology, Massachusetts General Hospital, Boston, MA 02114[1], Center for Agricultural Biotechnology, University of Maryland Biotechnology Institute, College Park, MD 20742[2], and Department of Molecular Genetics and Cell Biology, University of Chicago, Chicago, IL 60637[3], USA

Introduction

This chapter describes the use of cleaved amplified polymorphic sequences (CAPS) for genetic mapping and bacterial typing. As illustrated in Figure 1, the CAPS method utilizes amplified DNA fragments digested with a restriction endonuclease to display restriction site polymorphisms (Konieczny and Ausubel 1993,Tragoonrung et al. 1992, Weining and Langridge 1991, Williams et al. 1991). CAPS offer several advantages as molecular markers. First, because the sizes of the cleaved and uncleaved amplification products can be adjusted arbitrarily by the appropriate placement of the PCR primers, CAPS markers can be readily assayed using standard agarose gel electrophoresis. Second CAPS markers are co-dominant when used as genetic markers for mapping in eukaryotes. That is, when an organism is heterozygous for the cleaved and non-cleaved alleles of a particular CAPS marker, the gel electrophoresis pattern that is obtained is different from the patterns obtained when the CAPS marker is homozygous for either the cleaved or non-cleaved alleles. This means that relatively good map positions can be obtained using a relatively small number of F_2 recombinants, a major advantage for mapping mutations whose phenotypes are difficult to score. Third, only a small quantity of DNA is needed to assay CAPS markers. For example, DNA for CAPS mapping in plants can be prepared from a portion of a single leaf using a rapid DNA isolation procedure. Fourth, the CAPS mapping procedure is technically simple but the results are robust because an amplification product is always obtained. Fifth, CAPS markers can be assayed relatively quickly. For example, in the case of eukaryotic gene mapping, assuming that F_2 recombinants are available, a gene can be mapped by a single investigator in less than two weeks. Finally, as discussed below, the CAPS method is amenable to a variety of automation possibilities.

In the sections that follow, we describe two specific applications of CAPS markers. The first is the use of CAPS markers for genetic mapping in the model flowering plant *Arabidopsis thaliana* ("*Arabidopsis*" for simplicity). The second application is for subspecies typing of bacterial strains. In principle, the CAPS method could be adapted for use in genetic mapping in a variety of eukaryotic species, including humans, and for typing "strains" or individuals in any species.

Development of CAPS markers for mapping in Arabidopsis thaliana

Konieczny and Ausubel (1993) first adapted the CAPS procedure for genetic mapping by developing a set of CAPS markers for use with *Arabidopsis*. Mapping in *Arabidopsis* using DNA markers is accomplished by crossing two polymorphic ecotypes (wild accessions or isolates). Because *Arabidopsis* preferentially self-fertilizes, it is easy to maintain homozygous inbred ecotypes in the laboratory. Most mapping in *Arabidopsis* has been carried out with two particular inbred ecotypes, 'Columbia' and 'Landsberg *erecta*' which exhibit polymorphisms every 100-250 base pairs on average (Chang et al. 1988, Konieczny and Ausubel 1993). Starting with mapped and sequenced *Arabidopsis* DNA fragments, Konieczny and Ausubel (1993) designed 18 sets of primers which amplified products ranging in size from 0.316 to 1.728 kb from both the Columbia and Landsberg *erecta* ecotypes. These amplified products were digested with a panel of restriction endonucleases to identify restriction enzymes that generated ecotype-specific patterns. Using this set of CAPS markers, Konieczny and Ausubel (1993) showed that only 28 F_2 progeny plants derived from a single cross were required to unambiguously map an *Arabidopsis* gene to one of the ten *Arabidopsis* chromosome arms. Subsequently, many additional CAPS markers have been developed for *Arabidopsis* mapping in different laboratories. The availability of Columbia × Landsberg *erecta* recombinant inbred lines facilitates the mapping of newly developed CAPS markers that do not correspond to mapped genes or sequences (Lister and Dean 1993). Currently, a set of 74 *Arabidopsis* CAPS markers have been catalogued by one of us (E. D.) and can be accessed via the World Wide Web browser[1].

Table 1 lists a selection of suggested *Arabidopsis* CAPS markers evenly distributed throughout the five *Arabidopsis* chromosomes that could be used to map new genes. The selection of the markers listed on the table is based on their relatively accurate map position determined on the Columbia × Landsberg *erecta* recombinant inbred lines (Lister and Dean 1993), low cost of restriction enzymes used to detect polymorphic sites, feasibility of detection of polymorphic pattern in standard 1.5% agarose gel, and reproducibility of data generated for a particular marker. Alternative markers that use more expensive enzymes are listed in cases where not enough information is available to fill in gaps present for a particular chromosome (PAB5/Chromosome 1 and m246/Chromosome 2).

---

[1] http://weeds.mgh.harvard.edu/ausubel.htm/ or http://genome-www.standford.-edu/Arabidopsis/ (look for *Arabidopsis* information: CAPS marker information and tables for each chromosome).

Newly developed *Arabidopsis* CAPS markers can be submitted to this data base by sending an e-mail containing the relevant information to F. M. Ausubel or E. Drenkard at ausubel or drenkard@frodo.mgh.harvard.edu.

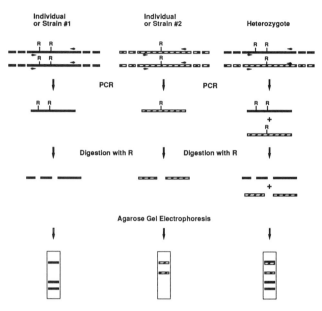

FIGURE 1. Assaying CAPS markers by agarose gel electrophoresis. The cartoon illustrates a CAPS marker which utilizes a diagnostic restriction endonuclease ("R") that cleaves the amplified fragment at either one or two sites. DNA isolated from individual or strain #1, individual or strain #2, or from an interspecific hybrid between #1 and #2, is used as a template for the PCR using synthetic oligonucleotides as primers (represented by short horizontal arrows) to amplify a DNA fragment. "R" represents the recognition sequence for restriction endonuclease R. The amplified fragment is subjected to digestion with restriction enzyme R and the digested fragments are subjected to agarose gel electrophoresis. The vertical rectangular boxes at the bottom of each panel represent an agarose gel and the short horizontal lines represent the patterns of bands observed following electrophoresis and staining with ethidium bromide.

Although the number of *Arabidopsis* CAPS markers continues to grow, the usefulness of the current set of CAPS markers is limited by the fact that a variety of different restriction endonucleases, some of which are relatively expensive, are required to display the polymorphisms. A large set of CAPS markers that all utilize the same restriction enzyme would greatly simplify the CAPS procedure and make it amenable to automation. In principle, developing such a set of CAPS markers is feasible for *Arabidopsis* because restriction site polymorphisms are abundant and are well distributed throughout the genome (Chang et al. 1988, Konieczny and Ausubel 1993). For example, it is possible to calculate that there are 5,000 to 13,000 polymorphic *Alu*I sites between the Columbia and Landsberg *erecta* ecotypes allowing the potential construction of very dense CAPS maps.

In the future by judiciously choosing the location of the CAPS primers, it will be possible to design CAPS markers with the following useful features. First, the primary PCR products for all CAPS markers could be approximately the same size. Second, the amplified products could be cleaved once in one ecotype and twice in the other as illustrated in Figure 1. This feature allows verification that the restriction endonuclease cleavage step has been successful. Third, in the case of the Columbia and Landsberg *erecta* ecotypes, which are the most commonly used for mapping, all of the Columbia CAPS alleles could be chosen to be the ones that are cleaved once; the corresponding Landsberg *erecta* alleles would be cleaved twice (or vice versa). With the above design features, the resulting simplicity of the restriction pattern that distinguishes the two parents and the heterozygote makes allele assignment very straightforward and facilitates automation using a gel scanning device.

An additional advantage of the design features described in the preceding paragraph is that it facilitates analysis of the data resulting from mapping strategies that involve pooling of DNAs from individual F_2 plants prior to amplification with the CAPS primers. This method is based on the principle of homozygosity mapping (Churchill et al. 1993 Lander and Botstein 1987, Michelmore et al. 1991). Briefly, in a group of F_2 mutants, loci that are not linked to the mutation will be represented equally by the Landsberg *erecta* and Columbia alleles. Alternatively, for any locus that is linked to the mutation, the progeny will be strongly biased. For example, if the mutation was originally isolated in a Landsberg *erecta* parent, then among the pooled mutants, amplification with closely linked CAPS markers should result in primarily a Landsberg *erecta* CAPS digestion pattern.

Using CAPS Markers for Typing Bacterial Strains

Calderwood et al. (1996) adapted the use of CAPS markers for typing strains of the human bacterial pathogen *Staphylococcus epidermidis*. Using an approach similar to the one used by Konieczny and Ausubel (1993) to identify CAPS markers for *Arabidopsis*, Calderwood et al. (1996) synthesized seven pairs of oligonucleotide primers corresponding to five previously sequenced *S. epidermidis* genes and then used these primers to amplify DNA sequences from 33 *S. epidermidis* strains. Using a panel of restriction endonucleases, seven restriction site polymorphisms were identified in five of the amplified products. Each fragment-enzyme combination that was polymorphic demonstrated only two alleles in the 35 *S. epidermidis* isolates studied, corresponding to the presence or absence of a single restriction site. Overall, five distinct combinations of CAPS alleles were detected. The categorization of the strains using CAPS correlated well with epidemiologic information as well as with categorization of the strains based on pulse field gel electrophoresis.

The use of CAPS for bacterial typing appears to be technically simpler, more reproducible and/or more discriminatory than most previous methods used for subspecies typing of bacteria including biotyping, antimicrobial sensitivity profiles, serotyping, bacteriophage and bacteriocin typing, multilocus electrophoresis, plasmid profiles, ribotyping, pulse field gel electrophoresis, and arbitrarily primed PCR analysis.

Alternative methods for bacterial typing using arbitrary primers and the amplification of repetitive sequences or previously amplified DNA is described in Chapters 10 and 21. As full genome sequences become available for a variety of bacterial species, it will become straightforward to develop sets of CAPS markers for different species based on polymorphisms in the recognition sites of a particular inexpensive and robust restriction endonuclease such as *Eco*RI. The CAPS typing method could be readily adapted to typing individuals or strains of any species for which substantial DNA sequence information is available.

The next section contains protocols for assaying CAPS markers in plants *(Arabidopsis)* and bacteria. It should be readily feasible to adapt these protocols for assaying CAPS markers in other organisms.

Materials

Plant DNA mini-preparation

1. Extraction buffer: 100 mM Tris, 50 mM EDTA, 500 mM NaCl, 10 mM β-mercaptoethanol, pH 8. If the β-mercaptoethanol is added just before use, this solution can be stored indefinitely at room temperature.
2. 3.0 M sodium acetate adjusted to pH 5.2 with acetic acid.
3. BTE: 50 mM Tris, 1.0 mM EDTA pH 8.
4. TE: 10 mM Tris, 1.0 mM EDTA pH 8.
5. RNAse A: 10 mg/mL DNAse-free RNAse A dissolved in water and stored in aliquots at -20°C.
6. 20% SDS.
7. 5.0 M potassium acetate.
8. Isopropanol.
9. Ethanol.
10. Liquid nitrogen.
11. Microcentrifuge.
12. Vortex mixer.
13. Water bath or incubator set at 65-70°C.
14. Ice.
15. Mortar and pestle.
16. 1.5 ml microcentrifuge tubes.

Isolation of bacterial DNA from colonies (boiling method)

1. Water bath or incubator set at 100°C.
2. Microcentrifuge.
3. 1.5 mL microcentrifuge tubes.

CAPS reaction

1. 2.5 mM dNTPs: 2.5 mM of each dATP, dCTP, dGTP and dTTP. Dilute commercial molecular biology grade 100 mM solutions and store in aliquots at -20°C.

2. Appropriate forward and reverse primers
3. *Taq* polymerase.
4. *Taq* polymerase 10× buffer (provided by the *Taq* polymerase supplier).
5. Thermal cycler.
6. Tubes that fit the thermal cycler.

Restriction enzyme digestions

1. Appropriate restriction enzyme.
2. 10× concentrated restriction enzyme buffer.
3. Water bath or incubator set at the appropriate temperature for the restriction enzyme used.

Experimental protocol

Mini-preparation of plant DNA

1. Harvest approximately 0.1 g of fresh plant tissue and freeze in a 1.5 mL microcentrifuge tube using liquid nitrogen or dry ice. The amount of tissue may vary between 0.05 and 0.2 g, may consist of a mixture of leaves, stems and flowers or seedlings, and may be stored frozen at -20°C for at least 6 moths before processing without adverse effects.
2. Grind the frozen tissue to a fine powder in a mortar and pestle.
3. Transfer frozen powder to a 1.5 mL microcentrifuge tube containing 0.5 mL of extraction buffer and vortex for 10 s. Keep tubes on ice while processing the rest of the samples.
4. Add 35 μL of 20% SDS and vortex briefly. Incubate 10 min at 65-70°C.
5. Add 130 μL of 5.0 M potassium acetate. Mix thoroughly by shaking tube back and forth and incubate on ice for 5 min.
6. Spin 5 min in a microcentrifuge at 14,000 rpm to remove undigested debris.
7. Transfer the supernatant to a fresh tube (~700 μL), and precipitate nucleic acids by adding 60 μL of 3 M sodium acetate and 640 μL isopropanol.
   From this step on, samples should be handled gently to decrease shearing of the DNA.
8. Spin 5 min in microcentrifuge at 14,000 rpm and discard supernatant. Resuspend pellet in 200 μL BTE and spin again at 14,000 rpm to discard insoluble material.
9. Transfer supernatant to a fresh tube and reprecipitate by adding 1/10 volumes of 3 M sodium acetate and 2 volumes of ethanol.
10. Resuspend pellet in 100 μL of TE and add 0.01 μg/μL RNase A. Incubate 1 h at 37°C or overnight at 4°C.
11. Reprecipitate DNA as in step 9 and centrifuge 10 min at 14,000 in microcentrifuge. Wash the DNA with 70% ethanol and centrifuge 5 min to obtain a DNA pellet. This second ethanol precipitation improves the amplification of the DNA in PCR reactions.
12. Carefully remove the supernatant and briefly air dry the final pellet. Resuspend it in 50 μL of TE buffer.

Boiling method for bacterial colonies

1. Transfer a large (1-2 mm) colony into a 1.5 mL microcentrifuge tube containing 20 μL water and boil for about 5 min. Agar carry-over should be minimized when picking colonies. The use of sterile toothpicks is preferred to transfers with wire loop.
2. Centrifuge 5 min in a microcentrifuge at 14,000 rpm to pellet debris and transfer supernatant to a fresh tube. Add 1 μL of supernatant to the CAPS reaction.

CAPS reaction

1. For each DNA sample to be tested mix the following reagents in a microcentrifuge tube:
   - 0.5 μL 2.5 mM dNTPs.
   - 1 μL 20 ng/μL forward primer.
   - 1 μL 20 ng/μL reverse primer.
   - 1 μL 10× *Taq* buffer (supplied with the enzyme).
   - 0.1 μL (0.5 units) *Taq* polymerase.
   - 5.5 μL water.
2. Dispense 9 μL aliquots into microcentrifuge tubes to be used for amplification reaction. Add 1 μL of DNA samples prepared by methods described above to each tube. The amount of DNA in 1 μL of a plant DNA mini-prep ranges from 10 ng to 100 ng, but the PCR reaction works well over this range. For cesium purified DNA samples 8 ng of DNA per reaction gives good results. The use of 20 ng of oligonucleotides per reaction gives good amplification, especially when the DNA samples are prepared using a mini-prep procedure.
3. Insert the tubes into a thermal cycler, and amplify using the following cycle: 2 min at 95°C, 50 cycles of 1 min at 95°C, 1 min at 56°C, and 3 min at 72°C, and 10 min at 72°C. If the thermal cycler is not equipped with a heated lid, a small amount of mineral oil must be added to the samples to prevent them from condensing on the lids of the tubes during the reaction. The recommended amplification program gives good results in standard tubes using water-cooled thermal cyclers or Peltier-effect machines. Investigators using sophisticated thermal cyclers and thin-walled tubes may find that the incubation times in the cycling reaction can be shortened considerably, and that the product yield can be sufficiently high that it is desirable to cleave only a portion of the product to reduce the amount of enzyme required.

Restriction enzyme digestion and analysis of amplified products

1. For each sample to be tested, mix the following:
   - 2 μL appropriate 10× restriction enzyme buffer.
   - 8 μL water.
   - 2-10 units appropriate restriction enzyme.
2. Add 10 μL of the premix (from step 1) to each sample tube containing the amplified product and incubate at a suitable temperature for 2 h.
3. Add 2 μL of DNA loading dye. Separate the products on an agarose gel. Many of the CAPS markers can be successfully scored after electrophoresis in a standard 1.5% agarose gel. Resolution of fragments less than 500 bp long is greatly improved by

using agarose gels consisting of 1% conventional agarose (for strength) and 2% NuSieve GTG agarose (supplied by American Bioanalytical, Natick, MA, USA). Other commercial agarose preparations that can be used at high concentrations should also give good results.

TABLE 1. Selected *Arabidopsis thaliana* CAPS markers

Marker	Chromosome no. (Map position) Fragment size (kb)	Enzyme: ecotype(s)[a], no. of cuts (size of products in kb)	Primers (Source[c] or Reference[d])
PAI1[b]	1 (9) 1.347	*FokI*: C, 3 (0.882, 0.368, 0.101, 0.097); L, 2 (0.979, 0.368, 0.101)	5'-GTAGAGGATTGAGCTTAAGGCAAG-3' 5'-AGCACGCGAACCATTACCATGAAG-3' (12)
m235[b]	1 (44.5) 0.534	*HindIII*: C=Nd, 1 (0.309, 0.225); L, 0 (0.534)	5'-GAATCTGTTTCGCCTAACGC-3' 5'-AGTCCACAACAATTGCAGCC-3' (5)
GAPB[b]	1 (86.2) 1.481	*DdeI*: C=CV=R, 3 (0.605, 0.284, 0.225, 0.174); L, 4 (0.35, 0.284, 0.255, 0.225, 0.174)	5'-TCTGATCAGTTGCAGCTATG-3' 5'-GGCACTATGTTCAGTGCTG-3' (1,6,11)
PAB5[b*]	1 (181.0) 0.885	*MboII*: C, 3 (0.360, 0.320, 0.165, 0.04); L, 2 (0.485, 0.36, 0.04)	5'-AGTATCATCAAAATCGAGAGATTG-3' 5'-GATGCAACCGCCGCAGCCATAGCGAT AAGA-3' (9)
ADH[b]	1 (187.1) 1.291	*XbaI*: C=CV, 0 (1.291); L=C24=R, 1 (1.097, 0.262)	5'-GCGTGACCATCAAGACTAAT-3' 5'-AAAAATGGCAACACTTTGAC-3' (1,2,6,11)
m246[b*]	2 (14.1) 1.354	*MaeIII*: C, 0 (1.354); L=C24=CV=N=R, 1 (1.122, 0.232)	5'-TGAAGAGCTATCCGAGATGG-3' 5'-GCTTGAACTCCTCCTCCTTC-3' (1,2,6,11)
GPA1[b]	2 (48.4) 1.594	*AflII*: C=C24=CV, 2 (0.705, 0.680, 0.209); L=R, 1 (1.385, 0.209)	5'-GGGATTTGATGAAGGAGAAC-3' 5'-ATTCCTTGGTCTCCATCATC-3' (1,2,6,11)
m429[b]	2 (80.0) 0.316	*ScrFI*: C=N, 0 (0.316); L=C24=CV=R, 1 (0.216, 0.100)	5'-TGGTAACATGTTGGCTCTATAATTG-3' 5'-GGCAGTTATTATGAATGTCTGCATG-3' (1,2,6,11)
GAPC[b]	3 (3.8) 1.148	*EcoRV*: C=CV=N=R, 1 (0.735, 0.713); L=C24, 2 (0.713, 0.39, 0.34)	5'-CTGTTATCGTTAGGATTCGG-3' 5'-ACGGAAAGACATTCCAGTC-3' (1,2,6,11)
g4711[b]	3 (48.6) 1,5	*HindIII*: C, 0 (1.5); L, 1 (1.0, 0.5)	5'-ACCAAATCTTCGTGGGGCTCAGCAG-3' 5'-CCTGTGAAAAACGACGTGCAGTTTC-3' (3)

GL1[b]	3 (64.6) 0.519	*TaqI*: C, 3 (0.298, 0.1, 0.074, 0.047); L=C24, 2 (0.372, 0.1, 0.047); CV=R, 1 (0.419, 0.12)	5'-ATATTGAGTACTGCCTTTAG-3' 5'-CCATGATCCGAAGAGACTAT-3' (1,2,6,11)
BGL1[b]	3 (114.1) 1.269	*RsaI*: C, 2 (0.785, 0.34, 0.105); L=CV=N=R, 1 (0.785, 0.485)	5'-TCTTCTCGGTCTATTCTTCG-3' 5'-TTATCACCATAACGTCTCCC-3' (1,6,11)
GA1[b]	4 (23.5) 1.196	*BsaBI*: C=CV=N=R, 1 (0.707, 0.527); L=C24, 0 (1.196)	5'-AAGCTTCGAACTCAAGGTTC-3' 5'-CCGGAGAATCGTACGGTAC-3' (1,2,6,11)
g4539[b]	4 (67.2) 0.6	*HindIII*: C=CV=R, 0 (0.6); L=Ws, 1 (0.48, 0.12)	5'-GGTCATCCGTTCCCAGGTAAAG-3' 5'-GGACGTAGAATCTGAGAGCTC-3' (1,6,7)
RPS2[b]	4 (86.2) 0.785	*Sau3A I*: C, 2 (0.18, 0.605); L, 3 (0.18, 0.251, 0.354)	5'-CTCAGAGTCTTGGACTTGTCG-3' 5'-TTCGACGGATGGACTCTCGTG-3' (4)
DHS1[b]	4 (124.8) 1.668	*BsaAI*: C, 1 (1.188, 0.48); L=C24=CV=N=R, 0 (1.668)	5'-CAAGTGACCTGAAGAGTATCG-3' 5'-AGAGAGAATGAGAAATGGAGG-3' (1,2,6,11)
ASA1[b]	5 (16.4) 1.728	*BclI*: C=C24=CV, 1 (1.042, 0.686); L=R, 2 (0.686, 0.553, 0.489)	5'-CTTACTCCTGTTCTTGCTTAC-3' 5'-CCTCTAGCCTGAATAACAGAAC-3' (1,2,6,11)
PAT1[b]	5 (approx. 25) 1.9	*DdeI*: C, 2 (1.0, 0.6, 0.3); L=W, 1 (1.0, 0.9)	5'-GTCGACGTGGTGCGGTGGGTTG-3' 5'-GTATGAGAACATAGTAACCCCATG-3' (10)
PHYC[b]	5 (78.5) approx. 2.0	*PstI*: C (1.7, pieces < 0.3); L, (0.8, 0.7, several pieces < 0.3)	5'-CCTAATGGAGAATCATTCGG-3' 5'-CTACAGAATCGTCCTCAACG-3' (8)
DFR[b]	5 (101.8) 1.143	*BsaAI*: C=C24=CV=N, 1 (0.609, 0.534); L=R, 2 (0.609, 0.318, 0.216)	5'-AGATCCTGAGGTGAGTTTTC-3' 5'-TGTTACATGGCTTCATACCA-3' (1,2,6,11)
LFY3[b]	5 (133.5) 1.33	*RsaI*: C, 5 (0.708, 0.236, 0.147, 0.126, 0.078, 0.035); L=C24=CV=N=R, 4 (0.855, 0.236, 0.126, 0.078, 0.035)	5'-TAACTTATCGGGCTTCTGC-3' 5'-GACGGCGTCTAGAAGATTC-3' (1,2,6,11)

[a]C=Columbia, CV=Cvi, L=Landsberg *erecta*, Nd=Niederzenz, N=NossenR=RLD, W=Wassilewskija

[b]Marker mapped on RI lines (Lister and Dean 1993).

[c]Source no. (unpublished): (1) Alonso Blanco, C. (CVi), (2) Altmann, T. (C24), J., (3) Diener, A., (4) Glazebrook, J. and Rogers, E. (5) Hardtke, C. and Berleth, T., (6) Ichikawa, H. (RLD), (7) Parker, J., (8) Poole, D.

[d]Reference no.: (9) Belostotsky and Meagher (1993), (10) Bender and Fink (1995), (11) Konieczny and Ausubel (1993), (12) Li et al. (1995).

*Note: The enzyme used to detect the polymorphism is very expensive.

Comments

Use of CAPS for genetic mapping in Arabidopsis

A minimum of 20 F_2 plants or F_3 families is required to establish linkage to one of the 21 CAPS markers listed in Table 1. Most of the markers in the table are less than 21.5 centiMorgans (cM) away from any arbitrary position, and calculations based on the binomial distribution show that linkage between a gene of interest and a marker less than 21.5 cM away could be detected with 95 % confidence using 20 plants that are homozygous for that gene. In the cases where weak linkage is detected when using one marker, other markers present in the vicinity of the linked one must be checked in order to get a more accurate position. Some scientists find it convenient to start with 25-30 plants in order to increase the probability of detecting linkage between one marker and the gene of interest.

A useful strategy to reduce the number of CAPS markers to be examined is to start testing CAPS markers near to the center of each chromosome and then proceed with markers approximately 50 cM away from the previous ones, until linkage is detected. Subsequent testing of other markers around the linked one will provide a more precise position for the gene of interest. Another consideration when using CAPS mapping is that it requires significant quantities of several restriction enzymes, which vary widely in price. A cost-effective strategy is to first test markers that depend on inexpensive enzymes, and that can be analyzed on 1.5 % agarose gels (less expensive than NuSieve agarose).

The use of the markers listed in Table 1 could prove to be useful during the first steps of a mapping strategy since they combine several of the characteristics mentioned above (low cost of enzyme, detection of polymorphism on 1.5% agarose gel) and additionally they are evenly distributed throughout the 5 chromosomes.

Estimation of genetic map distances using CAPS in Arabidopsis

To estimate the genetic map distance between a CAPS marker and a gene of interest when the population used for mapping is composed of F_2 plants that were derived from a cross between a homozygous mutant of ecotype A and ecotype B, that are homozygous for the mutation in the gene of interest:

1. Count the number of *chromosomes* in which the CAPS marker is either the ecotype A or B allele. Plants in which the CAPS marker is homozygous for the A allele count as two A chromosomes. The same consideration is applicable when the marker is homozygous for the B allele, and should be counted as two B chromosomes. Heterozygous plants count as one A chromosome and one B chromosome.
2. Calculate recombination frequency (r) between the gene of interest and the CAPS marker as the number of B chromosomes/total chromosomes.

In converting recombination frequency to map distance, it is necessary to consider that chromosomes in which two recombination events occurred between the markers are counted as having no recombination events, and that this source of error increases with increasing distance between the markers. Furthermore, recombination events can influence the probabilities of a second recombination event occurring in the vicinity, a phenomenon called interference. For *Arabidopsis*, a reasonable estimate of map distance is given by the Kosambi function:

$$D = 25 \ln \frac{100 + 2r}{100 - 2r}$$

where r is the recombination frequency expressed as a percentage, and D is the distance in cM (Koorneef and Stam 1992).

Analysis of patterns obtained using CAPS in bacterial strain typing

The basic criteria for interpreting data originating from the use of CAPS in bacterial strain typing is to attribute the different patterns of fragments generated by each restriction enzyme to allelic variations at the locus amplified with a particular primer combination. A CAPS type is defined as a unique multilocus genotype and consists of a particular combination of alleles (Calderwood et al. 1996). Genetic distances (D) between pairs of CAPS types are determined as the proportion of loci at which dissimilar alleles are present among isolates (Calderwood et al. 1996).

References

Belostotsky, D.A. and Meagher, R.B. (1993) *Proc Natl Acad Sci USA* 90:6686-6690.

Bender, J. and Fink, G.R. (1995) *Cell* 83:725-734.

Calderwood, S.B., Baker, M.A., Carroll, P.A., Michel, J.L., Arbeit, R.D. and Ausubel, F.M. (1996) *J Clin Microbiol*, in press.

Chang, C., Bowman, J.L., de John, A.W., Lander, E.S. and Meyerowitz, E.M. (1988). *Proc Natl Acad Sci USA* 85:6856-6860.

Churchill, G.A., Giovannoni, J.J. and Tanksley, S.D. (1993) *Proc Natl Acad Sci USA* 90:16-20.

Konieczny, A. and Ausubel, F.M. (1993) *Plant J* 4:403-410.

Koorneef, M. and Stam, P. (1992) Genetic Analysis. In Koncz, C., Chua, N.-H. and Schell, J. (eds.), Methods in Arabidopsis Research, World Scientific, Singapore, pp 83-99.

Lander, E.S. and Botstein, D. (1987) *Science* 236:1567-1570.

Li, J., Zhao, J., Rose, A.B., Schmidt, R., and Last, R.L. (1995) *Plant Cell* 7:447-461.

Lister, C. and Dean, C. (1993) *Plant J* 4:745-750.

Michelmore, R.W., Paran, I. and Kesseli, R.V. (1991) *Proc Natl Acad Sci USA* 88:9828-9832.

Tragoonrung, S., Kanizin, V., Hayes, P.M. and Blake, T.K. (1992) *Theor Appl Genet* 84:1002-1008.

Weining, S. and Langridge, P. (1991) *Theor Appl Genet* 82:209-216.

Williams, M.N.V., Pande, N., Nair, M., Mohan, M. and Bennet, J. (1991) *Theor Appl Genet* 82:489-498.

Genetic bit analysis: a solid-phase method for genotyping single nucleotide polymorphisms

Jennifer E. Reynolds, Steven R. Head, Tina C. McIntosh, Leslie P. Vrolijk and Michael T. Boyce-Jacino
Molecular Tool Inc., A GeneScreen Company, Alpha Center, Hopkins Bayview Research Campus, Baltimore, MD 21224, USA

Introduction

The most common genetic variant in the human genome is the single nucleotide polymorphism (SNP). Since SNP and insertion/deletions result in base changes that contribute to the majority of phenotypic diversity, methods to unequivocally genotype these sites are increasingly important in genetic research, gene mapping and clinical medicine. Studies that employ a variety of methods for the random identification of human DNA polymorphisms have led to the estimate that human haploid genomes differ from each other at approximately one in every 300 to 500 nucleotide positions (Botstein et al. 1980, Cooper et al. 1985, Nickerson et al. 1992) and that the majority of these variations are biallelic SNPs. In an analysis of random genomic fragments from the horse and human genomes, we found that SNPs are present approximately every 250 to 400 nucleotides.

Many techniques are currently employed in the typing of single nucleotide polymorphisms. A common technique is single-stranded conformational polymorphism analysis (SSCP; Orita et al. 1989, Weidner et al. 1994). This method is dependent upon electrophoretic resolution of DNA sequence conformation differences and is thus, difficult to carry out in an inexpensive and highly reproducible fashion. A multiplexed heteroduplex approach has also been applied to the detection of disease mutations (Prior et al. 1995), but this method is also difficult to standardize or automate. In allele-specific oligonucleotide hybridization (ASO; Saiki et al. 1989), the different thermal stabilities of hybrids formed between target DNA sequences and relatively short synthetic oligonucleotide probes complementary to one allele, but mismatched to another allele, are used to differentiate alleles. With this method, the conditions for each polymorphism have to be carefully optimized and small differences in hybridization or wash temperatures can significantly change the typing results. Another mutation detection technique, TaqMan ASO (Holland et al. 1991, Livak et al. 1995), takes advantage of the ability of *Taq* DNA polymerase to degrade oligonucleotide probes hybridized to target DNA during PCR, however, this method has been found to be somewhat impractical for genotyping diverse SNPs due to affinity requirements for all hybridizing oligonucleotides, regardless of allele. Other techniques, including Southern blot analysis (Southern 1975), denaturing gradient gel electrophoresis (Myers et al. 1987), amplification refractory mutation system (ARMS; Newton et al. 1989), and chemical cleavage methodologies (Cotton et al. 1988) are also commonly used to study single base variations, though these techniques are cumbersome and not amenable to automation. DNA fingerprinting technologies such as AFLP (Vos et al. 1995), random amplified polymorphic DNA (RAPD; Williams et al. 1990), DNA amplification fingerprinting (DAF; Caetano-Anollés et al. 1991) and arbitrarily-primed PCR (AP-PCR; Welsh and McClelland 1990) are useful for studying complex genomes, but not so for examining specific nucleotide changes in the genome, as is needed for specific mutation or polymorphism detection. In fact, direct fluorescence-based DNA sequencing (Hattori et al. 1993, Kwok et al. 1994) of a mutation is useful for determining the exact nucleotide variation in a sample, though of course, this technique is slow and rather expensive.

To address some of the limitations of gel-based technologies, solid-phase approaches to genotyping mutations and polymorphisms are being developed today. One such technique is the oligonucleotide ligation assay (OLA) that exploits the ability of DNA ligase to distinguish single nucleotide differences at positions complementary to the termini of co-terminal probing oligonucleotides (Nickerson et al. 1990). Despite the advantages of a solid-phase approach, this methodology has been noted to be sensitive to DNA hybridization conditions. A modification of this approach, coupled amplification and oligonucleotide ligation (CAL) analysis, has also been used for multiplexed genetic typing (Eggerding 1995; Eggerding et al. 1995). This procedure allows for the simultaneous amplification and genotyping of DNA in a single assay, however, because of this, the procedure requires highly optimized conditions and expensive equipment. Technologies have also been developed using solid-phase-bound arrays of oligonucleotides (Guo et al. 1994; Stimpson et al. 1995) designed to improve the capacity for larger-scale polymorphism detection. Of course, since these technologies are hybridization-based, they currently suffer from the same disadvantages as other allele-specific hybridization assays.

A number of groups have developed various solid-phase and gel-based methods designed to exploit the advantages of primer extension-based detection of SNP as a diagnostic tool (Syvanen et al. 1990, Sokolov 1990, Prezant and Fischel-Ghodsian 1992, Jalanko et al. 1992, Ugozzoli et al. 1992, Livak and Hainer 1994, Ihalainen et al. 1994), though most of these technologies remain dependent upon radioactivity and other cumbersome laboratory techniques. However, the addition of multiplexing and the use of fluorescent reagents (Tully et al. 1994) have improved the speed and simplicity of some of these primer-extension technologies.

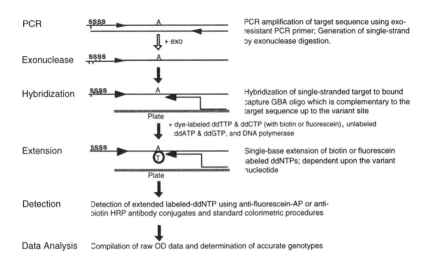

FIGURE 1. Genetic bit analysis (GBA).

In contrast, the solid-phase method GBA has distinct advantages in SNP genotyping since it relies upon the robust ability of DNA polymerase to distinguish single nucleotide differences in a format that is automatable and simple to execute (Nikiforov et al. 1994a, 1994b, 1995). The combination of common simple polymorphisms with solid-phase analysis also lends itself well to the automation of DNA manipulation and sequence analysis. Since the GBA test is performed in a microtiter plate format and exploits the highly developed enzyme-linked detection technology of ELISA, the operational advantages of the system are significant. The biochemistry can be automated using liquid handling robots, and the data acquisition and analysis of test results can be automated using readily available microtiter plate readers that transmit quantitative information to a computer database. This digitized data can then be automatically interpreted to provide customized reports and statistical information on genotype (or single-base sequencing) results. Standardization and reproducibility of GBA is facilitated by the stable attachment of capture and detection oligonucleotide primers to microtiter plates, which permits large scale batch preparation of the GBA plates, signal uniformity, and quality control of the test. Since the GBA methodology permits DNA polymorphic sequences to be analyzed under identical conditions, GBA is easy to apply to known DNA sequence polymorphisms and mutations. New mutations and polymorphisms can be rapidly incorporated either into existing tests or into new tests. These features also provide considerable flexibility to test design.

Molecular Tool's GBA system is a non-radioactive single-base sequencing method performed in microtiter plates and involves the following steps (Figure 1): (i) amplification by polymerase chain reaction of the target DNA sequence containing the variant base using one exonuclease-resistant primer per set of primer pairs; (ii) generation of a single-stranded template by exonuclease digestion; (iii) hybridization-capture of the target strand to a pre-synthesized GBA primer attached in the well of a microtiter plate.; (iv) extension of the GBA primer with haptenated (biotinylated or fluoresceinated) ddNTPs and the Klenow fragment of DNA-polymerase I; (v) detection of the extension product using ELISA (enzyme-linked immunoassay) colorimetry; and (vi) analysis of the colorimetric data to determine sample genotypes.

Materials

Enzymes

All enzymes used in GBA technology are available commercially. *Taq* DNA polymerase may be obtained from Perkin-Elmer. *Escherichia coli* DNA polymerase, Klenow fragment (exonuclease-free) and T7 gene 6 exonuclease are obtained from United States Biochemical.

Oligonucleotide synthesis

All oligonucleotides used in GBA are synthesized using standard phosphoramidite chemistry on an Applied Biosystems 392/394 DNA synthesizer, using reagents obtained from Glen Research. For the synthesis of phosphorothioate primers, tetraethylthiuram disulfide (TETD; Applied Biosystems) is recommended by the manufacturer. All oligonucleotides are deprotected with concentrated ammonia and desalted using NAP 5 (0.2 μmol scale synthesis) or NAP 25 (1 μmol) gel filtration columns (Pharmacia).

Labeled dideoxynucleoside triphosphates

Haptenated chain-terminating 2',3'-dideoxynucleoside triphosphates used in the single nucleotide extension reaction may be purchased from DuPont NEN, Amersham Lifesciences and Enzo Diagnostics. The DuPont compounds are derivatives of amino-propynyl-substituted 2',3'-dideoxypyrimidines or 2',3'-dideoxy-7-deazapurines. The chemistry of these chain terminators and their use in DNA sequencing have been previously described (Prober et al. 1987).

Solutions

1. 0.5 M and 50 mM N,N-dimethyloctylamine hydrochloride (ODA-HCl) stock: Mix 10.26 mL 95% N,N-dimethyloctylamine (Aldrich) with 80 mL water, adjust to pH 7.0 with concentrated HCl, and then bring up the volume to 100 mL. A 50 mM stock of ODA-HCl can then be prepared using standard dilution calculations.
2. 1× TNTw buffer: 10 mM Tris-HCl pH 7.5, 150 mM NaCl, 0.05% Tween 20.
3. 1× PCR buffer: 1.5 mM MgCl$_2$, 50 mM KCl, 10 mM Tris-HCl pH 8.5, 400 μM each dNTP.
4. 1× exonuclease buffer: 50 mM Tris-HCl pH 7.5, 1 mM DTT and 100 μg/mL acetylated BSA.

5. 3× hybridization salts: 4.5 M NaCl, 30 mM EDTA and 3 mM cetyltrimethyl-ammonium bromide (CTAB).
6. Polymerase extension mix: 1.5 μM of ddATP, ddCTP, ddGTP and ddUTP, 10 mM manganese-15 mM isocitrate, 10 mM MgCl$_2$, 25 mM NaCl, 20 mM Tris-HCl pH 8.0 and 0.015 units/μL of exonuclease-free Klenow polymerase. Two of ddNTPs are labeled separately with biotin and fluorescein depending on the bases to be assayed by GBA.

Experimental protocol

PCR and GBA primer design and qualification are described in brief in the "Comments" section of this paper, and several examples given. Updates to the biochemical protocol for GBA previously described by Nikiforov and colleagues (1994a, 1994b, and 1995) are given below. Figure 1 schematically illustrates the general steps involved in GBA.

Immobilization of GBA oligonucleotides onto 96-well plates

1. GBA primers (10pmol/well) are immobilized to microplates (Immulon 4; Dynatech) by overnight incubation, covered, at 37°C with 50 μL/well of 50 mM ODA-HCl.
2. After incubation, plates are washed three times with 1× TNTw to remove unbound oligonucleotide.

NOTE: This attachment method is non-covalent. Plates can be used immediately after incubation, refrigerated while still damp from the TNTw wash for up to one month, or dried under low heat (50°C) and stored in a desiccated environment before use. If stored in sealed pouches with desiccant, the plates are stable for at least 1 year.

PCR amplification

1. DNA samples (at 1-10 μg/mL) are amplified in 30 μL reactions containing 0.5 μM of each PCR primer and 0.025 units/μL *Taq* DNA polymerase in 1× PCR buffer. One PCR primer of each pair is phosphorothioate-protected at the 5' end as described below: Following an initial 2 min denaturation step at 95°C, reactions are amplified for 35 cycles of denaturation (30 s at 95°C), annealing (110 s at 55°C) and extension (155 s at 72°C) in a BioOven III thermocycler (Sun Bioscience). Successful PCR should yield >70 fmoles/ μl of PCR product.

NOTE: All post-PCR biochemical steps take place at room temperature. Use of dispensing instruments such as repeat pipetters, multichannel pipetters and robots such as the 96-tip Quadra (Tomtec), the 8-channel Multidrop dispenser (Titertek), and the Biomek robot (Beckman Instruments) provide for quick and accurate additions of reagents throughout PCR and GBA steps.

Preparation of single-stranded PCR fragments

To facilitate the hybridization reaction of the PCR product to the solid phase-bound GBA primer, the PCR product is selectively digested to a single strand through the use of a 5'-3' exonuclease, T7 gene 6 exonuclease. The hybridizing strand (complementary to the surface-bound oligo) is protected through the introduction of 4 phosphorothioate bonds at the 5' end of its PCR primer during oligonucleotide synthesis. The phosphorothioate modification is available through major oligonucleotide manufacturers.

1. Just before use, the exonuclease enzyme is diluted to a concentration of 3 units/μl in exonuclease buffer.
2. For a 30 μL PCR reaction, 6 μL of the diluted exonuclease (3 units/μL) are added to give a final concentration of 0.5 units/μL. The reaction is then incubated at room temperature for 1 h or at 37°C for 30 min. The reaction can be assayed for loss of double-stranded PCR product after incubation as observed by gel electrophoresis.
3. Following incubation, 18 μL of hybridization salts are added to the 36 μL of exonuclease-treated PCR products (total volume of 54 μL).

Hybridization

1. Following addition of hybridization salts, 25 μL of the single-stranded template mixture is added to the 96-well plate containing the immobilized GBA primer (see above) and is allowed to hybridize for 1 h at room temperature. The 54 μL final volume following the PCR, allows two well testing such as for duplicates or duplexed PCR. Periodic shaking of the plate can improve the hybridization rate.
2. After incubation, the plate is washed 3 times with 1× TNTw to remove any non-hybridized template. Wash solution can be removed after the third wash to prepare for extension or left in wells for storage at 4°C for up to 24 h.

NOTE: Usually, 20-30 μl of the exonuclease-treated PCR product in hybridization salts is sufficient for strong GBA signals.

Extension

1. Following the hybridization step, 30 μL of polymerase extension mix are added to each well and incubated at room temperature for 15 min. Labeled bases are selected based upon the sequence composition of the target. For example, a C to T mutation would be assayed using a mix of ddATP, fluorescein-ddCTP, ddGTP, and biotin-ddUTP; a G to A mutation using a mix of fluorescein-ddATP, ddCTP, biotin-ddGTP, and ddUTP, and so forth. Each of the different labeled bases can be assayed for independently as described below. Alternatively, a single labeled base can be used in two separate extension reactions. The protocol below assumes 2 base differential labeling in a single well.
2. After incubation, the extension reaction is quenched with 30 μL per well of 0.2 M EDTA pH 8.0, and the plates washed 3 times with 1x TNTw.
3. The plates are then washed once with 0.1 N NaOH to remove the hybridized template, followed by a final 3 washes with 1× TNTw in preparation for the detection steps. Plates can again be stored in wash solution at this point for up to 24 h prior to detection of the extension product.

Detection

The detection phase of the GBA involves two separate antibody additions and colorimetric detections. The fluorescein-labeled ddNTP is detected with an anti-fluorescein alkaline phosphatase (AP) conjugate (Boeringer-Mannheim; cat. no. 1426338), and the biotin-labeled ddNTP is detected with an anti-biotin horse-radish peroxidase (HRP) conjugate (Baxter, cat. no. 03-3720).

1. The anti-fluorescein Ab conjugate is diluted 1:1500 in 1× TNTw and 1% fraction V BSA (Sigma). A volume of 30 μL of this antibody solution is added per well followed by incubation at room temperature for 30 min. Note that antibody lots may vary in the optimal dilution but can be assayed by GBA to give endpoint readings as described below of >1.5 OD units.
2. After incubation, the 96-well plate is washed 6 times with 1× TNTw.
3. The AP substrate p-nitrophenyl phosphate (PNPP) (NPPD; Moss) is added, 100 μL per well, and the OD at 405 nm is recorded after 24 min.
4. The PNPP is then washed off with 3 washes of 1× TNTw. The plate is then ready for detection of the biotin-labeled ddNTP.
5. The anti-biotin-HRP Ab conjugate is diluted 1:500 in 1× TNTw and 1% fraction V BSA, and 30 μL are added to each well of a 96-well plate.
6. After incubation at room temperature for 30 min, the 96-well plate is washed 6 times with 1× TNTw.
7. The substrate tetramethylbenzidine (TMB) (TMBE; Moss) is then added, 100 μl per well, and the OD at 620nm recorded after 24 min.

Genotype determination

1. Genotypes may be determined directly from the raw OD data. Positive OD signals are generally considered those with OD values greater than 1.0, and negative OD signals are considered those less than 0.3. GBA system noise (either in the form of template independent or template dependent noise) is typically quantified by OD signals between 0.3 and 1.0. Samples displaying this noise in an otherwise noise-free GBA analysis are subsequently dropped from the genotype analysis. These threshold criteria may vary slightly depending upon the locus and the biochemical protocol employed.
2. Raw OD data from the GBA analysis is tabulated for each sample, and the absolute ratio of the signals (greater signal/ lesser signal) is calculated. In general, 405:620 nm signal ratios of at least 5:1 are considered acceptable for accurate genotype determination for homozygous signals assuming the positive signal is >1.0, and signal ratios of approximately 1:1 are expected for heterozygous samples (with OD values of greater than 1.0 for each base assayed). The more the signal ratios for homozygotes increase above 5:1, the more confident the genotype call. Likewise, the more the signal ratios deviate from 1:1 for the heterozygous samples, the less confident the genotype call.
3. Following the guidelines in steps 1 and 2, genotypes can be determined by examining both the calculated ratios of the two signals and the absolute signal values for each base assayed (X and Y signals). An "XX" homozygote is called if the X signal is at least 5 times greater than the Y signal, and the Y signal is below 0.3 (the threshold for noise in the system). A "YY" homozygote is called if the Y signal is at least 5 times greater than the X signal, and the X signal is below 0.3. An "XY" heterozygote is called if the X and Y signals are both greater than 1.0, and the ratio of the two signals

does not deviate from 1.0 ± 0.3. If the two signals are below 0.3, then the result is recorded as a "no signal" or "NS". These guidelines are conservative so as to ensure that only highly confident genotype calls are made. This is despite the fact that some of the "failed" genotypes might appear evident, as is the case in the example below for samples 5 (probable YY homozygote) and 6 (probable XY heterozygote).

Sample	X signal	Y signal	Ratio	Genotype	Comments
1	2.00	0.05	40:1	XX	–
2	0.10	1.65	16.5:1	YY	–
3	1.50	1.75	1.2:1	XY	–
4	0.06	0.10	1.7:1	NS	X and Y <0.3
5	0.33	1.98	6:1	fail	X signal >0.3
6	1.12	0.87	1.3:1	fail	Y signal <1.0

4. Since the data generated by GBA for a biallelic locus is binary, the results of a GBA can also be plotted on an x/y graph with x corresponding to the absorbance reading of the first base analyzed and y to the absorbance reading of the second base. Thus, a scatter-plot can be generated where each point represents the plot of X:Y signals for a particular sample. In a scatter-plot, homozygotes of Base 1 would lie close to the y axis (high signal in y, low in x), homozygotes for Base 2 would lie along the x axis, and heterozygotes would give signals for both bases and therefore, would lie along the diagonal. Using the scatter-plot format, both the raw OD values and the ratio of signals can be visually displayed for analysis.

Comments

PCR and GBA primer design and qualification

Following identification of a polymorphic site, PCR primers are designed using the software program Primer version 1.0 (Whitehead Institute). This software program allows the input of parameters such as product length, primer melting temperature, product melting temperature and optimum product annealing temperature. Typically, each primer pair will be qualified on two DNA control samples and a negative water control under standardized PCR conditions. The samples are then run on 15% non-denaturing polyacrylamide gels at 40W for 40 min. Amplification products can be quantified by comparison with multiple dilutions of a mass marker (Gibco BRL) and qualification of the primer pair is based on observation of the predicted fragment size and sufficient yield (greater than 70 fmoles/μL). Overall, GBA has been successfully performed on PCR products ranging from 50 to 450 bp in length, and our experience has shown that short products yield more robust results.

It is also important when designing PCR primers that their sequences do not overlap the GBA primer sequence. The GBA primers are designed to capture the single-stranded PCR product by hybridization and are then used to interrogate the polymorphic site by single-base polymerase-mediated extension. GBA primers are designed to be complementary to the single-stranded template, ending one base short of the polymorphic site (at its 3' end). Note that either DNA strand can be the target for the GBA primer as long as the GBA primer is complementary to the phosphorothioate-protected PCR strand. Selection of which strand to assay is based on evaluation of the candidate GBA primer for potential false-priming to other regions of the target (template dependent noise), potential self-priming (template independent noise), and basic features such as melting temperature. Analyses of potential noise in GBA primers are done by using the Oligo 4.0 program (National Biosciences). This program is used to determine whether the oligonucleotides are capable of forming stable secondary structures or self-annealing which may interfere with template hybridization. Self-priming reactions usually occur when structures with thermal energies (DG values) of greater than -6 Kcal are predicted involving complementarity of the 3' end of the GBA primer. In practice, we have found that a signal to noise ratio of 5 to 1 is acceptable, but signal to noise ratios of up to 100 to 1 can be achieved once the primer chemistry and reaction biochemistry have been fully optimized.

TABLE 1. PCR and GBA Primers

Locus	Primer	Pth[a]	5'-3' Primer Sequences	Primer size	Product size
ALDOB	PCR-U	–	AGATGCTGCCACCTCTTATCTACTT	25	147
	PCR-L	x	TACTGACCTCTACTGCCACATTTTC	25	
	GBA	–	*CTCGGATAAGCATTGTTTTCTT*	22	
CA2	PCR-U	–	AGAGTGCTGACTTCACTAACTTCGA	25	115
	PCR-L	x	ATCCAGGTCACACATTCCAGAAGAG	25	
	GBA	–	*TCCTCGTGGCCTCCTTCCTGAATCC*	25	
CST5	PCR-U	–	ACTCATGCCACAGACCTCAATGAC	24	81
	PCR-L	x	GCGGCTGTAGTACTCATCCTTATTA	25	
	GBA	–	*AGACCTCAATGACAAGAGTGTGCAG*	25	

[a]The phosphorothioated (Pth) or protected primer is indicated by an x in this column; The primer and product sizes are given in base pairs; GBA capture primers are listed in italics; PCR-U indicates the upper (sense) PCR primer, and PCR-L indicates the lower (antisense) PCR primer.

GBA analysis of three SNP markers

To illustrate the robustness of the GBA methodology for genotyping single nucleotide polymorphisms, we describe below various details on GBA genotyping of three human SNP markers, ALDOB, CA2, and CST5. These SNP loci were identified through literature review. Genbank accession numbers and references for the polymorphisms are as follows: ALDOB (M15657; Brooks and Tolan, 1993), CA2 (M77176; Venta and Tashian, 1991), and CST5 (X70377; Balbin et al., 1993). Table 1 lists the PCR and GBA primer sequences for these SNP loci. All primers were designed as described above. Figures 2 shows the design of GBA primers for loci CA2 and CST5. Both of these GBA primers interrogate the lower strand and are 25 bp in length.

Locus CA2:
```
       5' T  CCTCGTGGCC  TCCTTCCTGA  ATCC                           GBA PRIMER
ATTGAAGCTA  GGAGCACCGG  AGGAAGGACT  TAGGAACCTA  ATGACCTGGA          TEMPLATE
131         141         151         161  |     171
                                         A/G polymorphic site (either a T or C will
                                         incorporate at 3' site of GBA primer)
```
Locus CST5:
```
       5' AG  ACCTCAATGA  CAAGAGTGTG  CAG                           GBA PRIMER
GTACGGTGTC  TGGAGTTACT  GTTCTCACAC  GTCACACGGG  ACCTGAAACG          TEMPLATE
591         601         611         621  |     622
                                         A/G polymorphic site (either a T or C will
                                         incorporate at 3' site of GBA primer)
```

FIGURE 2. GBA primer design

Following the design and testing of the PCR and GBA primers for these loci, population testing of individual samples for the SNPs was possible. In each case, roughly 75 samples plus controls were tested for each SNP. A subset of data from testing at the ALDOB locus is given in Table 2 as an example.

TABLE 2: Raw OD data and genotype determination for ALDOB

Sample	C	T	Ratio	Genotype
1	0.18	2.22	12.3	TT
2	0.18	2.40	13.3	TT
3	0.19	2.36	12.4	TT
4	0.21	2.47	11.8	TT
5	0.23	2.65	11.5	TT
6	0.28	2.14	7.6	TT
7	0.28	2.26	8.1	TT
8	0.30	0.05	6.0	fail
9	0.35	2.43	6.9	TT
10	1.33	0.06	22.2	CC
11	1.76	1.90	1.1	TC
12	1.77	2.03	1.1	TC
13	1.80	1.93	1.1	TC
14	1.86	1.51	1.2	TC
15	2.00	2.18	1.1	TC
16	2.03	1.51	1.3	TC
17	0.21	0.07	3.0	NS
18	2.09	1.81	1.2	TC
19	2.17	0.27	8.0	CC
20	2.27	1.84	1.2	TC
21	2.33	0.24	9.7	CC
22	2.48	0.12	20.7	CC
23	2.58	0.30	8.6	CC
24	2.65	0.25	10.6	CC
25	2.67	0.12	22.3	CC

From: Process Group 7940006, Locus 700002, Plate 7600000134, Batch 11061

This table shows the raw OD values for each base assayed (a T/C polymorphism), the ratio of signals, and the genotype call using our conservative criterion. For ALDOB, the ddUTP nucleotide was labeled with biotin and ddCTP labeled with fluorescein.

An XY scatter-plot was used to visually access the raw data for ALDOB using the entire dataset, including positive and negative controls (Figure 3). Raw OD values for 405 nm were plotted on the x-axis, and raw 620nm OD values were plotted on the y-axis. As can be seen in Figure 3, three distinct sample populations were seen for this locus. Homozygotes of the 405 nm type (CC) lie along the x-axis, TT homozygotes lie on the y-axis, and heterozygotes (TC) lie along the diagonal. Negative PCR and GBA controls lie at the origin.

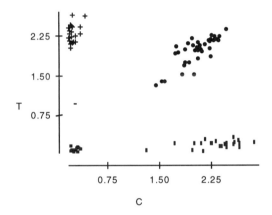

FIGURE 3: Scatter-plot for ALDOB. '+' symbols represent the TT homozygotes, '|' symbols represent the CC homozygotes, and the '●' symbols represent TC heterozygotes; Negative control samples at the origin are "no signals" and the '-' sample along the Y axis is a failure; Data is from Process Group 7940006, Locus 700002, Plate 7600000134, Batch 11061.

Limitations

The most common source of noise is caused by complementarity between the 3' end of the GBA primer with itself (template independent noise) or other sequences found on the template strand. Modification of the GBA primer can solve many of these problems. Occasionally the PCR primers left over from the PCR step will hybridize to the GBA primer and interfere with template hybridization. More stringent hybridization conditions (250 mM NaCl), as well as redesign of PCR primers, can help solve this problem. On rare occasions, cross reactivity between certain sequences of the GBA primer and the antibody-conjugates are seen. This can be determined by applying the antibody-conjugate to unextended GBA primer and measuring the signal.

Overview

The advantages of a solid-phase primer-extension genotyping technology such as GBA are many fold. In contrast to gel-based testing, solid-phase arrays can be manufactured in a standardized way with high quality control, thereby ensuring that variation in performance of the test is more a factor of input DNA quality and less of operator expertise. The solid-phase format also provides advantages in processing, since reagents can be added by hand at small scales, or by robots on a larger scale, without changes to the test. Furthermore, once tests are developed, new mutations and/or polymorphisms can be readily added to the test format with minimal effort. The primer-extension technology itself offers the advantages of an enzyme-mediated single-base sequencing reaction following the hybridization step, for a two-tiered level of specificity.

Molecular Tool Inc. has successfully developed robust genetic tests based on GBA technology for many different diallelic DNA marker systems. Since most of these biallelisms have generally occurred at widely spaced genetic positions, the GBA tests developed for them have been based upon single-site discrimination (only the two relevant bases are tested), which is well suited to the 96-well microplate, indirect colorimetric detection format. These GBA marker systems have been developed from a wide variety of sources including humans, maize, horses, and infectious agents such as *M. tuberculosis*. Clearly, standardization afforded by a simple diallelic genotyping system such as GBA has obvious advantages for paternity and identity applications, and in fact, a GBA-based paternity test using SNPs is currently nearing commercial implementation. GBA-based genotyping of SNPs has direct utility for clinical diagnostic applications, as well. For example, we have developed prototype tests for known non-random SNP mutations for disease genes such as BRCA1, ApoE and CFTR, among others.

In addition to the development of successful single-site GBA tests, tests based upon multiple-site GBA using a modified GBA procedure called nested-GBA (or N-GBA) have been prototyped. Using this method, the exact base sequence of a string of bases can be determined. Application of such an approach to microchip arrays of oligonucleotides will greatly expand the utility of SNP genotyping for applications such as human identity and gene mapping. Since the exact genotype of a site is determined by GBA, the information derived from testing remains useful as the test format is miniaturized and simplified. Solid-phase primer-extension assays address the many limitations of other genotyping approaches and enable the development of truly massively parallel genotyping strategies needed to handle the increasing demands of research and clinical genotyping.

References

Balbin M, Freije JP, Abrahamson M, Velasco G, Grubb A and Lopez-Otin C (1993) *Hum Genet* 90:668-690.
Brooks CC and Tolan DR (1993) *Am J Hum Genet* 52:835-40.
Botstein D, White RL, Skolnick M and Davis RW (1980) *Am J Hum Genet* 32:314-31.
Caetano-Anollés G, Bassam BJ and Gresshoff PM (1991) *BioTechnology* 9:553-557.

Cooper DN, Smith BA, Cooke HJ, Niemann S and Schmidtke J (1985) *Hum Genet* 69:201-205.
Cotton RGH, Rodrigues NR and Campbell RD (1988) *Proc Natl Acad Sci* USA 85: 4397-4401.
Eggerding FA (1995) *PCR Methods Appl* 4:337-345.
Eggerding FA, Iovannisci DM, Brinson E, Grossman P and Winn-Deen ES (1995) *Hum Mutat* 5:153-165.
Guo Z, Guilfoyle RA, Thiel AJ, Wang R and Smith LA (1994) *Nucleic Acids Res* 22:5456-5465.
Hattori M, Shibata A, Yoshioka K and Sakaki Y (1993) *Genomics* 15:415-417.
Holland PM, Abramson RD, Watson R and Gelfand DH (1991) *Proc Natl Acad Sci USA* 88:7276-7280.
Ihalainen J, Siitari H, Laine S, Syvanen A-C and Palotie A (1994) *BioTechniques* 16(5):938-943.
Jalanko A, Kere J, Savilahti E, Schwartz M, Syvanen AC, Ranki M and Soderlund H (1992) *Clin Chem* 38:39-43.
Kwok P-Y, Carlson C, Yager TD, Ankener W and Nickerson DA (1994) *Genomics* 23:138-144.
Livak KJ, Flood SJ, Marmaro J, Giusti W and Deetz K (1995) *PCR Methods Applic* 4:357-362.
Livak KJ and Hainer JW (1994) *Hum Mutat* 3:379-385.
Myers RM, Maniatis T and Lerman LS (1987) *Methods Enzymol* 155:501-527.
Newton CR, Graham A, Heptinstall LE, Powell SJ, Summers C, Kalsheker N, Smith CJ and Markham AF (1989) *Nucleic Acids Res* 17:2503-16.
Nickerson DA, Kaiser R, Lappin S, Stewart J, Hood L and Landegren U (1990) *Proc Natl Acad Sci USA* 87:8923-8927.
Nickerson DA, Whitehurst C, Boysen C, Charmley P, Kaiser R and Hood L (1992) *Genomics* 12:377-87.
Nikiforov TT, Rendle RB, Goelet P, Rogers Y-H, Kotewicz ML, Anderson S, Trainor GL and Knapp MR (1994a) *Nuc Acids Res* 22(20):4167-4175.
Nikiforov TT, Rendle RB, Kotewicz ML and Rogers YH (1994b) *PCR Methods Appl* 3:285-291.
Nikiforov TT and Rogers YH (1995) *Anal Biochem* 227:201-209.
Orita M, Iwahana H, Kanazawa H, Hayashi K and Sekiya T (1989) *Proc Natl Acad Sci USA* 86:2766-70.
Prezant TR and Fischel-Ghodsian N (1992) *Hum Mutat* 1:159-64.
Prior TW, Wenger GD, Papp AC, Snyder PJ, Sedra MS, Bartolo C, Moore JW and Highsmith WE (1995) *Hum Mutat* 5:263-268.
Prober JM, Trainor GL, Dam RD, Hobbs FW, Robertson CW, Zagurski RJ, Cocuzza AJ, Jensen M and Baumeister K (1987) *Science* 238:336-341.
Saiki RK, Walsh PS, Levenson CH and Erlich HA (1989) *Proc Natl Acad Sci USA* 86:6230-6234.
Sokolov BP (1990) *Nucleic Acids Res* 18:3671.
Southern E (1975) *J Mol Biol* 98:503-517.
Stimpson DI, Hoijer JV, Hsieh WT, Jou C, Gordon J, Theriault T, Gamble R and Baldeschwieler JD (1995) *Proc Natl Acad Sci USA* 92:6379-6383.
Syvänen AC, Aalto-Setala K, Harju L, Kontula K and Soderlund H (1990) *Genomics* 8:684-692.
Tully G, Sullivan KM, Nixon P, Stones RE and Gill P (1996) *Genomics* 34:107-113.
Ugozzoli L, Wahlquist JM, Ehsani A, Kaplan BE and Wallace RB (1992) *Genet Anal Tech Appl* 9:107-112.
Venta PJ and Tashian RE (1990) *Nucleic Acids Res* 19:4795.
Vos P, Hogers R, Bleeker M, Reijans M, van de Lee T, Hornes M, Frijters A, Pot J, Peleman J, Kuiper M and Zabeau M (1995) *Nucleic Acids Res* 23:4407-4414.
Weidner J, Eigel A, Horst J and Kohnlein W (1994) *Hum Mutat* 4:55-6.
Welsh J and McClelland M (1990) *Nucleic Acids Res* 18:7213-7218.
Williams JGK, Kubelik AR, Livak KJ, Rafalski JA, Tingey SV (1990) *Nucleic Acids Res* 18:6531-6535.

Genomic fingerprinting using oligonucleotide arrays

Kenneth L. Beattie
Health Sciences Research Division, Oak Ridge National Laboratory, Oak Ridge, TN 37631, USA

Introduction

Genosensors, also termed "DNA chips", are miniature arrays of surface-tethered DNA probes (typically oligonucleotides) to which a nucleic acid sample (the "target" sequence) is hybridized. The target strands will bind to the oligonucleotide array according to specific base pairing rules to produce a hybridization fingerprint that reflects the nucleotide sequence of the target strands. The emerging field of sequencing by hybridization (SBH) aims to develop microfabricated genosensors for ultrafast nucleic acid sequence analysis. Full sequence determination, employing all 4^n oligonucleotide probes of length n, is the most futuristic and ambitious implementation of SBH (Bains and Smith 1988, Southern 1988, Drmanac et al. 1989, Khrapko et al. 1989, 1991, Bains 1991, Drmanac et al. 1993, Lysov et al. 1994, Broude et al. 1994, Hoheisel 1994, Parinov et al. 1996). Diagnostic sequence comparison, using smaller numbers of probes, will be more easily implemented for analysis of specific mutations or polymorphisms, identification of microbial species (Southern et al., 1992, Maskos and Southern 1992, Pease et al. 1994, Mirzabekov 1994, Beattie et al. 1995a,b, Yershov et al. 1996, Doktycz and Beattie 1996) or profiling of gene expression (Beattie et al., 1996a). The genosensor concept is depicted in Figure 1.

Microfabricated arrays of oligonucleotides or longer DNA probes can be used to rapidly generate digital hybridization fingerprints that are interpreted by computer. Implementation of hybridization arrays in a microfabricated format should enable rapid acquisition of comprehensive sequence information about a nucleic acid sample, using an array of hundreds to thousands of DNA probes. Alternatively, smaller sets of probes, duplicated in subarrays across the chip, can be used to interrogate numerous samples in parallel, for a smaller set of sequence features.

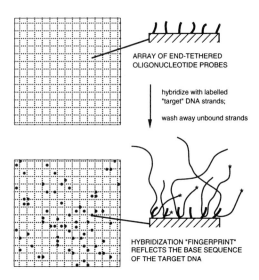

ARRAY OF END-TETHERED
OLIGONUCLEOTIDE PROBES

hybridize with labelled
"target" DNA strands;

wash away unbound strands

HYBRIDIZATION "FINGERPRINT"
REFLECTS THE BASE SEQUENCE
OF THE TARGET DNA

FIGURE 1. The genosensor concept.

This chapter presents an overview of oligonucleotide arrays and how they may be used for high throughput DNA marker analysis and messenger RNA profiling. Simple procedures for preparation of oligonucleotide arrays and hybridization of nucleic acids to genosensor arrays will first be given, then some general strategies for the use of oligonucleotide arrays in DNA fingerprinting and mRNA profiling will be discussed. It is hoped that the information provided in this chapter will enable any laboratory to adopt oligonucleotide fingerprinting as a productivity-enhancing research tool.

Preparation and use of oligonucleotide arrays

Within the SBH community two approaches are being developed to create arrays of synthetic oligonucleotides for genosensor chips: (i) *in situ* synthesis of numerous sequences directly onto a support surface; and (ii) attachment of presynthesized oligonucleotides at each site in the array. Both strategies use the well established phosphoramidite method of solid phase chemical synthesis of oligonucleotides (Matteucci and Caruthers 1981). One *in situ* approach employed at Affymetrix (Fodor et al. 1991, Jacobs and Fodor 1994, Pease et al. 1994, Pirrung and Bradley 1995), utilizes photolithographic deprotection of 5'-OH groups at specific sites on the chip to direct specific base addition at those sites during chemical synthesis by the phosphoramidite method. The Southern laboratory, in partnership with Beckman Instruments, uses a physical masking technique to direct the coupling of the four phosphoramidites within parallel channels formed between two flat glass or polypropylene sheets. By orienting the channels perpendicularly in two successive coupling cycles a 4×4 matix of the 4^2 dinucleotides is synthesized, then by continuing this process and dividing the channel width by four after each pair of coupling cycles, a grid of 4^n sequences is eventually synthesized (Southern, 1988, Southern et al. 1992, Maskos and Southern 1992, Matson et al. 1994). In another *in situ* strategy, being developed at Combion Inc. (Therialt 1996), miniature "ink jets" are used to deliver tiny droplets of phosphoramidite solutions to precise locations on a surface and to synthesize an array of numerous sequences onto the surface.

The post-synthesis attachment strategy of oligonucleotide array fabrication, accessable to ordinary laboratories and described below, can be implemented using commercially available equipment and materials. The desired set of oligonucleotides is first synthesized, then chemically immobilized at specific sites on the surface of the chip (Beattie et al. 1992, Lamture et al. 1994, Beattie et al. 1995a,b, Beattie 1996, Beattie et al. 1996a,b). A single small scale synthesis yields sufficient oligonucleotide for 10^5-10^6 chips. Delivery of small DNA probe aliquots to a microscope slide can be performed manually, or accomplished robotically using a Hamilton MicroLab 2200 fluid dispensing system. The Hamilton robot can simultaneously dispense up to 8 droplets of as little as 10 nL to a flat surface at 1 mm spacing (Beattie et al. 1995a,b), enabling about 1,000 probes to be placed onto an ordinary microscope slide.

More advanced options are available for preparation of higher density arrays. Using rows of tiny piezoelectric "ink jet" tips, Microfab Technologies Inc. (Plano, TX) can deliver tiny droplets (each tens to hundreds of picoliters in volume) to precise locations onto a chip (Eggers et al. 1994). Two-dimensional arrays of microfabricated piezoelectric microjets are under development at Accelerator Technology Corp. (Bryan, TX), which are designed to simultaneously "print" large numbers of probes onto the genosensor surface (McIntyre 1996). The Mirzabekov group has developed a miniature robotic multi-pin printing device (Yershov et al. 1996) to simultaneously deliver numerous minute droplets of probe solutions to thin (10-20 μm) slabs of polyacrylamide gel gridded on a glass support (Khrapko et al. 1991, Mirzabekov 1994).

For covalent linkage of oligonucleotide probes to silicon dioxide surfaces, we previously reported a procedure involving specific condensation of a primary amine on the 5'- or 3'-terminus of the oligonucleotide with an epoxysilane group on the glass (Lamture et al. 1994, Beattie et al. 1995a,b). More recently, we developed a much simpler and more reliable procedure involving direct coupling of 3'-aminopropanol-derivatized oligonucleotides to unmodified SiO_2 surfaces (Doktycz and Beattie 1996, Beattie 1996, Beattie et al. 1996a,b). The presumed reaction is shown in Figure 2.

The following evidence supports the above reaction scheme (Beattie et al. 1996b). The linkage is: (i) stable in hot water, enabling multiple cycles of hybridization; (ii) stable in mild acid but labile in mild base (favoring the ester linkage over the amide linkage); (iii) not formed with 5'-hexylamine-derivatized oligonucleotides (primary amine alone is insufficient); (iv) inhibited by pretreatment of glass with propanolamine but not propylamine; and (v) blocked by acetylation of primary amine on oligonucleotide. The attachment reaction proceeds rapidly in aqueous solution at room temperature and gives a lower background of nonspecific binding of target DNA to the surface, compared with the previous epoxy-amine linkage method. The attachment density (10^{10}-10^{11} molecules/mm^2) is similar for both methods discussed above.

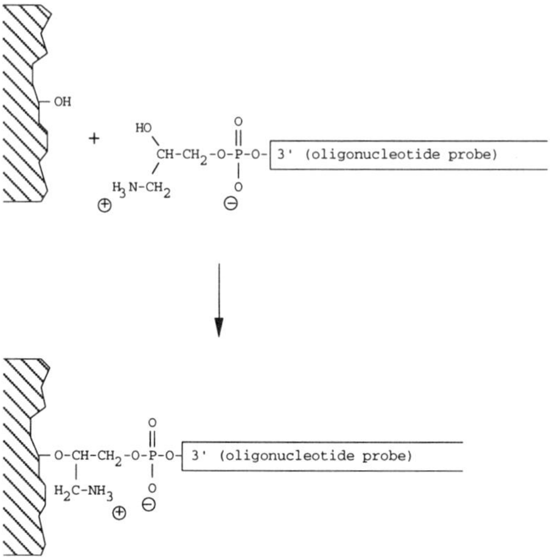

FIGURE 2. Simplified attachment chemistry by direct coupling of 3'-aminopropanol-derivatized oligonucleotides to unmodified SiO$_2$ surfaces.

The following procedure is carried out for attachment of oligonucleotides to glass surfaces using the new direct coupling chemistry. Oligonucleotides are chemically synthesized using the 3'-Amino-Modifier C3 CPG support (Glen Research; cat. no. 20-2950) with the standard phosphoramidite chemistry (Matteucci and Caruthers 1981). During cleavage of the oligonucleotides from the support the C3 amino group (actually a propanolamine function) is created at the 3'-end. Custom oligonucleotides with this 3'-propanolamine modification are available from Genosys Biotechnologies (The Woodlands, TX). Oligonucleotides are dissolved in water at a concentration of 10-20 μM. Glass microscope slides are cleaned by rinsing with acetone and ethanol, and dried in an 80°C oven. Droplets of oligonucleotide solution (typically 50-250 nL) are placed onto the clean, dry slide, incubated at room temperature for 5-15 min, then rinsed with water, air-dried and stored dessicated at room temperature. The attachment reaction occurs rapidly, and if some of the droplets dry during the application of all oligonucleotides in an array, the slide should be held at room temperature until all droplets dry before washing with water (the reaction is apparently complete upon drying). If droplets are applied manually, the slide can be placed above a printed template to guide the placement of droplets. A commercially available robotic fluid dispensing system (Hamilton MicroLab 2200 system equipped with 21G needles and 50 μL syringes) is capable of robotically dispensing droplets as small as 10 nL onto a glass slide at 1 mm center-to-center spacing (Beattie et al. 1995a,b).

For hybridization of target strands to nonamer oligonucleotides attached to microscope slides the following standard procedure (Beattie et al. 1995a,b) can be used. Slides are "prehybridized" by soaking for 1 h at room temperature in a "blocking solution" followed by a brief water wash. We have found that 10-20 mM tripolyphosphate is an effective and economical blocking solution for minimizing the nonspecific binding of ^{32}P-labeled target strands to the glass slide (Beattie et al. 1995b). Target DNA (typically, PCR product) is dissolved in (or added to) hybridization buffer (either 6× SSC or 3.3 M tetramethylammonium chloride in 50 mM Tris-HCl pH 8, 2 mM EDTA, 0.1% SDS and 10% polyethylene glycol 8000) at a concentration of 10-50 fmol strands/μL (10-50 nM). If the target strands are labeled with ^{32}P, a minimum of 2,000 cpm/μL is used in the hybridization mixture, and prior to addition of labeled DNA to the hybridization mixture, unincorporated label is removed by loading the DNA onto a Microcon-3 microconcentrator (Amicon) and washed three times with water. Furthermore, if PCR is used to amplify the target, the PCR product is processed with a Ultrafree spin-filter (30,000 molecular weight cut-off; Millipore) to remove excess PCR primers prior to hybridization. An aliquot of target DNA in hybridization buffer is pipetted onto the microscope slide (20 μL for an array occupying a third of the slide or 60 μL for the entire slide) and covered with a glass cover slip. The slide is incubated at 6°C for 2 h to overnight, and then washed at room temperature for at least 1 h with hybridization buffer without polyethylene glycol. For hybridization of immobilized probes of different lengths, variations in the hybridization and temperature should be explored to optimize the hybridization with respect to signal intensity and mismatch discrimination. We have found that hybridization of 12-mer arrays can be conveniently carried out at room temperature in the above hybridization buffer. If target strands are labeled with ^{32}P, hybridization can generally be quantitated within a few minutes using a phosphorimager (Beattie et al. 1995a,b), although overnight exposure against X-ray film is adequate for autoradiographic detection.

Strategies for genomic fingerprinting

High throughput polymorphic marker analysis is one of the most important potential uses of oligonucleotide array hybridization. This section discusses experimental strategies that can be followed to achieve rapid genotyping by oligonucleotide fingerprinting. As will be discussed, similar strategies may apply to transcriptional profiling by oligonucleotide fingerprinting.

A "targeted" approach can be used if the nucleotide sequences of the DNA sequence polymorphisms are known. A specific oligonucleotide probe would be included in the genosensor array for each allele of each DNA marker. The hybridization pattern produced with the appropriate PCR fragments (containing the DNA markers of interest) would quickly reveal the genotype. This approach is most applicable to DNA markers consisting of point mutations (small base substitutions, insertions and deletions).

Different sectors of the oligonucleotide array could be hybridized with different amplicon mixtures (produced by multiplex PCR), for simultaneous analysis of hundreds or even thousands of markers.

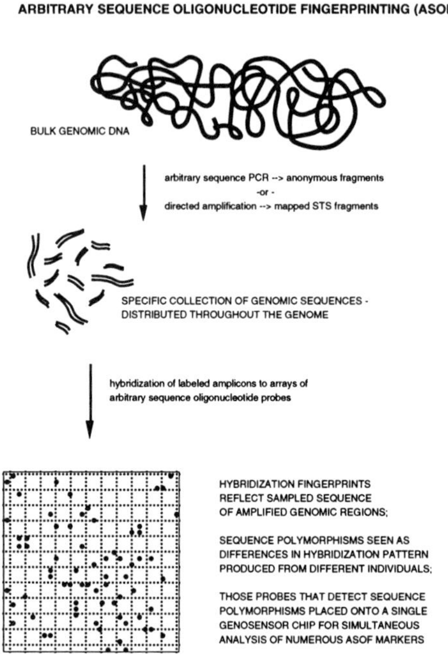

ARBITRARY SEQUENCE OLIGONUCLEOTIDE FINGERPRINTING (ASOF)

BULK GENOMIC DNA

arbitrary sequence PCR --> anonymous fragments
-or -
directed amplification --> mapped STS fragments

SPECIFIC COLLECTION OF GENOMIC SEQUENCES -
DISTRIBUTED THROUGHOUT THE GENOME

hybridization of labeled amplicons to arrays of
arbitrary sequence oligonucleotide probes

HYBRIDIZATION FINGERPRINTS
REFLECT SAMPLED SEQUENCE
OF AMPLIFIED GENOMIC REGIONS;

SEQUENCE POLYMORPHISMS SEEN AS
DIFFERENCES IN HYBRIDIZATION PATTERN
PRODUCED FROM DIFFERENT INDIVIDUALS;

THOSE PROBES THAT DETECT SEQUENCE
POLYMORPHISMS PLACED ONTO A SINGLE
GENOSENSOR CHIP FOR SIMULTANEOUS
ANALYSIS OF NUMEROUS ASOF MARKERS

FIGURE 3. Arbitrary sequence oligonucleotide fingerprinting (ASOF).

A "nontargeted" hybridization strategy, termed arbitrary sequence oligonucleotide fingerprinting (ASOF), was recently proposed as a new approach for high throughput DNA marker analysis (Beattie 1995, Beattie et al. 1996a). The ASOF strategy is schematically represented in Figure 3. Genomic DNA is first subjected to PCR to provide a specific sampling of fragments scattered across the genome. The subset of genomic sequences can be generated by amplification using a single short primer of arbitrary sequence, as in the RAPD (Williams et al. 1990), AP-PCR (Welsh and McClelland 1991) and DAF (Caetano-Anollés 1993, Caetano-Anollés and Bassam 1993) techniques described elsewhere in this volume. Alternatively, the genomic sampling step could be achieved by use of targeted multiplex PCR to prepare a collection of sequence tagged site (STS) fragments (Olson et al. 1989) selected to be more or less evenly distributed across the genome. The PCR-generated, labeled genomic fragments are hybridized to a genosensor array containing several hundred or a few thousand oligonucleotide probes of arbitrary sequence. DNA sequence polymorphisms within the amplified genomic fragments will be seen as differences in the hybridization fingerprint produced using genomic DNA from different individuals. Differences in the hybridization fingerprint can result from sequence variations within the priming sites (to make the existence of a PCR fragment variable), or from sequence variations within amplified sequences that interact with arrayed probes.

The ASOF approach differs from traditional DAF, AP-PCR or RAPD marker analysis in that the genomic fragments are analyzed by hybridization fingerprinting rather than by gel electrophoresis. Statistical considerations lead to the prediction that the number of polymorphisms revealed by hybridization fingerprinting will be much greater than by gel electrophoresis (Beattie 1995, Beattie et al. 1996a). After the ASOF strategy is used to discover which arbitrary sequence probes will reproducibly detect polymorphisms within a given collection of PCR fragments, a genosensor can then be fabricated using only the informative probes, enabling the analysis of a large number of ASOF markers in a single experiment.

The results of a preliminary experiment on the use of ASOF for comparing genomic DNA of different individuals are shown in Figure 4. These results are presented not as a final recommended ASOF protocol, since additional optimization is needed, but as preliminary indication of the feasibility of ASOF. PCR was carried out using a single 5'-^{32}P-labeled 9-mer primer of sequence, 5'-GTGTCGATG-3', and human genomic DNA from three unrelated humans. The PCR reactions contained (in 100 μL) 100 ng template DNA, 200 μM each dNTP, 0.2 μM 5'-[^{32}P]-labeled primer, 5 units Taq polymerase, and standard PCR buffer. Tubes were held 5 min at 95°C, subjected to 30 thermal cycles of 1 min at 90°C, 1 min at 30°C and 2 min at 72°C were conducted, brought to 95°C for 5 min and another 2.5 units Taq polymerase was added, and 30 more cycles of PCR were conducted as above. PCR mixtures were centrifuged through Ultrafree-30,000 spin filters (Millipore) to remove free primer, then suspended in 100 μL hybridization buffer [3.3 M tetramethylammonium chloride in 50 mM Tris-HCl pH 8, 2 mM EDTA, 0.1% SDS] with addition of 10% polyethylene glycol-8000. Half of the PCR mixture was placed onto each of two microscope slides containing an array of 200 different 9-mers of arbitrary sequence (Beattie, 1995; Beattie et al., 1996a), and a cover slip was placed onto each slide. After incubation overnight at 6°C, slides were washed with hybridization buffer at room temperature for 1 h, then covered with plastic film and placed against a phosphorimager screen overnight. Shown in Figure 4 are regions of the 9-mer arrays containing about 100 probes. As seen in Panel A, the hybridization pattern is very similar between two DNA samples, but there are positions that give different relative intensities or absence of a signal with one sample compared with the other. Panel B shows another example, using DNA samples from three different individuals and a different set of about 100 9-mer probes. As in Panel A, apparent differences were seen in the hybridization fingerprints obtained with genomic DNA from different individuals. These data, although very preliminary, suggest that ASOF can be be used to define specific combinations of oligonucleotide probes and PCR primers that may be exploited for high throughput genosensor-based polymorphic marker analysis.

A

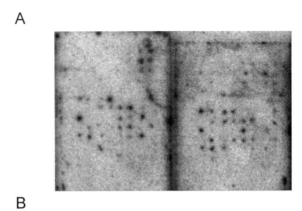

B

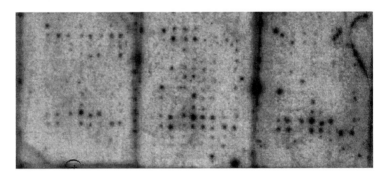

FIGURE 4. Preliminary ASOF results obtained using human genomic DNA. Experiments were performed as described in the text, using human DNA from two (Panel A) and three (Panel B) individuals.

In the ASOF approach an important consideration is the number and length of oligonucleotide probes represented in the genosensor array. Although the ultimate implementation of ASOF will be based upon experimentally optimized parameters of PCR primer and oligonucleotide probe composition, length and number, it is useful to define a starting point for these studies, based on theoretical considerations. The use of short (octamer-to-decamer) primers with genomic DNA of plants or animals typically yields 50-100 bands in a gel electrophoretic assay, in the size range of a few hundred to a few thousand base pairs. For a given total length of sampled genomic sequences it is possible to use statistical predictions to help specify the appropriate set of probes that should be included in the array. Let us assume that the total length of amplified target sequence that will be hybridized to the oligonucleotide array is 50,000 bp (100,000 bases). For a probe of length p, the average number of occurrences n, of the probe within the target sequences is represented by $n = 100,000/4^p$ and from this value we can predict the average number of hybridization signals that would be produced with a given composition (number and length) of DNA probe array. Table 1 summarizes these calculations for DNA probes of various length.

TABLE 1. Expected hybridization signals in ASOF.

Probe length (p)	Ave. no. of occurrences (n) within the amplified target	Ave. number of hybridization signals for a given size of array:	
		20×20 array	50×50 array
8-mer	1.53	(610)	(3,815)
9-mer	0.38	154	954
10-mer	0.10	38.1	238

From the above table it appears that an array of 9-mer probes may be most appropriate for use in the ASOF procedure, in agreement with the preliminary results shown above in Figure 4. If the arrayed probes are too short, they will all bind one or more amplified genomic fragments, whereas if the probes are too long, too few of them will capture a target strand. Now, let's consider the question, what is the approximate number of useful ASOF markers that could be discovered in a single hybridization with a 50×50 probe array? If we take an estimate of 0.005 for the average frequency of useful single base polymorphism per base pair (minor allele occurring at a frequency >0.1) (Botstein et al. 1980), and assume that 50% of single base changes will be detectable at either the level of hybridization or priming, then we predict that for short single primer PCR, only about three PCR fragments (among the expected 50-100 amplicons) will be affected by single nucleotide polymorphism. If we expect an average of 954 hybridization signals using a 50×50 array of 9-mers (Table 1), then the total length of hybridizing 9-mers will be about 8,586 bases, and an average of 21 hybridization signals should be disrupted by single nucleotide polymorphism. Thus, the detectability of polymorphisms is predicted to be greater at the level of hybridization of PCR products to the probe array than at the level of amplification, and about 20 new ASOF markers are expected to be defined in a single hybridization experiment with the 9-mer array.

The potential power of ASOF is fully realized in moving from the "discovery" mode to the "screening" mode. If the analysis is repeated with additional primers and probes and if the oligonucleotides that reproducibly detect ASOF polymorphism are then represented in a "screening chip", it should be possible to simultaneously analyze hundreds or thousands of ASOF markers in a single experiment. Although the ASOF markers will be diallelic and anonymous, the prospect of quickly screening large numbers of these markers may justify the approach. Finally, a subtle but notable feature of ASOF is that the technique is insensitive to the problem of mismatch discrimination. Even if a fraction of the hybridization signals represent imperfect hybrids, as long as sequence variation has a reproducible affect on the hybridization pattern, ASOF markers can be reliably analyzed. DNA polymorphisms that cannot be reliably detected can be ignored, and only the reliable markers need be utilized for high throughput screening.

Another very important application envisioned for oligonucleotide array fingerprinting is analysis of gene expression, or transcriptional profiling (Beattie 1995, Beattie et al. 1996a). Again, two approaches may be used. First, the "directed" approach employs immobilized probes corresponding to known gene sequences. In the near future it will be possible to represent all 100,000 or so human genes in a series of genosensors bearing unique probes for each coding sequence. For this application, the probes may be longer than in ASOF, enabling higher stringency hybridization, but in the case of gene families careful attention to related sequences must be given to avoid problems of cross hybridization. In the latter case, probes of 15-20 nucleotides in length may provide an optimal balance between sequence uniqueness and mismatch discrimination.

The ASOF procedure may be applied in a "nontargeted" universal approach to transcriptional profiling. Indeed, mRNA profiling may be the most significant use of ASOF. Using labeled PCR fragments derived from bulk mRNA or from cDNA, the expressed sequence-derived fragments may be hybridized to an array of arbitrary sequence probes to quickly generate a hybridization fingerprint in which each mRNA species binds to one or a few specific sites within the array. The latter approach corresponds to "differential display on a chip". A powerful feature of oligonucleotide fingerprinting for mRNA profiling is that strands bound at sites within the array showing differential hybridization intensity (under different physiological states) may be eluted from the chip, cloned and/or sequenced, to discover the genes that are involved in any given biological response. The considerations of oligonucleotide array design, discussed above for ASOF marker analysis, apply equally to the transcriptional profiling task. Adjustments in oligonucleotide length and PCR conditions can be made to yield meaningful hybridization fingerprints, wherein a reasonable fraction of the arrayed probes form a hybrid, while most hybridization signals involve a single mRNA species.

A new type of hybridization support consisting of microchannel glass (Tonucci et al. 1992) or porous silicon (Lehmann 1993) has recently been introduced for use in a flowthrough genosensor (Beattie et al. 1995a). In the flowthrough design, hybridization occurs within three-dimensional volumes of porous silicon dioxide or channel array glass, rather than on a two-dimensional surface. The approximately 100-fold greater surface area per unit cross section, compared with the flat surface design, greatly increases the binding capacity per hybridization cell, providing improved detection sensitivity. The flowthrough configuration has other important advantages over flat surface genosensors (Doktycz and Beattie 1996, Beattie 1996, Beattie et al. 1996a,c), which are particularly relevant in the use of ASOF for DNA marker analysis and mRNA profiling: (i) greatly improved hybridization kinetics with dilute nucleic acid samples; (ii) improved sample throughput achieved through successive cycles of hybridization (bound target strands are removed by hot water flushes); (iii) direct analysis of dilute, heat-denatured PCR fragments without physically isolating one of the two strands; (iv) analysis of a large-volume sample by flowing the solution through the device; and (v) ability to elute bound materials from individual hybridization cells for further analysis. Further development of the flowthrough genosensor should speed the implementation of genomic fingerprinting strategies for analysis of DNA markers and gene expression.

Acknowledgments

This work was carried out in the DNA Technology Laboratory at the Houston Advanced Research Center, under NIH Grant P20 HG00665, with additional support from the ICSC World Laboratory. The author's current address is Health Sciences Research Division, Oak Ridge National Laboratory, P.O. Box 2008, Oak Ridge, TN 37831-6123.

References

Bains W (1991) *Genomics* 11:294-301.
Bains W and Smith GC (1988) *J Theor Biol* 135:303-307.
Beattie KL (1995) US Patent, pending.
Beattie KL (1996) In: Sayler GS (ed), *Biotechnology in the Sustainable Environment*. Plenum Publishing Corp, New York. In press.
Beattie K, Beattie W, Meng L, Turner S, Bishop C, Dao D, Coral R, Smith D and McIntyre P (1995a) *Clin Chem* 41:700-706.
Beattie KL, Eggers MD, Shumaker JM, Hogan ME, Varma RS, Lamture JB, Hollis MA, Ehrlich DJ and Rathman D (1992) *Clin Chem* 39:719-722.
Beattie KL, Zhang B, Pratt JD and Beattie WG (1996c) *Nucleic Acids Res*, mns. submitted.
Beattie KL, Zhang B, Tovar-Rojo F and Beattie WG (1996a) In: Schlegel J (ed), *Pharmacogenetics: Bridging the Gap between Basic Science and Clinical Application.* IBC Biomedical Library, Southborough, MA. In press.
Beattie WG, Meng L, Turner S, Varma RS, Dao DD and Beattie KL (1995b) *Mol Biotechnol* 4:213-225.
Beattie KL, Raia GC, Pratt JD, Budowsky EI, Kumar S, Varma RS and Beattie KL (1996b) *Nuceicl Acids Res*, mns. submitted.
Botstein D, White RL, Skolnick M and Davis RW (1980) *Am J Human Genet* 32:314-331.
Broude NE, Sano T, Smith CL and Cantor CR (1994) *Proc Natl Acad Sci USA* 91:3072-3076.
Caetano-Anollés G (1993) *PCR Methods Applic* 3:85-94.
Caetano-Anollés G and Bassam BJ (1993) *Appl Biochem Biotechnol* 42:189-200.
Doktycz MJ and Beattie KL (1996) In: Beugelsdiik A (ed), *Automated Technologies for Genome Characterization*. John Wiley & Sons, New York. In press.
Drmanac R, Drmanac S, Strezoska Z, Paunesku T, Labat I, Zeremski M, Snoddy J, Funkhouser WK, Koop B and Hood L (1993) *Science* 260:1649-1652.
Drmanac R, Labat I, Brukner I and Crkvenjakov R (1989) *Genomics* 4:114-128.
Eggers M, Hogan M, Reich RK, Lamture J, Ehrlich D, Hollis M, Kosicki B, Powdrill T, Beattie K, Smith S, Varma R, Gangadharan R, Mallik A, Burke B and Wallace D (1994) *BioTechniques* 17:516-525.
Fodor SPA, Read JL , Pirrung MC, Stryer L, Lu AT and Solas D (1991) *Science* 251:767-773.
Hoheisel JD (1994) *Trends Genet* 10:79-83.
Jacobs JW and Fodor SPA (1994) *Trends Biotechnol* 12:19-26.
Khrapko KR, Lysov YP, Khorlin AA, Ivanov IB, Yershov GM, Vasilenko SK, Florentiev VL and Mirzabekov AD (1991) *DNA Sequence* 1:375-388.
Khrapko KR, Lysov YP, Khorlyn AA, Shick VV, Florentiev VL and Mirzabekov AD (1989) *FEBS Lett* 256:118-122.
Lehmann V (1993) *J Electrochem Soc* 140:2836-2843.
Lamture JB, Beattie KL, Burke BE, Eggers MD, Ehrlich DJ, Fowler R, Hollis MA, Kosicki BB, Reich RK, Smith SR, Varma RS and Hogan ME (1994) *Nucleic Acids Res* 22:2121-2125.
Lysov YP, Chernyi AA, Balaeff AA, Beattie KL, Florentiev VL and Mirzabekov AD (1994) *J Biomolec Struct Dynam* 11:797-812.
McIntyre P (1996) *IBC Conference on Biochip Array Technologies*, Marina del Rey, CA.
Maskos U and Southern EM (1992) *Nucleic Acids Res* 20:1675-1678.
Matson RS, Rampal JB and Coassin PJ (1994) *Anal Biochem* 217:306-310.

Matteucci MD and Caruthers MH (1981) *J Am Chem Soc* 103:3185-91.

Mirzabekov AD (1994) *Trends Biotechnol* 12:27-32.

Olson M, Hood L, Cantor C and Botstein D (1989) *Science* 245:1434-1435.

Parinov S, Barsky V, Yershov G, Kirillov E, Timofeev E, Belgovskiy A and Mirzabekov A (1996) *Nucleic Acids Res* 15:2998-3004.

Pease AC, Solas D, Sullivan EJ, Cronin MT, Holmes CP and Fodor SPA (1994) *Proc Natl Acad Sci USA* 91:5022-5026.

Pirrung MC and Bradley JC (1995) *J Org Chem* 60:6270.

Southern EM (1988) International patent application PCT GB 89/00460.

Southern EM, Maskos U and Elder JK (1992) *Genomics* 13:1008-1017.

Therialt T (1996) *IBC Conference on Biochip Array Technologies*, Marina del Rey, CA.

Tonucci RJ, Justus BL, Campillo AJ and Ford CE (1992) *Science* 258:783-785.

Welsh J and McClelland M (1991) *Nucleic Acids Res* 18:7213-7218.

Williams JGK, Kubelik AR, Livak KJ, Rafalski JA and Tingey SV (1990) *Nucleic Acids Res* 18:6531-6535.

Yershov G, Barsky V, Belgovskiy A, Kirillov E, Kreindlin E, Ivanov I, Parinov S, Guschin D, Drobishev A, Dubiley S and Mirzabekov A (1996) *Proc Natl Acad Sci USA* 93:4913-4918.

Differential display of RNA

José E. Galindo, Ghislaine M.-C. Poirier, Hongqing Guo, Arne Huvar, Pamela C. Wagaman, Jessica Zhu, Jennifer Tench, Jackson S. Wan, and Mark G. Erlander
R.W. Johnson Pharmaceutical Research Institute, San Diego, CA 92121, USA

Introduction

Differential gene expression occurs in all phases of life including development, maintenance, injury and death of an organism. Being able to identify these genes will help understand not only gene function but also the underlying molecular mechanisms of a particular biological system. Differential display (DD; Liang and Pardee 1992, Welsh et al. 1992) is a common method currently used by investigators for this purpose. Since the introduction of DD, over one hundred reports have been published which document either successful applications or method improvements. However, some users have experienced frustration and failure with most of the criticisms being: (i) a high false positive rate; (ii) questioned ability of DD to identify both abundant and rare mRNA; (iii) coding regions of mRNA are usually not cloned; and (iv) verification process is time consuming and usually requires a fair amount of RNA (Bertioli et al. 1995, Debouck 1995). Below is a detailed protocol for DD as well as for reverse Northern analysis for the verification of putative positives. Included in the reverse Northern protocol is a method for amplifying RNA; this is helpful for verifying putative positives when only small amounts of RNA are available. In addition, in the "overview" section we address the above criticisms for DD.

Materials

Differential display

A) cDNA synthesis
  1. SuperScript preamplification system (Gibco BRL; cat. no. 18089-011):
     • 10× PCR buffer: 200 mM Tris-HCl pH 8.4, 500 mM KCl.
     • 25 mM $MgCl_2$.

- 10 mM dNTP mix: 10 mM each dATP, dGTP, dCTP, dTTP.
- 0.1 M DTT.
- SuperScript II reverse transcriptase (RT; 200 units/μL).
- RNaseH (2 units/μL).
- H_2O (DEPC treated).
2. Degenerate anchor primers (1 μg/μL): $T_{12}VG$, $T_{12}VA$, $T_{12}VT$, and $T_{12}VC$ where V represents an equal molar mixture of G, A and C.
3. RNA template: mRNA (100 ng/μL) or total RNA (1 μg/μL).

B) Differential display PCR
1. Degenerate anchor primers (25 μM) (*see* cDNA synthesis).
2. 5' primer (10-mer) (25 μM) (*see* Note 1).
3. 10× PCR buffer (Perkin Elmer; cat. no. N808-0006): 100mM Tris-HCl pH 8.3 @ 25°C, 500 mM KCl, 15 mM $MgCl_2$, 0.01% [w/v] gelatin (autoclaved).
4. 20 μM dNTP mix: 20 μM each of dATP, dGTP, dCTP, dTTP.
5. 200 μM dNTP mix: 200 μM each of dATP, dGTP, dCTP, dTTP.
6. α-^{35}S-dATP (1000Ci/mmol) (Amersham).
7. AmpliTaq 5U/μL (Perkin Elmer; cat. no. N801-0060).
8. Sequencing loading dye: 95% formamide, 20mM EDTA, 0.05% bromophenol blue, 0.05% xylene cyanol FF.
9. X-ray film: Kodak XAR 5.
10. TA cloning kit (Invitrogen; cat. no. K2000-01):
    - 10× ligation buffer: 60 mM Tris-HCl pH 7.5, 60 mM $MgCl_2$, 50 mM NaCl, 1 mg/mL bovine serum albumin, 70 mM β-mercaptoethanol, 1 mM ATP, 20 mM dithiothreitol, 10 mM spermidine.
    - pCR^{TM}2.1 vector (25 ng/μL).
    - Sterile water.
    - T4 DNA ligase (4.0 Weiss units/μL).
    - SOC medium: 2% tryptone, 0.5% yeast extract, 10 mM NaCl, 2.5 mM KCl, 10 mM $MgCl_2$, 10 mM $MgSO_4$, 20 mM glucose (dextrose).
    - 0.5 M β-mercaptoethanol.
    - INVαF' cells.
11. LB plates containing 50 μg/mL ampicillin and spread with 40 μL of 40 mg/mL X-Gal.

Reverse Northern

A) cDNA probe synthesis
1. SuperScript preamplification system (Gibco BRL; cat.no. 18089-011).
   - Oligo dT$_{12-18}$ (0.5 μg/μL).
   - 10× PCR buffer: 200 mM Tris-HCl pH 8.4, 500 mM KCl.
   - 25 mM $MgCl_2$.
   - 0.1 M DTT.
   - SuperScript II RT (200 units/μL).
   - H_2O (DEPC treated).
2. Random hexamers 7.4 mg/mL (Pharmacia).

3. RNA template: 0.4 $\mu g/\mu L$ mRNA or 0.4 $\mu g/\mu L$ amplified RNA (aRNA)
4. d(GAT)TP mix: 20 mM each of dGTP, dATP, dTTP.
5. dCTP (120 μM).
6. α-^{32}P-dCTP (3000Ci/mmol) (Amersham).
7. 3 N NaOH (make fresh).
8. 1 M Tris-HCl pH 7.4.
9. 2 N HCl.
10. G-50 Sephadex column (Boehreinger Mannheim).

B) Dot blot
1. Nylon membranes: maximum strength Nytran 0.45 mm (Schleicher & Schuell).
2. Whatman chromatography paper.
3. Denaturing solution: 1.5 M NaCl, 0.5 M NaOH.
4. Neutralizing solution: 1.5 M NaCl, 1 M Tris-HCl pH 7.4.
5. 2× SSC.
6. UV-Stratalinker 2400 (Stratagene).
7. Rotating hybridization oven (Bellco).
8. 2× Southern pre-hybridization buffer (5 prime-3 prime; cat. no. 5302-820800): 10× SSC pH 7.0, 10× Denhardt's solution, 0.1 M sodium phosphate pH 6.8, 0.2% SDS and 0.01 M EDTA.
9. 2× Southern hybridization buffer (5 prime-3 prime; cat. no. 5302-500197): 10× SSC pH7.0, 2× Denhardt's solution, 0.04 M sodium phosphate pH 6.8, 0.4% SDS and 0.01 M EDTA.
10. Sheared salmon sperm DNA (10 mg/mL).
11. Formamide.
12. Pre-hybridization solution: 1 part 2× Southern pre-hybridization buffer, 1 part formamide, and 0.1 mg/mL salmon sperm DNA boiled for 5 min.
13. Hybridization solution: 1 part 2× Southern hybridization buffer, 1 part formamide, and 0.1 mg/mL salmon sperm DNA boiled for 5 min.
14. Wash solution I: 2× SSC, 0.1% SDS.
15. Wash solution II: 0.2× SSC, 0.1% SDS.
16. Wash solution III: 0.1× SSC, 0.1% SDS.
17. Phosphorimager 445SI (Molecular Dynamics).

C) RNA amplification
1. T7(dT)$_{15}$ primer: 5′ -AAACGACGGCCAGTGAATTGTAATACGACTCACTATAGGCGCT$_{15}$- 3′ .
2. SuperScript plasmid system for cDNA synthesis (Gibco BRL; cat. no. 8248).
   - RNase-free H$_2$O .
   - 5× first-strand buffer: 250 mM Tris-HCl pH 8.3, 375 mM KCl, 15 mM MgCl.
   - 0.1 M DTT.
   - 10 mM dNTP mix.
   - SuperScript II RNase H⁻ reverse transcriptase.
   - 5× second strand synthesis buffer: 125 mM Tris-HCl, 500 mM KCl, 25 mM MgCl$_2$, 50 mM (NH$_4$)$_2$SO$_4$, 0.75 mM β-NAD$^+$.
   - DNA Ligase (10 U/μL).

- DNA Polymerase I (10 U/μL).
- RNase H (2 U/μL).
3. 7.5 M ammonium acetate.
4. RNase-free H$_2$O.
5. Ampliscribe T7 transcription kit (Epicentre; cat. no. AS2607).

Experimental protocol

Differential display

A) cDNA synthesis

For each RNA condition to be compared, the following reaction in duplicates is performed for each of the 4 degenerate anchor primers. In addition, a control reaction (without SuperScript II RT) is performed. Thus there is a total of 12 reactions per RNA condition. If more than a few bands are seen on the DD gel from the reaction without SuperScript II RT, then your RNA may be contaminated by chromosomal DNA. You may want to treat the samples with DNase I before cDNA synthesis.

1. Mix 1 μL of RNA template with 1 μL of an anchor primer and 10 μL H$_2$O in a 1.5 mL tube, incubate at 70°C for 10 min, then place on ice for 1 min.
2. Add 2 μL 10× PCR buffer, 2 μL MgCl$_2$, 1 μL dNTP mix and 2 μL DTT, incubate at 42°C for 5 min to anneal and add 1 μL SuperScript II RT. Incubate at 42°C for 50 min for cDNA synthesis, terminate the reaction at 70°C for 15 min, then place on ice.
3. Add 1 μL RNaseH and incubate at 37°C for 20 min.
4. The cDNAs can be store at -20°C until use.
5. Before use, dilute the cDNA to 1:20 with H$_2$O. You may need to dilute more if you see a smear on the sequencing gel after PCR.

B) Differential display PCR

To minimize the redundancy associated with differential display, each of the 4 degenerate anchor primers T$_{12}$VG, T$_{12}$VA, T$_{12}$VT, and T$_{12}$VC is paired with only 1/4 of the 256 possible 10-mers (*see* Note 1).

1. Mix 1 μL cDNA template (from cDNA synthesis), 1 μL degenerate anchor primer (use the corresponding primer that was used during cDNA synthesis), 1 μL 5' primer, 2 μL 10× PCR buffer, 2 μl 20 μM dNTP mix, 1 μL α-^{35}S-dATP, 0.5 μL AmpliTaq and 11.5 μL H$_2$O in a 0.5 mL thin wall PCR tube. Perform PCR using the following program: 40 cycles of 20 s at 94°C, 20 s at 42°C, and 30 s at 72°C, followed by 10 min at 72°C. We use the GeneAmp PCR system 9600 machine (Perkin Elmer).
2. Mix 1 μL PCR product with 1 μL sequencing loading dye, heat to 80°C for 5 min and load on a 6% acrylamide-8 M urea sequencing gel (load the duplicate reactions for each RNA condition in adjacent lanes) (*see* Note 2). Run at 55 W for about 3.5 h. Stop the gel after the top dye (xylene cyanol FF) runs off the bottom. To determine the size of the bands, also load a molecular size marker such as the M13 DNA sequencing reaction control from the Sequenase DNA sequencing kit (USB; cat. no. US70770) in a separate lane (*see* Figure 1 for example of DD gel).

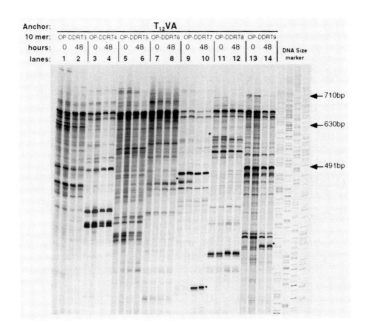

FIGURE 1. An example of a differential display gel. HeLa cell was untreated (0 hours) or treated with interferon-γ (IFN-γ) for 48 h prior to mRNA isolation and PCR as described in the text. The reactions were performed in duplicates and loaded adjacent to each other. $T_{12}VA$ was used during cDNA synthesis and in all DD PCRs. The 10-mers used are marked above the lanes. The asterisks mark some of the bands representing mRNA with putative increase or decrease in expression after treatment.

3. Transfer gel onto Whatman chromatography paper, dry without fixing in a vacuum gel dryer for 2 h, expose to X-ray film from overnight to 2 nights. Be sure to use some kind of fluorescent sticker on the gel so that you can align it to the film.

4. Align the film to the gel, use tape to secure the position, excise out the differentially expressed bands by cutting through the film, select only bands that are present in both lanes derived from the duplicate cDNA reactions (you may want to expose again after this step to see if the correct bands were excised).

5. Add 100 μL H_2O to the excised band in a 1.5 mL tube, sit at room temperature for 10 min, boil for 15 min, spin down and transfer the liquid to a new tube, use directly as template for PCR amplification.

6. Mix 2 μL template (from previous step) (do not use more than 3 μL because the urea may inhibit the reaction), 2 μL degenerate anchor primer, 2 μL 5' primer, 4 μL 10× PCR buffer, 25.6 μL H_2O, 4 μL of 200 μM dNTP solution and 0.4 μL AmpliTaq in a 0.5 mL thin wall PCR tube. Perform PCR using the following program: 40 cycles of 20 s at 94°C, 20 s at 42°C, 30 s at 72°C, followed by 10 min at 72°C.

7. Check the size of the PCR product by running 5 μL of the reaction on a 1.5% agarose gel. Then use 2 μL of the reaction directly for cloning *via* the TA cloning kit (Invitrogen; cat. no. K2000-01).

8. Mix 2 μL of PCR product, 1 μL 10× ligation buffer, 2 μL pCR™2.1 vector, 4 μL sterile water and 1 μL T4 DNA ligase in a 1.5 mL tube, incubate at 14°C overnight.

9. Transform into INVαF' cell following the Invitrogen protocol or into equivalent cell, spread onto LB plate, incubate at 37°C overnight.

10. Pick and sequence 4-6 white colonies for each band. You may need to sequence more than one because the band may be composed of multiple DNA species.

Reverse Northern

Reverse Northern is a quick and easy way to verify induced mRNA identified by differential display. However, the major difficulty with this technique is the detection of low abundant mRNA. We have shown that by using PCR amplified fragments of cloned cDNA instead of miniprep DNA, mRNA with abundance level as low as 1 in 40,000 can be detected. Furthermore, amplified RNA (aRNA) (Poirier et al., mns. submitted, Van Gelder et al. 1990) can be used for probe synthesis, thus reducing the amount of RNA required for the verification procedure.

A) cDNA probe synthesis
Repeat the following for each RNA condition to be compared:

1. Mix 5 μL mRNA and 2 μL oligo dT_{12-18} (or 5 μL aRNA and 740 ng of random hexamers in a total of 7 μL H_2O) in a 1.5 mL tube, incubate at 70°C for 10 min, then place on ice for at least 1 min.

2. Add 2.5 μL 10× PCR buffer, 2.5 μL $MgCl_2$, 2.5 μL DTT, 1 μL d(GAT)TP mix, 1 μL dCTP (cold) and 7.5 μL α-^{32}P- dCTP. The α-^{32}P-dCTP is reduced from 50 μL (500 μCi) to 7.5 μL by drying in a speedvac. Do not let it dry completely because it will stick to the plastic side. Incubate at 42°C for 5 min.

3. Add 1 μL SuperScript II RT, incubate at 42°C for 50 min.

4. Add 3 μL NaOH, incubate at 68°C for 30 min to remove the RNA.

5. Add a mixture of 10 μL Tris-HCl, 3 μL HCl and 9 μL H_2O to neutralize.

6. Take 1 μL out to measure counts per minute.

7. Spin through a G-50 sephadex column to remove the unincorporated nucleotides.

8. Take 1 μl from the flow through to measure counts per minute and compare to that from step 6. You should get 20% to 30% incorporation. On average, 10^8 cpm are obtained from 2 μg of mRNA.

B) Dot blot
1. Isolate miniprep DNA from a 1 mL overnight culture for each clone to be tested. Alternatively, PCR amplify the insert from each of the clones.
2. Spot 0.5 μL DNA (1/20 dilution of miniprep; ~1 μg) from each clone (or 1 μg of each PCR product) in a grid-like pattern onto duplicate nylon membranes; one membrane for each RNA condition (see Figure 4).
3. Air dry for 20-30 min, denature for 5 min at room temperature by placing membrane (DNA side up) on Whatman paper pre-soaked with denaturing solution, wash twice in neutralizing solution for 5 min at room temperature, and rinse in 2× SSC.
4. UV cross-link once (Stratalinker auto cross-link setting).
5. Place the duplicate membranes each in a separate hybridization tubes, add 10 mL pre-hybridization solution (for a 15 × 15 cm membrane) and incubate in a rotating hybridization oven at 42°C overnight.
6. For each RNA condition, replace the pre-hybridization solution with a mixture of 10 mL hybridization solution and equal amount (cpm) of boiled (5 min) probe (1-2.10^7 cpm/mL of hybridization solution), incubate with rotation at 42°C overnight.
7. Wash membrane twice in wash solution I for 5 min at room temperature, wash twice in wash solution II for 15 min at 65°C, and expose membrane on X-ray film (keep membrane moist).
8. You may want to wash the membrane again in wash solution III for 15-30 min at 65°C to reduce background, then expose again.

C) RNA amplification; protocol modified from Van Gelder et al. (1990).
1. Add 2 μg of total RNA, 2 μL of T7(dT) primer (50 ng/μL) to RNase-free H$_2$O for a total volume of 11 μL. Incubate at 70°C for 10 min. Quench on ice for 1 min.
2. To this heat denatured RNA add 4 μL of 5× first strand buffer, 2 μL 0.1 M DTT, 1 μL 10 mM dNTP. Incubate at 37°C for 2 min. Then add 2 μL of SuperScript reverse transcriptase. Mix and incubate at 37°C for 1 h.
3. Add to the first-strand reaction 91 μL H$_2$O, 30 μL 5× second strand buffer, 10 mM dNTP mix, 1 μL DNA ligase, 4 μL DNA Polymerase I and 1 μL RNase H for a final volume of 150 μL.
4. Precipitate the resulting double stranded cDNA by adding 0.5 volumes of 7.5 M ammonium acetate and 2.5 volumes of 100% ethanol. Wash pellet with 70% ethanol. Resuspend cDNA pellet with 8 μL RNase-free H$_2$O.
5. To amplify RNA, use Ampliscribe T7 kit (Epicentre) and add all reagents at room temperature to prevent DNA precipitation. In order, add 2 μL of 10× buffer, 6 μL of 100 mM NTP mix, 2 μL DTT, 8 μL cDNA template and 2 μL of T7 RNA polymerase. Mix and incubate at 37°C for 4 h.
6. Add 10 μL of 7.5 M ammonium acetate, mix, chill on ice for 15 min, spin at 14,000 rpm in microcentrifuge for 45 min (at 4°C) to pellet cRNA. Wash pellet with 500 μL 70% ethanol, spin at 14,000 rpm for 10 min (at 4°C) to pellet cRNA. Remove ethanol and air dry at room temperature. Resuspend cRNA in 10 μL RNase-free H$_2$O. Quantitate cRNA and run small aliquot (0.5 to 1 μg cRNA) on a 1.5% TBE agarose gel. The expected result is a smear from 400 bp to 4,000 bp.

Notes

1. The arbitrary 10-mers used in DD-PCR should contain 50% GC content. Since the site of annealing is highly depended on the 3' end of the 10-mer, the last 4 bases of the oligo should be as different as possible. The 10-mer performs best when the most 3' base is a G or C; however, to increase the diversity we have also used A or T with the caveat that the band intensity decreases slightly. Therefore, a total of 256 different 10-mers can be used.

2. To minimize false positives it is essential that the bands are sharp and well separated on the gel. We found that standard flat well comb produces better bands than sharkstooth comb. Also, the DNA should be loaded as close to the bottom of the well as possible; this will produce a sharp flat band and eliminate streaks appearing in the center of the lane.

Overview

Previous kinetic studies (Axel et al. 1976, Bishop et al. 1974) indicate that ~98% of all mRNA species within the cell have a prevalence ranging from 1/10,000 to 1/100,000, and thus are considered rare. Furthermore, the remaining ~2% of mRNA species (about 200-500) contribute to over 50% of the total mRNA mass (Axel et al. 1976, Bishop et al. 1974). Therefore, it is relatively easy to identify a few very abundant differentially expressed mRNA (i.e. using plus/minus screening); but if the quest is to identify the majority of differentially expressed mRNA species, then a method that is not sensitive to mRNA abundance is required. To evaluate how well DD can identify abundant as well as rare differentially expressed mRNA, we used DD to find differentially expressed mRNA within HeLa cells in response to interferon-γ (IFN-γ). Described below is a summary of our results (original report: Wan et al., mns. submitted).

To identify differentially expressed genes in HeLa cells in response to IFN-γ, we surveyed approximately 1300 mRNA species *via* DD with 72 primer sets. An example gel is shown in Figure 1. In total, 23 differentially expressed genes were identified by DD. The median mRNA abundance of the genes identified was approximately 1/20,000 and ranged from 1/1000 to 1/200,000. Thus, DD is able to identify abundant and rare mRNA.

The hit rate (true positives) of DD was 1 in 2. However, the hit rate can vary depending on the criteria for selecting bands for evaluation. We have observed that bands which are intense, sharp, and appear in a all-or-none fashion are more likely to be true positives.

Redundancy (i.e. cloning the same cDNA more than once) was observed. This redundancy is primer sequence-dependent and can be observed in at least four ways. First, we as well as others, have shown that the 4 most 3' bases of the 10-mer are the most critical in determining annealing specificity (Wan et al., mns. submitted, Bauer et al. 1993). By comparing the sequence of a 10-mer with its site of annealing on the cDNA, we have found that in 9 out of 10 cases, the mismatches occurred within the 6 bases at the 5' end of the 10-mer and not in the 4 most 3' bases. However, 10-mers identical at the last

4 bases do not necessary produce similar band patterns (Figure 2, lanes 2 and 3), indicating that although the 4 most 3' bases determine annealing specificity, the 5' end of the primer probably plays a role in stabilizing or destabilizing a particular primer-cDNA complex.

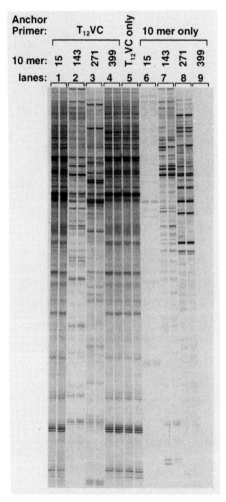

FIGURE 2. Differential display redundancy may be caused by "weak" 10-mers. The reactions were performed in duplicates and loaded adjacent to each other. $T_{12}VC$ was used during cDNA synthesis. For the PCR, 1.25 μM $T_{12}VC$ and of each 10-mers [15 (TACGCAGTAC), 143 (TAACGCGTAC), 271 (AGTAGGGTAC), 399 (GATCGTGTAC)] was used in lanes 1-4 respectively; 1.25 μM $T_{12}VC$ alone was used in lane 5; 1.25 μM of each 10-mer alone was used in lanes 6-9.

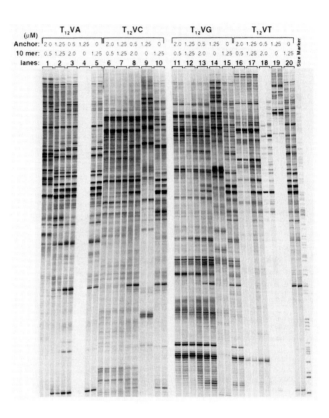

FIGURE 3. Using the same 10-mer in combination with the 4 anchor primer may cause redundancy in differential display. The reactions were performed in duplicates and loaded adjacent to each other. $T_{12}VA$, $T_{12}VC$, $T_{12}VG$ and $T_{12}VT$ was used for cDNA synthesis for lanes 1-5, 6-10, 11-15, 16-20 respectively. 10-mer #7 (CATTCAGCAC) was used in all reactions except for lanes 4, 9, 14, and 19. The primers and concentrations used for the PCR are marked above each lane. $T_{12}VA$ (lane 4) and $T_{12}VT$ (lane 19) alone consistently produce patterns with low intensity.

Secondly, redundancy can be caused by "weak" 10-mers. We define "weak" primers here as 10-mers which produce low intensity bands or blank lanes when used alone in a DD PCR (no $T_{12}VX$ primer, i.e. anchor primer, is used in the reaction; see Figure 2, lanes 6 and 9). With a weak 10-mer, one observes the same pattern of amplified cDNA whether one uses 10-mer and anchor primer or anchor primer only (see Figure 2, and compare lanes 1 and 4 with 5). Thus, weak 10-mers do not contribute to the amplification process. We recommend that each anchor primer be used as a single primer in a DD PCR so that the anchor only pattern can be compared with anchor plus 10-mer patterns for the process of eliminating weak primers from the 10-mer repertoire.

Thirdly, redundancy is contributed by the 3' anchor primers as well. We have observed that the 4 anchor primers ($T_{12}VG$, $T_{12}VA$, $T_{12}VT$, $T_{12}VC$) do not completely separate the mRNA population into 4 subpopulations when these anchor primers are used as first-strand synthesis primers for cDNA synthesis. This is illustrated in Figure 3 (lanes 5, 10, 15, and 20), which demonstrates that a 10-mer used alone in a DD-PCR amplifies similar cDNA/patterns from all four first-strand synthesis preparations.

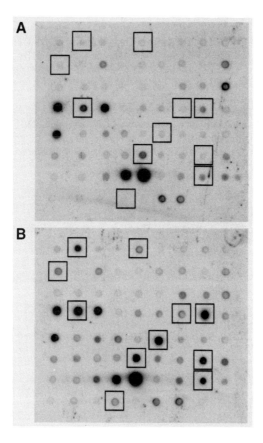

FIGURE 4. Reverse Northern showing cDNAs induced by IFN-γ treatment. 71 cDNA clones were spotted onto duplicate nylon membranes (A) and (B) at the same relative positions. The membranes were then probed with labeled cDNA from untreated HeLa cells (A) and from HeLa cells treated with IFN-γ for 48 h (B). Clones in boxes were confirmed by Northerns to be IFN-γ induced.

Finally, the molar ratio between the 10-mer and the anchor primer plays a role in redundancy. A high 10-mer/anchor primer ratio will produce patterns similar to that of 10-mer alone while a low ratio will produce patterns similar to that of using only the anchor primer (Figure 3). Therefore, to minimize all types of observed redundancy we recommend the following: (i) the last 4 bases of the 10-mer should be as diverse as possible; (ii) the anchor primers should be used alone as a control to eliminate weak 10-mers; (iii) a different set of 10-mers should be used for each anchor primer; and (iv) an equal molar ratio of 10-mers and anchor primers should be used.

Because cDNAs obtained from DD are usually short and at the non-coding region (3' untranslated region), we found that the Merck EST database can be helpful for identification of human mRNAs because each cDNA in the Merck EST database is sequenced from both 5' and 3' orientation. Thus, a match at the 3' end can often lead to coding information at the 5' end. This problem should eventually go away when full length cDNA contig databases are available when all human EST sequencing have been completed.

To verify putative positives we use a reverse Northern approach. An example of this is shown in Figure 4. When only small amounts of RNA are available (2-5 μg of total), we have demonstrated that amplified RNA (aRNA) (Poirier et al., mns. submitted, Van Gelder et al. 1990) is representative of the original mRNA population and thus can be used as template for screening putative positives.

References

Axel, R., Feigelson, P., and Schutz, G. (1976) *Cell* 7:247-254.

Bauer, D., Muller, H., Reich, J., Riedel, H., Ahrenkiel, V., Warthoe, P., and Strauss, M. (1993) *Nucleic Acids Res* 21:4272-4280.

Bertioli, D. J., Schlichter, U. H., Adams, M. J., Burrows, P. R., Steinbiss, H. H., and Antoniw, J. F. (1995) *Nucleic Acids Res* 23:4520-4523.

Bishop, J. O., Morton, J. G., Rosbash, M., and Richardson, M. (1974) *Nature* 250:199-204.

Debouck, C. (1995) *Current Opinion in Biotechnology* 6:597-599.

Liang, P., and Pardee, A. B. (1992) *Science* 257:967-971.

Van Gelder, R. N., von Zastrow, M. E., Yool, A., Dement, W. C., Barchas, J. D., and Eberwine, J. H. (1990) *Proc Natl Acad Sci USA.* 87:1663-1667.

Welsh, J., Chada, K., Dalal, S. S., Cheng, R., Ralph, D., and McClelland, M. (1992) *Nucleic Acids Res* 20:4965-4970.

DNA markers: Protocols, Applications and Overviews
ed. by G. Caetano-Anollés & P.M. Gresshoff
© 1997 John Wiley & Sons, New York

Molecular markers and forest trees

David M. O'Malley and Ross Whetten

Forest Biotechnology Group, North Carolina State University, Raleigh, NC 27695-8008, USA

Introduction

Forests cover some 4 billion hectares of the earth's surface. These forests are ecological and economic resources that are threatened by over-exploitation and the possibility of rapid environmental change due to global warming. A better understanding of the genetic mechanisms that influence tree adaption and productivity is critically needed for management of the world's forest resources (Burley 1996). Intensive culture of genetically improved trees in plantations could help meet the increasing demand for wood products. Forest tree species present some unique biological problems and are at the earliest stages of domestication. Genetic analysis using new molecular marker technologies provides a powerful approach to understanding the organization and distribution of genetic resources in natural and managed populations of forest trees. In this review, the application of molecular markers to forest trees is discussed, emphasizing applications in genomic mapping and population studies.

The role that genetic marker technology plays in forest genetics could now mature from an ancillary technique into a practical tool to achieve breeding objectives and to study fundamental questions about the biology of forest trees. Forest geneticists adopted allozyme methods during the 1970's and used these genetic markers during the 1980's for population studies (reviewed in Adams et al. 1992). The "neutral" theory postulated that the extensive genetic variation detected for allozymes could be explained if the magnitudes of selection coefficients for allozymes were generally very small, similar to the mutation rate. Population processes of interest to forest geneticists, such as migration or outcrossing rate, could be estimated from the distribution of marker genotypes among individuals and populations based on an assumption of neutrality. However, correlations between allozyme genotypes and phenotypic traits involving growth and fitness are not uncommon in outbred populations, and in some cases, the weight of the biological evidence strongly supports an allozyme locus as the explanation for phenotypic variation

(Watt 1994). An alternative explanation for these correlations is linkage disequilibrium between a neutral marker locus (the allozyme) and a "hidden" quantitative trait locus (QTL) that also segregates in the population (Ruiz and Barbadilla 1995). Considering the large number of genes in eukaryotes (perhaps 10^4–10^5), the hypothesis that important variation is not at the marker locus seems more appropriate as a null hypothesis. Now, molecular marker technologies have enabled a genomic mapping approach where phenotypic effects can be correlated with segregating neutral molecular markers within families to reveal QTLs as locations on a genomic map.

Ten years ago in his review of quantitative genetics in forestry, van Buijtenen (1988) identified the interface of quantitative genetics and molecular genetics as an emerging field that promised opportunity because genes that contributed to the variance of polygenic traits could be identified. Nevertheless, "entirely new methods" were needed. QTL mapping with molecular markers has provided a *high resolution* extension of the genetic analysis that was already possible in experimental organisms (e.g. *Arabidopsis*) that are easily bred over multiple generations. In contrast, the scope and power of Mendelian analysis in forest trees has been extremely restricted because breeding studies are hindered by long generation times and severe inbreeding depression. Examples of useful, simply inherited polymorphisms were rare in forest trees (e.g. a simply inherited disease resistance gene in *Pinus lambertiana*; Kinloch et al. 1970). Most insight on the genetic control of economic traits in forest trees has been obtained from heritability estimates based on family mean differences in common garden studies (e.g. Namkoong and Kang 1990). In forest trees and other undomesticated organisms, *complex trait dissection* using molecular markers (Lander and Schork 1994) has provided *a new paradigm* for genetic analysis. The impact of complex trait dissection on tree breeding could be profound because it enables within family genetic analysis. Complex trait dissection attributes a portion of the phenotypic variation to one or more loci segregating within a family, and can be carried out in an opportunistic fashion in arbitrary pedigrees and collections of outbred families. At the population level, marker trait associations can be established based on linkage disequilibrium and using a sib pair approach. Complex trait dissection and experimental investigations in population genetics depend on the ability to generate large numbers of *informative* genetic markers.

Markers and DNA sequence

Molecular genetic markers are readily assayed phenotypes that have a direct 1:1 correspondence with DNA sequence variation at a specific location in the genome. In principle, the assay for a genetic marker is not affected by environmental factors (i.e. the heritability of a marker is 1.0). Genetic markers are a "short-cut" to DNA sequence information. The increased efficiency in screening is offset by the limited scope of the information that is obtained. Sequence differences can be detected by direct sequencing of DNA fragments that have been PCR amplified from diploid DNA, based on the sensitivity of sequencing methods to distinguish heterozygous sites from homozygous sites (e.g. Phelps et al. 1995, Chadwick et al. 1996). Heterozygous sites can also be screened using a mismatch detection approach (e.g. Mashal et al. 1995, Youil et al. 1996). However, nucleotide diversity (i.e. the proportion of heterozygous sites on two homologous DNA fragments drawn randomly from a population; Nei and Li 1979) is low, approximately 0.0005 to 0.02, depending on the nature of the sequence (coding or noncoding), on the gene, and on the species (e.g. Innan et al. 1996, Moriyama and Powell 1996). This range of nucleotide diversity corresponds with approximately 0.5 to 20 differences per kb of DNA sequence.

Knowledge of DNA sequence allows markers to be identified by exploiting specific differences between alleles through several approaches. For example, Voo et al. (1995) exploited a cDNA sequence difference between two alleles of the gene coding for 4-coumarate co-A ligase (*4CL*) so that PCR amplified fragments from the genomic DNA would contain a polymorphism for a restriction site. The DNA fragments amplified from the two alleles could be distinguished by cleavage of the product with a restriction enzyme (Konieczny and Ausubel 1993). DNA sequence differences can be detected in PCR amplified DNA fragments from specific genes without direct knowledge of DNA sequence variation as single strand conformation polymorphisms (SSCPs; Orita et al. 1989). PCR amplified fragments can also reveal differences in fragment size due to insertions/deletions, and can be allele-specific (e.g. Paran and Michelmore 1993, Gu et al. 1995). Microsatellite markers exploit the variable repeat number for simple sequence repeats occur within genes or in non-coding regions of the genome (e.g. Powell et al. 1996a).

Restriction fragment length polymorphisms (RFLPs) are extensively used as genetic markers in trees (e.g. Bradshaw et al. 1994, Devey et al. 1994, Byrne et al. 1995, Mukai et al. 1995). RFLPs generated by a cloned cDNA hybridization probe could recognize variation in and near a specific gene when stringent hybridization conditions are used. Low copy number, anonymous genomic sequences (non coding) can also be used as hybridization probes. Most RFLP variation in plants is due to insertion/deletion differences that are located between restriction enzyme recognition sites, and a small portion of the variation can be due to sequence variation within restriction sites. RFLP markers are especially effective when used in combination with some DNA sequence information, but except for the identity of the probe, prior knowledge of DNA sequence information is not essential.

Little DNA sequence information is available for most forest tree species, but anonymous markers can be generated by PCR based methods that do not require prior knowledge of DNA sequence. Amplification using a single small primer reveals extensive DNA fragment polymorphisms that are often simply inherited and useful as molecular markers [e.g. randomly amplified polymorphic DNA (RAPD; Williams et al. 1990), arbitrarily primed PCR (AP-PCR; Welsh and McClelland 1990), DNA amplification fingerprinting (DAF; Caetano-Anollés et al. 1991)]. Another anonymous PCR based marker system is amplified fragment length polymorphism (AFLP) analysis (Vos et al. 1995). AFLPs are produced by cleaving genomic DNA using a pair of restriction endonucleases, then ligating adapters to the ends of the DNA fragments. A subset of the fragments is selectively amplified from these ligated fragments. These DNA fragments can be resolved on DNA sequencing gels and have a higher multiplex ratio than RAPDs, often 20-40 markers per gel lane.

Markers and comparative mapping

Integration of genomic maps within and among taxa is an important goal for genomic mapping in forest trees. Assembly of consensus maps is facilitated by marker systems with a high level of variability, a codominant mode of inheritance, and unambiguous homology. Homologous genes have a common ancestry (reviewed by Hillis et al. 1996). The relationship of genes whose common ancestry is traced through a gene duplication event is paralogy. Paralogous DNA sequences could arise from tandem duplications, polyploidy, or other mechanisms (e.g. retropseudogenes). The relationship of genes whose common ancestry can be traced back through speciation events is orthology. Synteny is the association of markers caused by their location on the same chromosome. Synteny is usually conserved among closely related species and chromosomes that are homologous among closely related species are called homeologs. Even when there have been extensive chromosomal rearrangements during speciation and evolution, the colinearity of marker genetic maps can be highly conserved. The correct identification of orthologous markers is the basis for inferring the phylogeny of species from the phylogeny of genes. For the purpose of constructing genomic maps, unrecognized paralogy could cause confusion with respect to the map location of markers (i.e. the *same* marker could *appear* to map to different locations in different crosses or species). The map position of a marker locus relative can help to infer orthology and paralogy.

O'Brien (1991) suggested a classification scheme for markers based on their utility for comparative mapping. Markers based on coding sequences that are highly conserved and readily recognized in diverse taxa were classified as Type I anchor loci. Many allozymes could be considered Type I markers. Type I marker loci tend to have low levels of variability and limited utility for mapping and complex trait dissection within species. Type II anchor loci are highly variable markers useful for pedigree analysis within species but are less useful for comparative mapping among divergent taxa (e.g. microsatellites). RFLPs have been useful markers for comparative mapping in some groups of plants. In wheat, Devos et al. (1993) recognized *homeologous* probes that hybridize with loci on all three wheat genomes and *non-homeologous* probes that tend to be genome specific. In related species, homology is generally assumed for restriction fragments that have identical size in different individuals using a hybridization probe for single or low copy number sequences (Hillis et al. 1996), but orthology can be difficult to infer when gene families are large.

Only a few genes from trees have been sufficiently well characterized at the level of DNA sequence to be considered candidates for Type I anchor loci. Due to large genome size and a large copy number for gene families recognized by hybridization probes (Neale et al. 1994), the RFLP approach in conifers could be less useful to identify orthologous markers than in some other groups of organisms. PCR amplification of orthologous conifer gene sequences based on angiosperm DNA sequences is impeded by the large genome size and phylogenetic divergence of conifers. Type I marker development for comparative mapping in conifers would be greatly facilitated by the availability of cDNA sequences (e.g. Kinlaw et al. 1996). Some conifer genes that could be candidates for Type I markers include *cad* and *4cl* (Voo et al. 1995, MacKay et al. 1995).

Comparative mapping could provide a way for forest geneticists to use genomic information developed in experimental organisms and more readily manipulated plants, as well as in related tree species. In the Poaceae (grass family), the different taxa can be treated as a single genetic system, where the map from one species provides a good guide for the homeologous map position of important genes in another species, and markers are practically interchangeable among the species (e.g. Ahn et al. 1993, Bennetzen et al. 1993, Devos et al. 1995, Moore et al. 1995). Comparative mapping in the cereal crops has shown close correspondence of QTLs for several agronomic traits (large seeds, reduced mature inflorescence disarticulation, day-length insensitivity) that were selected independently during domestication of rice, sorghum, and maize (Paterson et al. 1995). Thus, important trait genes isolated in one species could be recognized in another species, and candidate genes for a QTL mapped in a poorly characterized species could be identified by reference to the map of a better characterized species. Recent work has shown that genome organization is highly conserved among vertebrate animals (Lyons et al. 1994). Comparative mapping of mammals and of fish could provide insights on the genome structure and organization of the vertebrate ancestor prior to divergence of the tetrapod lineage some 400 million years ago (Morizot 1994). Genomic information from fish can provide insight for genomic mapping in humans (e.g. Elgar et al. 1996, Trower et al. 1996). The gymnosperm lineage can be traced to approximately 350 million years ago, suggesting that the gymnosperm-angiosperm divergence is almost as old as the fish-tetrapod divergence. Perhaps the coalescence of blocks of synteny and common elements of chromosomal organization among extant seed plants will provide insight on the ancestral genome of the earliest seed plants and a comprehensive way to use genomic information in all higher plants of economic importance.

Markers and genomic maps

Genetic maps are constructed by analysis of the cosegregation of genetic markers in the products of meiosis, usually in a family of diploid progeny derived from one or two parents. Markers for different genes that are physically located on different chromosomes are expected to assort independently, but markers for genes located on the same chromosome are transmitted together unless the parental combinations are broken up by recombination. A map with nearly complete coverage of the genome could require several hundred markers, and the number of possible orders of these markers is very large (*n*! for *n* loci). A large family size is important for determining locus order because several recombination events are needed to obtain statistical support for a local order. Interval support is usually high for markers ≥5 cM apart in genomic maps based on 1×10^2 meioses, but 1×10^3 meioses could be required for interval support for a map with 0.5 cM resolution. For finer scale resolution of locus order (< 1 cM), construction of physical maps is more feasible than construction of genetic maps based on recombination and linkage.

Genomic map construction from marker segregation data in outbred families is a complicated task compared with map construction for inbred F_2 families obtained by selfing an F_1 hybrid. In an F_2 family, *all* codominant markers segregate in a 1:2:1 ratio (3:1 for dominant markers). BC_1 families are obtained by backcrossing the F_1 to an inbred parental line, and all markers segregate 1:1. The parental linkage arrangements of the markers (coupling and repulsion) are known with certainty from the parental lines, and only 2 alleles are segregating at each marker locus. Codominant markers segregating in F_2 families provide more linkage information and are more efficient for mapping than BC_1 families. Outbred families are more complex. Up to 4 alleles can segregate at one locus because there are two potentially heterozygous parents and they could have different heterozygote genotypes. To estimate recombination fraction for a pair of marker loci, at least one parent must be heterozygous for both markers. The recombination fraction between two marker loci can be estimated from 21 different parental genotypic configurations (Ritter and Salamini 1996). The marker genotypes of the grandparents (i.e. a 3-generation outbred pedigree) are useful to infer linkage phase of markers. In principle, the linkage information for diverse mating patterns with a variety of different segregation ratios for several different loci can be combined using likelihood methods, but the statistical support for locus order could be low. Most linkage information for outbred families comes from 1:1 and 1:2:1 segregation ratios.

Following Sewell and Neale (1995), a codominant marker that segregates 1:1 in an outbred, full-sib family can be classified as maternally informative for linkage, MI, if the maternal parent is heterozygous, or paternally informative, PI, if the paternal parent is heterozygous. A codominant marker could be informative for both parents, BI (1:2:1 ratio), or fully informative, FI, if the 2 parents have different heterozygous genotypes (1:1:1:1 ratio). Separate paternal parent and maternal parent maps can be constructed from the PI and MI markers. The FI and BI markers provide common reference points (anchors) that allow the two maps to be integrated (e.g. Ritter and Salamini 1996). Otherwise, two parent specific homolog maps will be made for each pair of chromosomes, and the homolog maps that belong to the same chromosome will be unrecognized. FI and BI segregations can only be obtained with codominant markers and are needed to characterize the mode of gene action (dominant vs codominant) for QTLs that segregate in outbred families (e.g. Schäfer-Pregl et al. 1996).

The ideal situation for linkage analysis in outbred full sib families is to have large numbers of codominant markers that are heterozygous in every parent so that a map can be constructed that integrates information from both parents. Heterozygosity is defined as $h_e = 1 - S_i p_i^2$, where p_i is the frequency of the ith allele of a marker locus. If $h_e = 0.25$, then approximately 1/4 of parents are expected to be heterozygotes for that locus. Assuming independence, the 2 parents of a full-sib family will both be heterozygotes in only 1/16 of families for a marker with $h_e = 0.25$. There is some choice involved in screening primers or probes for markers useful in mapping studies, so estimates of heterozygosity intended to be representative of genetic variability of a species are not necessarily representative of markers chosen for mapping. We define a conditional probability of both parents being informative for a marker (PBI) as the probability that a marker segregates in both parents, given that the marker segregates in at least one of them. For a locus with $h_e = 0.25$, the

PBI is $(1/4)^2/(1/4 + 1/4 - 1/16) = 1/7 = 0.14$, assuming independence. In contrast, the PBI is 0.60 for a marker with $h_e = 0.75$. Many mapping studies provide the summary data needed to obtain an estimate of PBI that roughly characterizes the level of heterozygosity of the set of markers actually used for mapping.

Anonymous markers and individual tree maps

Although most RAPDs have a dominant mode of inheritance (i.e. presence vs absence of a band, and heterozygous phenotypes are identical to the homozygote band present phenotype), these markers have been widely used for genetic analysis in outbred species (e.g. Tulsieram et al. 1992, Nelson et al. 1993, Binelli and Bucci 1994, Bradshaw et al. 1994, Grattapaglia and Sederoff 1994, Byrne et al. 1995, Kubisiak et al 1995, Mukai et al. 1995, Kaya and Neale 1995, Plomion et al. 1995, Yazdani et al. 1995, Devey et al. 1996a, O'Malley et al. 1996). PCR amplification of anonymous DNA sequences provided a less laborious alternative to RFLP markers. In gymnosperms, megagametophyte tissue from open pollinated seeds collected from an open-pollinated tree can provide a haploid "family" for mapping allozymes or PCR based genetic markers. The DNA that can be extracted from a megagametophyte (approximately one microgram) is enough template for hundreds of PCR reactions, but not enough to resolve RFLPs. Dominant markers present no interpretation problem in haploid segregations obtained from megagametophytes from individual trees, where all markers are expected to segregate 1:1. RAPD markers that segregate 1:1 are also abundant in outbred full-sib families (e.g. Grattapaglia and Sederoff 1994), and are readily identified by screening the parents and a small number of progeny. These markers are in a pseudotestcross configuration, and could be MI or PI depending on which parent is heterozygous and which parent is homozygous for the null allele. For a pair of parents of a full-sib family analyzed as a pseudotestcross, PBI = 0, and separate maps are developed for each parent from the segregation data.

PCR amplification of RAPD markers from diploid genomic DNA shows that a small portion of RAPD markers have a codominant mode of inheritance. In some cases, codominant RAPD markers are found in progeny that are not present in band profiles from either parent. These bands could be caused by heteroduplex formation, where noncomplementary but similar strands anneal (i.e. products of different alleles), and migrate more slowly on gels than the parental DNA fragments (e.g. Davis et al. 1995, Antolin et al. 1996, Novy and Vorsa 1996). Heteroduplex formation has implications for the genetic interpretation of RAPDs for population and inheritance studies, but could be exploited using SSCP methods to provide more codominant markers from RAPDs. Alternatively, RAPD fragments can be cloned and sequenced to convert them to sequence tag sites (STS) that are PCR amplified using a pair of longer, more specific primers. Some RAPD STS are allele-specific, some are codominant, and some will be polymorphic when analyzed using SSCP methods (Paran and Michelmore 1993, Gu et al. 1994, Bradshaw et al. 1994, Bodenes et al. 1996).

Integrating RAPD marker maps from individual trees within a species (i.e. recognizing syntenic marker groups for several parents) depends on identification of homologous markers in different families. The homology of RAPD markers in different families or species can be tested through precise estimation of fragment size, cross-hybridization of amplified bands, and restriction digestion of reamplified RAPD bands to demonstrate identical restriction fragment sizes. Reiseberg (1996) used this kind of rigorous approach to test homology of RAPD fragments and showed that 90% of markers in three closely related species of sunflower map to the same relative positions in each species (i.e. the fragments were mostly orthologous). In *Pinus*, Furman et al. (1997) used reamplification and restriction digestion of RAPD bands to test putative orthologous markers that were used in a phylogenetic analysis of 6 closely related species. Kaya and Neale (1995) mapped a set of RAPD markers in 4 families of *Pinus brutia* using megagametophytes. The observed heterozygosity for RAPD markers in this small sample was 0.37. They concluded that it was difficult to compare linkage groups in different trees because most RAPD markers segregating in one tree were not segregating in others. The problem, finding markers that are both segregating in two "parents", is analogous to finding FI or BI markers in a full-sib family with RFLPs. Averaging over the 6 possible pairs of the 4 trees, the PBI was 0.19, a little less than the expected 0.23 for $h_e = 0.37$. For *Picea mariana* and *Picea abies* (Isabel et al. 1995, Bucci and Menozzi 1995), the heterozygosity estimates were $h_e = 0.33$ (28 loci), and $h_e = 0.34$ (20 loci), respectively, tested by segregation in megagametophytes. The heterozygosity of RAPDs could be smaller than RFLPs, but given the relative ease of obtaining large numbers of RAPD markers in different trees, syntenic groups of RAPD markers can be readily assembled by comparison of pairs of several "parents". Similarly, the integration of pseudotestcross maps from diploid families requires comparisons of at least two families (i.e. four parents).

The major disadvantage of RAPD markers is that they are difficult to reproduce among different laboratories. The PCR amplification process that produces RAPDs involves competition among fragments that can be affected by small variations in the reaction conditions. Fastidious attention to reagents and amplification conditions is required to ensure that the same markers resolved today can be recovered in the future in the same or in a different laboratory. Skroch and Nienhuis (1995) evaluated the quality of RAPD genetic data. They reported 2% scoring errors and 76% reproducibility (i.e. getting the same band back in a replicate sample run on a different day in the same laboratory). Reproducibility increased with band intensity, the quality of the data from any band that could be scored was high (low % scoring errors). Intensive screening can identify the polymorphisms that have intense bands on gels, a high level of repeatability, Mendelian inheritance, and therefore utility as *genetic markers* (e.g. Plomion et al. 1995). RAPD methods do reveal a large amount of polymorphism, but not all of this variation makes suitable genetic markers for genomic mapping. Heun and Helentjaris (1993) reported that 37% of RAPD bands among maize F_1 hybrids were unambiguously polymorphic and 95% of these bands showed a dominant pattern of inheritance. The major benefit of RAPD methods is the relative ease of obtaining markers, and their utility in map construction is enhanced when used in combination with codominant markers. AFLPs (Vos et al. 1995) are an advance in technology to resolve anonymous markers that has a higher multiplex ratio, perhaps higher reproducibility, and has more potential for automation than RAPDs by using DNA sequencing equipment.

RFLPs and genome organization

Restriction fragment length polymorphisms (RFLPs) were the first molecular markers that were widely used for genomic mapping and population studies. The mode of inheritance of RFLP markers is codominant. RFLP maps have been constructed for several tree species (e.g. Bradshaw et al. 1994, Devey et al. 1994, Byrne et al. 1995, Mukai et al. 1995, Devey et al. 1996a). Most of the hybridization probes in *Populus, Eucalyptus* and *Citrus* detected single copy sequences and the RFLPs were readily interpreted as single loci (Bradshaw et al. 1994, Byrne et al. 1994, Liou et al. 1996). In *Eucalyptus*, 210 RFLP loci were identified and 39% of the marker segregations were FI or BI (i.e. the PBI was 0.39). In both *Eucalyptus* and Citrus, multiple loci detected by the same hybridization probe were more likely to map to the same linkage group than to a different linkage group. In contrast with the angiosperm genera, hybridization probes based on conifer cDNA often yield complex patterns of bands from restriction digested genomic DNA, suggesting large gene families (e.g. Baker and White 1996, Kinlaw et al. 1994). Sewell and Neale (1995) presented a consensus map for *Pinus taeda* (loblolly pine) based on 320 RFLP markers that segregated in 2 outbred pedigrees. The PBI for one map was 0.33 and for the other map was 0.21. A small portion of the RFLP markers (12%) were scored in both RFLP maps. Devey et al. (1994) reported a map for a 3-generation pedigree of *Pinus radiata*, based on 165 RFLP markers detected using cDNA hybridization probes developed for *Pinus taeda*, and 43 other markers. In this *Pinus radiata* family, the PBI was 0.23, and most markers (73%) segregated in a 1:1 ratio, transmitted by one parent or the other. Considering two *P. taeda* and one *P. radiata* families, Devey et al. (1996) interpreted 96 marker loci to be homologous among two of the three mapping families, and used 44 of these markers to anchor linkage groups in the different families. Neale et al. (1994) reported that multiple loci detected by the same probe in pine are more likely to map to different linkage groups than to the same group.

The low heterozygosity of RFLP markers limits their usefulness as Type II markers. The PBI values for pine suggest that h_e is approximately 0.33 to 0.50 for RFLPs used as markers in mapping families. Even though RFLP markers have a codominant mode of inheritance, only 1/5 to 1/3 of the markers used for map construction in pine families are expected to segregate in both parents. These markers can provide anchors to orient parent-specific maps; these anchor points will be sparse unless a large number of markers is generated. In hybrid eucalyptus, the h_e for RFLPs could be larger than 0.50. However, paralogy could interfere with the interpretation of complex RFLP patterns, such as those found in pine. Focusing on PCR amplification of genomic sequences that encode expressed genes could greatly simplify the task of identifying orthologous markers in the presence of pseudogenes (Voo et al. 1995, Hillis et al. 1996, Perry and Furnier 1996).

The complexity of RFLP band patterns in the Pinaceae raises questions about conifer genome organization and evolution (Neale et al. 1994). Almost all species have 1n = 12 chromosomes, and chromosomal evolution and speciation in the Pinaceae are believed to occur at a slow rate (Levin and Wilson 1976). However, the genome size of conifers is large compared with other plants. For pine, 2C = 38-53 pg (O'Brien et al., 1996), which is approximately 8 times larger than maize, 50 times larger than rice, and 200 times larger than *Arabidopsis*. Most of the pine genome is composed of repetitive DNA (Rake et al. 1980), but the average copy number of expressed conifer genes is not known. One mechanism that could have increased the copy number of gene sequences in conifers is ancient polyploidy, as is suspected in maize where two copies are found for many single copy probes from rice (e.g. Devos et al. 1995). The cereal grain species provide a model for this kind of chromosomal evolution (e.g. Ahn et al. 1993, Kurata et al. 1994). However, inheritance studies show most allozyme loci are encoded by single loci, and colinearity of allozyme linkage groups is conserved in the Pinaceae (Conkle 1981). Alternatively, gene sequence copy number could be increased through mechanisms that result in the dispersal of gene families throughout the genome (e.g. retrotransposons; Kvarnheden et al. 1995) or through mechanisms that cause tandem duplications. The duplicated sequences could be transcribed or non functional pseudogenes.

An understanding of the structure and organization of conifer genomes is important for genomic mapping as well as for genetic engineering, but little is currently known about conifer genomes. Bobola et al. (1992) characterized rDNA repeat units of spruce, and found as many as 5 units that were 2-3 fold larger than typical repeat units for angiosperms, and could constitute approximately 4% of genomic DNA. The variation among individuals in rDNA copy concentration varied as much as 6-fold within black spruce. Brown et al. (1993) found 7 rDNA sites in white spruce by FISH (fluorescent in situ hybridization), one the highest numbers of rDNA sites reported for a plant. Karvonen and Savolainen (1993) found segregating rDNA variants for the IGS and ITS regions of Scots pine, which has 8 rDNA loci. Karvonen et al. (1994) found length variants (0.5 kb) in the highly conserved ITS regions of rDNA repeat units of spruce. These variants occurred within individuals and within different species of the genus *Picea*, suggesting past interspecific hybridization. Abundant copia-like retrotransposon sequences have been identified in the Pinaceae (Kam et al. 1996, Kossack 1989). These sequences are dispersed throughout the genome, have a size of approximately 5 kb, and could constitute about 2% of the genomic DNA. In maize, 10 retroelement families account for at least half of the maize nuclear DNA (San Miguel et al. 1996).

Only a few studies of pine genes have identified the genomic sequences corresponding with cDNAs. For example, Voo et al. (1995) identified two classes of *4CL* cDNA in *Pinus taeda* xylem based on DNA sequence, then showed that the corresponding genomic fragments amplified from megagametophytes were allelic. Detailed studies of the protein, cDNA, mRNA expression, and genomic sequences of cinnamyl alcohol dehydrogenase, *Cad* , in *Pinus taeda* yielded evidence for only a single expressed *Cad* locus (MacKay et al. 1995). Direct repeat sequences were frequent in the 5' region of *Cad* and the number and sequence of the repeats varied among four alleles sampled at this locus (MacKay, personal communication). Kvarnheden et al. (1995) reported a *cdc2* gene family in *Picea*, and showed that 6 of 10 genomic clones were processed retropseudogenes. These gene sequences were 86-92% identical with the cDNA, lacked introns, and contained deletions, insertions and/or non-silent point mutations. Perry and Furnier (1996) described 7 alcohol dehydrogenase loci (*Adh*) loci from genomic DNA sequences in *Pinus banksiana*, that corresponded with expressed cDNA sequences. These *Adh* loci were clustered in 2 groups only 3 cM apart. They also reported that there were frequent and often large (>38 bp) direct repeats in the untranslated regions of the *Adh* genomic sequences. Distinguishing among the alleles and loci was facilitated by the variation in the number and kind of repeated sequences. Although the *Adh2* allozyme locus of conifer seeds mapped to one of the clusters, the cDNA and genomic sequence that corresponded with this locus was not identified.

Microsatellites and forest trees

Hypervariable microsatellite markers have become the marker system of choice for trait mapping in human and animal genetics, and their application in plant genetics is increasing (e.g. Wu and Tanksley 1993, Akkaya et al. 1995, Kresovich et al. 1995, Chin et al. 1996). Microsatellite loci contain a simple sequence (mono-, di-, or trinucleotide) that is repeated (reviewed by Powell et al. 1996). The repeat number of microsatellites can be highly variable, so that most individuals are heterozygous and the proportion of individuals that have the same marker genotype is small. The PBI can be very high for some microsatellite markers. Thus, these markers segregate in most parents and allow the same genomic region to be "tagged" in many different individuals. Furthermore, microsatellite markers can be resolved and scored using semi-automated DNA sequencing equipment. However, the frequency of microsatellite sequences in plant DNA could be an order of magnitude lower than for the human genome. Perhaps the development of more efficient methods to clone microsatellites will make these markers more available for plants.

The application of microsatellites to genomic mapping in forest trees has thus far been limited by the development costs required to obtain a large set of microsatellite markers, and by the small proportion of markers that can be readily transferred from one species to other species within a genus. Morgante et al. (1996) reported that the proportion of candidate microsatellite clones from a *Picea* genomic library that yield a single locus polymorphism pattern is low relative to other plants, and suggested that this could be due to the high level of sequence complexity and large genome size of spruce. There have been relatively few microsatellites characterized for conifers, compared to the hundreds of markers needed for linkage map construction (e.g. *Picea sitchensis*:van de Ven and McNicol 1996; *Picea abies*: Morgante et al. 1996; *Pinus radiata*: Smith and Devey 1994). Byrne et al. (1996) characterized variation in 4 microsatellites from *Eucalyptus nitens* in six species. They found that the microsatellite loci were conserved within subgenus, but only half of the microsatellite sequences could be PCR amplified from two other subgenera, and none could be PCR amplified from one subgenus. Microsatellite markers can also be detected in searches of DNA sequence databases (e.g. Chin et al. 1996), but DNA sequencing projects for tree species are just beginning (e.g. Kinlaw et al. 1996).

Microsatellite markers have special advantages for the determination of parentage, hence could be useful for paternity or gene flow studies in managed or natural populations. The advantage of microsatellite markers is that the proportion of the population that can be excluded as a potential father increases with the number of alleles that are present at a locus, and the possible parents in a population decreases. Analysis of parentage was the motivation for the first work on microsatellite markers in tropical trees (Condit and Hubbell 1991). Chase et al. (1996) described several microsatellites for a tropical tree species, *Pithecellobium elegans* and compared microsatellite marker paternity exclusion probabilities to the exclusion probabilities for allozymes. Dow et al. (1996) used microsatellite markers to characterize seed dispersal and parentage of juveniles for *Quercus macrocarpa*. Steinkellner et al. (1996) characterized 5 microsatellites from *Quercus robur* and *Quercus petraea* in similar studies. The number of microsatellite markers needed for these population studies are perhaps an order of magnitude less than for genomic mapping. For example, Blouin et al. (1996) assessed the utility of 20 microsatellite markers to classify individuals by relatedness for a population of mice derived from the wild, and resolved family relationships (full-sibs vs half-sibs).

Molecular markers for phylogenetic and population studies

Nucleotide sequence variation in organelle genomes has provided a source of genetic markers that differ from nuclear DNA markers in their properties (Hipkins et al. 1994). The inheritance of mitochondrial and chloroplast genomes is uniparental and from the seed parent in angiosperms, but chloroplasts show paternal inheritance in gymnosperms, and both mitochondria and chloroplasts have paternal inheritance in some species (e.g. Neale et al. 1991). The chloroplast genome evolves more slowly than nuclear genes, which makes chloroplast DNA polymorphisms especially useful for systematic studies in plants (e.g. Tsumura et al. 1995). The rate of evolution for trees could be even slower than for herbaceous plants (Bousquet et al. 1992). cpDNA variation can provide insights on biogeography. For example, there is floristic similarity for forest trees in the US and Asia, and pairs of sister taxa are recognized for *Magnolia*. Qiu et al. (1995a, 1995b) addressed speculation concerning past migrations by showing that *M. tripetela* from the southern US had the lowest level of divergence with two Asian taxa, suggesting separation during the late Miocene or early Pliocene, and migration to North America near that time. Smith and Doyle (1995) used cpDNA polymorphisms to gain new perspectives on the phylogeny of the Juglandaceae. They found that cladistic analysis of morphological traits and cpDNA polymorphisms yielded the same topology, but this phylogenetic tree did not agree with the fossil record regarding the basal-most lineages versus the more terminal lineages, suggesting a problem with the rooting of the phylogenetic trees.

Organelle genome differences have been used for studies of interspecific hybridization and introgression in forest trees. For example, Dawson et al. (1995) used RAPDs, RFLPs, and SSCPs from PCR amplified mitochondrial DNA sequences to diagnose interspecific hybridization of *Gliricidia sepium* and *G. maculata*. The geographic distribution of the species and the distribution of the RAPD and RFLP markers suggested three areas of hybridization. Mitochondrial SSCPs showed a high degree of divergence among populations and there were diagnostic haplotypes for the two species. This genetic data is important for the conservation and utilization of *Gliricidia* germplasm in breeding efforts. Sutton et al. (1991) used chloroplast and mitochondrial DNA polymorphisms resolved as RFLPs to identify introgressed populations of *Picea sitchensis* and *Picea engelmanii*. Markers for *P. sitchensis* were found to the east of the coastal watersheds where it commonly occurs. The individuals in introgressed populations usually had a chloroplast marker from one species and a mitochondrial marker from the other.

Genetic markers from organelle genomes have been used at the intraspecifc level to provide new information on genetic differentiation of populations, despite the typically low level of DNA sequence variation. Ponoy et al. (1994) identified 16 cpDNA haplotypes using RFLPs for *Pseudotsuga menzesii*. They found high levels of haplotype diversity within populations and a low level of differentiation among populations, and concluded that the patterns of variation were concordant with the patterns observed with allozymes. Chloroplasts in conifers such as *Pseudotsuga* are transmitted by pollen, hence have the potential for wider dispersal by wind than mitochondria, which are inherited only from the seed parent. Mitochondrial DNA polymorphisms show strong patterns of differentiation in several species of pine (e.g. Dong and Wagner 1993, Strauss et al. 1993). Aagard et al. (1995) reported that nearly half of the RAPD markers detected in study of racial differentiation of *Pseudotsuga menzesii* could have been amplified from mitochondrial DNA sequences. Some of these markers showed maternal inheritance among hybrids and backcrosses between races and had a much higher level of differentiation than allozymes.

Intraspecific variation in organelle genomes can be used to study biogeography through the comparison of phylogenetic trees based on DNA sequence variation with the geographic distribution of populations (i.e. phylogeography; Avise 1996). Phylogeographic studies provide insight on population history as well as genetic differentiation. While phylogeographic approaches based on mitochondrial DNA polymorphisms have been used extensively to study animal populations, the low level of intraspecific sequence variation in plant organelle genomes, and the complexity of mitochondrial genomes has limited application of similar approaches in plants. One example of biogeographic structure revealed by cpDNA polymorphisms is *Tiarella trifoliata*, where Soltis et al. (1992) found two major clades in the western US (Alaska to northern California) that differed by a single restriction site mutation. One clade was distributed in the northern portion of the species range, except for disjunct populations to the south which occur at high elevations in the mountains of Oregon and northern California. The southern clade was mainly distributed in northern California and Oregon, but had disjunct populations located in Washington. The geographic structure of the cpDNA variation for *T. trifoliata* (and two other plant species) support biogeographic data that suggest distinct refugia existed during the glacial episodes of the Pleistocene and contributed to intraspecific genetic differentiation. In an example from forest trees, Sewell et al. (1996) found a northern and a southern cpDNA haplotype involving five restriction site changes for *Liriodendron tulipifera*. The southern haplotype was restricted to three populations in central Florida, while the northern haplotype was widespread among upland sites. This phylogeographic pattern is similar to those observed for animal species in the same region (Avise 1996) and suggests differentiation caused by isolation during periods with high ocean levels between glacial episodes. Ferris et al. (1995) found a cpDNA haplotype that was locally abundant in southern Britain, whose limited distribution could have arisen during the migration of forest species northward as the last ice age ended. In another example, El Mousadik and Petit (1996) used 11 cpDNA haplotypes to study differentiation in *Argania spinosa*. They detected two lineages that divided the range of the species.

Phylogeographic analysis in forest trees could become more common through the development of "universal" plant organelle primer sequences, which can PCR amplify cpDNA and mtDNA fragments from many different species (Dumolin et al. 1995, Demesure et al. 1996), coupled with the SSCP technique which can reveal single nucleotide differences among PCR products from different sources (e.g. Watano et al. 1995). For pine and other conifers, Vendramin et al. (1996) provided a list of 20 primer pairs that enable PCR amplification cpDNA microsatellite sequences located throughout the 120 kb chloroplast genome of pine, based on the cpDNA sequence of *Pinus thunbergii*. Microsatellite variation from chloroplasts has been reported for pine (Powell et al. 1995, Morgante et al. 1996, Cato and Richardson 1996). These microsatellites could have a high level of variability to better detect and define haplotypes.

Genetic markers have been used extensively in forest trees to obtain insights on the genetics and biology because the long generation times and difficulties with breeding trees make traditional approaches to genetic analysis difficult. Allozymes have been extensively used to assess genetic diversity and estimate levels of heterozygosity representative of species. However, allozyme methods do not detect all of the variation in an enzyme gene. Recently, Barbadilla et al. (1996) proposed an alternative charge-state model for the interpretation of allozyme variation that accounts for intragenic recombination and does not assume equilibrium. Some implications of their model are that all parameters measuring genetic diversity using allozymes will be underestimated and biased, tests based on the infinite allele model are not applicable to allozyme data, and the power of disequilibrium and fitness studies will be reduced by intra-allozyme heterogeneity. Genetic differentiation revealed by molecular methods could be much greater than inferred from allozymes, suggesting that gene flow could have been overestimated in some organisms (e.g. Pogson et al. 1995, Furnier and Stein 1995). Never the less, allozyme variation has provided a useful population genetic perspective and valuable insights for forest geneticists. The newer molecular marker methods may prove more powerful to address population questions.

Lynch (1996) argued that the connection between molecular estimates of genetic diversity and adaptation will be weak, an argument often supported in empirical studies. For example, Karhue et al. (1996) looked for associations between molecular markers and patterns of differentiation for adaptive traits in *Pinus sylvestris*. They concluded that molecular markers do not reflect the patterns of differentiation of quantitative traits. An alternative approach is to use molecular markers to infer phylogenetic divergence of species and populations. These patterns are more likely to reflect patterns of feature diversity than estimates of heterozygosity or correlations with markers (e.g. Faith 1992, Furman et al. 1996). This approach has been used extensively with mtDNA in animals, but phylogeographic patterns inferred from organelle DNA may not be representative of differentiation of the nuclear genome. The application of phylogenetic analysis to intraspecific variation could provide some valuable insights for conservation biology (Rand 1996).

Complex trait dissection and forest trees

There are two very different models for genetic analysis in trees using genetic markers. One is the experimental plant model, based on inbred lines and well-differentiated varieties or long-term selection studies. Similar to other plants, it is possible to obtain large numbers of progeny from specific crosses, but there are no varieties, domestication is just beginning, inbreeding depression is severe, and a prolonged juvenile stage and large size of trees makes experimental approaches difficult. The extreme alternative is the example provided by human genetics where controlled crosses are not used and there is a strong reliance on naturally occurring variation and family structures. Lander and Schork (1994) described 4 ways to carry out complex trait dissection: (i) segregation analysis of markers and phenotypes in experimental crosses among inbred lines, (ii) linkage analysis in outbred pedigrees, (iii) association studies in populations, and (iv) allele sharing methods. Mapping QTLs in experimental crosses of inbred lines involves relatively straightforward statistical models with only 2 alleles per locus, and both marker loci and and putative QTLs are either homozygous or heterozygous. Furthermore, the experimental lines are genetically divergent, either through generations of artificial selection, or they originate from distinct varieties or different species. In contrast, statistical analysis in outbred full-sib families of forest trees is more complicated (Ritter and Salamini 1996, Schäefer-Pregl et al. 1996). Varieties fixed for genetic differences are unavailable for trees because domestication has begun only recently, but some breeding programs do exploit the genetic divergence of species through hybridization (Bradshaw and Grattapaglia 1994). Within forest tree species, QTLs are likely to be segregating within populations and outbred parents are likely to be heterozygous for genes controlling economic traits. The power to detect a QTL segregating in a population is lower for a single full-sib family than for a group of families because the one family analyzed could be homozygous for the QTL (Muranty 1996). Allocation of experimental effort to samples within and among families is difficult to optimize when so little is known about the distribution and frequencies of QTLs in populations. The population approach has been used extensively in human genetics, where the association of markers with disease phenotypes in a collection of samples from small, essentially unrelated families has been analyzed using linkage disequilibrium or a sib-pair approach. The problem of detection and mapping is exacerbated by the low level of heterozygosity for many marker loci. One approach is to use several markers simultaneously to infer haplotypes, although the markers used will vary from family to family especially for dominant markers. However, the four haplotypes from two parents can often be distinguished and some level of statistical analysis can be carried out.

Genetic markers and breeding

Forest trees generally have an extended juvenile growth phase that lasts several years and it is this feature that makes genetic markers especially useful for tree breeding. For many species, the correlation between juvenile expression of trait with mature expression of the trait is low (e.g. height growth). Some traits are difficult to assay, such as disease resistance, and a marker that reliably predicts resistance and traces the inheritance of the gene through generations of breeding would also have value. Some traits are expressed only in mature trees. Lehner et al. (1995) identified RAPD markers linked to the gene, *P*, that controls the *pendula* phenotype of Norway spruce (*Picea abies* f. *pendula*). The pendula gene is responsible for narrow crowns and strong apical dominance in older trees. Gender is sometimes under the control of a single gene in dioecious species. *Pistachia vera* is a small dioecious tree species that produces nuts when the trees are approximately 7 years old. Hormaza et al. (1994) identified a RAPD marker closely associated with gender in *P. vera* that allows breeders to distinguish which seedlings will become nut producers, and which will become pollen producers. Similarly, Sondur et al. (1996) found RAPD markers linked to the gene that determines gender in *Carica papaya*. Even though these traits were known to be controlled by single genes, markers for these traits have clear value for breeding because they predict the characteristics of mature trees several years before they are expressed.

The genetic control of disease resistance in forest trees has become a focus for complex trait dissection experiments (e.g. Villar et al. 1996, Newcombe et al. 1996, Cervera et al. 1996, Benet et al. 1995, Wilcox et al. 1996). DNA pooling methods (e.g. Giovannoni et al. 1991, Michelmore et al. 1991, Churchill et al. 1993, Darvasi and Soller 1994, Shalom et al. 1996) facilitate identification of markers linked to known trait genes (e.g. Devey et al. 1995, Hormaza et al. 1994, Lehner et al. 1995). Pooling methods can also identify markers linked to unknown major genes whose phenotypes do not segregate in Mendelian ratios due to environmental factors (i.e. heritability <1). The mapping of a fusiform rust resistance locus in loblolly pine provides a good example of complex trait dissection using DNA pooling methods (Wilcox et al. 1996). The genetic model that best explained the cosegregation of a marker with the presence or absence of disease phenotypes had two parameters, the recombination fraction between a disease resistance gene and the marker, and the frequency of "escapes". This model assumes that not every susceptible plant that was inoculated with the disease became infected for "environmental" reasons, and the estimate for the frequency of "escapes" was 22%. Further experiments that increased inoculum density reduced the frequency of escapes, providing support for the original model (H.V. Amerson, NCSU, personal communication). It is not surprising that major disease resistance loci were not previously known for loblolly pine or other tree species, given the environmental effects that obscured observation of Mendelian inheritance. Evidence for major resistance loci was generally lacking in forest trees before mapping studies due to the difficulties in experimental manipulation of both highly variable host plants and diverse strains of pathogens. Another difficulty is that some diseases use alternate hosts during their life cycle. The biotic interaction of herbivorous insects and their plant hosts is another complex trait that could be analyzed using genetic markers. Rausher (1996) argues for diffuse coevolution of plants and their natural enemies, but the mapping tools are only now in hand to look for evidence of pairwise interactions that would be the basis for gene-for-gene coevolution in natural populations. Identifying the resistance genes and determining their role in the interaction of trees with pests and pathogens in nature and in disease management programs are the next challenges.

Polygenic inheritance of quantitative traits was generally assumed before genomic mapping in interspecific F_2 families showed that much of the phenotypic variation within family could be explained by segregating markers (e.g. Paterson et al. 1991). Bradshaw and Stettler (1995) mapped trait effects in an interspecific F_2 family of *Populus* and reported that QTLs accounted for a large portion of the phenotypic variation in every trait studied. Grattapaglia et al. (1996) reported QTLs that accounted for a large portion of the phenotypic variation in productivity and wood traits in a half-sib family from a *Eucalyptus grandis* seed parent at maturity. Plomion et al. (1996) reported evidence for QTLs in early height growth of *Pinus pinaster* that suggested different QTLs could be important at different stages in development for forest trees. These results suggest that the low juvenile-to-mature correlation for pine could involve progressive changes in genetic control of traits such as height during maturation. Groover et al. (1994) reported QTLs for wood traits in *Pinus taeda*, and evidence for more than 3 alleles at a QTL locus segregating in an outbred full-sib family. Mapping quantitative traits in these four studies demonstrates the potential to track important amounts of genetic variation for quantitative traits using molecular markers.

Marker Assisted Selection (MAS) could be applied to quantitative traits in forest trees to increase productivity (e.g. O'Malley and McKeand 1994, Kerr et al., 1996). Forest trees show high levels of genetic variation within and among families for economic traits. If markers can be found that predict superior mature performance in the best families, seedlings could be chosen for propagation based on their marker composition. The ability to propagate many tree species declines with maturation, so that when the value of an individual genotype is expressed, it could be too late to propagate the individual. This approach could result in substantial gains even if the vegetative propagation systems do not have a very large multiplication factor. The integration of MAS into breeding programs poses some challenges and there are questions about the efficiency of MAS relative to conventional selection based on phenotypes. Kerr et al. (1996) assessed methods for MAS using computer simulations and evaluated its economic benefits. They concluded that MAS could be exploited for breeding quantitative traits in forest trees. One important issue for MAS is the identification of QTLs in the diverse germplasm and family structures available in mature test plots of forest trees. Better quantitative approaches are needed to identify the families that could be segregating for major gene effects, hence that are good candidates for mapping.

References

Aagaard JE, Vollmer SS, Sorensen FC and Strauss SH (1995) *Mol Ecol* 4:441-447.
Adams WT, Strauss SH, Copes DL, Griffin AR (eds) (1992) *Special Issue: Population Genetics of Forest Trees*, New Forests 6:1-427.
Ahn SJ, Anderson A, Sorrells ME and Tanksley SD (1993) *Mol Gen Genet* 241:483-490.
Akkaya MS, Shoemaker RC, Specht JE, Bhagwat AA, Cregan P (1995) *Crop Sci* 35:1439-1445.
Antolin MF, Bosio CF, Cotton J, Sweeney W, Strand MR and Black WC IV (1996) *Genetics* 143:1727-1738.
Avise JC (1996) In: Avise JC and Hamrick JL (eds), *Conservation Genetics, Case Histories from Nature*. Chapman & Hall, New York. pp 431-470.
Baker SM and White EE (1996) *Theor Appl Genet* 92:827-831.
Barbadilla A, King LM and Lewontin RC (1996) *Mol Biol Evol* 13(2):427-432.
Benet H, Guries RP, Boury S and Smalley EB (1995) *Theor Appl Genet* 90:7-8.

Bennetzen JL, Hulbert SH, San-Miguel PJ, Carter-Peek C, Chiu J and Handa AK (1993) *Trends Genet* 9:259-261.

Binelli G and Bucci G (1994) *Theor Appl Genet* 88:3-4.

Blouin MS, Parsons M, Lacaille V and Lotz S (1996) *Mol Ecol* 5:393-401.

Bodenes C, Pradere S and Kremer A (1996) In: Ahuja MR, Boerjan W and Neale DB (eds) *Somatic Cell Geneticsand Molecular Genetics of Trees*. Kluwer Academic Publishers, Netherlands, pp 239-248.

Bousquet J, Strauss SH, Doerksen AH and Price RA (1992) *Proc Natl Acad Sci USA* 89:7844-7848.

Bradshaw HD Jr and Stettler RF (1995) *Genetics* 139:963-973.

Bradshaw HD Jr and Grattapaglia D (1994) *Genetics* 1:191-196.

Bradshaw HD Jr, Villar M, Watson BD, Otto KG, Stewart S and Stettler RF (1994) *Theor Appl Genet* 89:2-3.

Brown GR, Amarasinghe V, Kiss G and Carlson JE (1993) *Genome* 36, 310-316.

Bucci G and Menozzi P (1995) *Heredity* 75:188-197.

Buijtenen JP van (1988) In: Weir BS, Eisen EJ, Goodman MM and Namkoong G (eds), *Proc Second International Conference on Quantitative Genetics* (May 31-June 5 1987), Raleigh, NC; Sinauer Associates Inc, Sunderland, Mass. pp 549-554.

Burley J (1996) In: Dieters, MJ, Matheson AC, Nikles DG, Harwood CE and Walker SM (eds), *Tree Improvement for Sustainable Tropical Forestry, Proc QFRI-IUFRO Conf*, Caloundra, Queensland, Australia, pp 1-5.

Byrne M, Murrell JC, Allen B and Moran GF (1995) *Theor Appl Genet* 91:6-7.

Byrne M, Marquezgarcia MI, Uren T, Smith DS and Moran GF (1996) *Australian J Bot* 44:331-341.

Caetano-Anollés G, Bassam BJ and Gresshoff PM (1991) *Bio/Technology* 9:553-557.

Cato SA and Richardson TE (1996) *Theor Appl Genet* 93:587-592.

Cervera MT, Gusmao J, Steenackers M, Peleman J, Storme V, Vandenbroeck A, van Montagu M and Boerjan W (1996) *Theor Appl Genet* 93:733-737.

Chadwick RB, Conrad MP, McGinnis MD, Johnston Dow L, Spurgeon SL and Kronick MN (1996) *Biotechniques* 20:676-83.

Chase M, Kesseli R and Bawa K (1996) *Am J Bot* 83:51-57.

Chin ECL, Senior ML, Shu H, and Smith JSC (1996) *Genome* 39:866-873.

Churchill GA, Giovannoni JJ, Tanksley SD (1993) *Proc Natl Acad Sci USA* 90:16-20.

Condit R and Hubbell SP (1991) *Genome* 34:66-71.

Conkle MT (1981) In: Conkle, MT, (ed) *Proc Symp Isozymes of North American Forest Trees and Forest Insects* (July 27 1979), Berkeley, California; USDA Forest Service Pacific Southwest Forest and Range Experiment Station, pp 11-17.

Darvasi A and Soller M (1994) *Genetics* 138:1365-1373.

Davis TM, Yu H, Haigis KM and McGowan PJ (1995) *Theor Appl Genet* 91:582-588.

Dawson, IK, A J Simons, R Waugh and W Powell (1996) *Mol Ecol* 5:89-98.

Demesure B, Sodzi N and Petit RJ (1995) *Mol Ecol* 4:129-131.

Devey ME, Fiddler TA, Liu B-H, Knapp SJ and Neale DB (1994) *Theor Appl Genet* 88:273-278.

Devey ME, Delfino-Mix A, Kinloch BB Jr and Neale DB (1995) *Proc Natl Acad Sci USA* 92:2066-2070.

Devey ME, Bell JC, Smith DN, Neale DB and Moran GF (1996a) *Theor Appl Genet* 92:673-679

Devey ME, Sewell MM and Neale DB (1996b) In: Dieters MJ, Matheson AC, Nikles DG, Harwood CE and Walker SM (eds), *Tree Improvement for Sustainable Tropical Forestry, Proc QFRI-IUFRO Conf*, Caloundra, Queensland, Australia, pp 478-480.

Devos KM, Moore G and Gale MD (1995) *Euphytica* 85:367-372.

Dong J and Wagner DB (1993) *Theor Appl Genet* 86:573-578.

Dow BD, Ashley MV and Howe HF (1995) *Theor Appl Genet* 91:137-141.

Dumolin S, Demesure B and Petit RJ (1995) *Theor Appl Genet* 91:1253-1256.

Elgar G (1996) *Hum Mol Genet* 25:1437-1442.

El Mousadik A and Petit RJ (1996) *Mol Ecol* 5:547-555.

Faith DP (1992) *Cladistics* 8:361-373.
Ferris C, Oliver RP, Davy AJ and Hewitt GM (1995) *Mol Ecol* 4:731-738.
Furman BJ, Dvorak WS, Sederoff RR, O'Malley DM (1996) In: Dieters MJ, Matheson AC, Nikles DG, Harwood CE and Walker SM (eds), *Tree Improvement for Sustainable Tropical Forestry, Proc QFRI-IUFRO Conf*, Caloundra, Queensland, Australia, pp 485-491.
Furman BJ, GrattapagliaD, Dvorak WS and O'Malley DM (1997) *Mol Ecol* 6:in press.
Furnier GR and Stine M (1995) *Can J For Res* 25:736-742.
Giovannoni JJ, Wing RA, Ganal MW, Tanksley SD (1991) *Nucleic Acids Res* 19:6553-6558.
Grattapaglia D and Sederoff RR (1994) *Genetics* 137:1121-1137
Grattapaglia,D, F Bertolucci, R Penchel and R R Sederoff (1996) *Genetics*144:1205-1214.
Groover A, Devey M, Fiddler T, Lee J, Megraw R, Mitchel-Olds T, Sherman B, Vujcic S, Williams C and Neale DB (1994) *Genetics*138:1293-1300.
Gu WK, Weeden NF, Yu J, Wallace DH (1995) *Theor Appl Genet* 91:465-470.
Heun M and Helentjaris T (1993) *Theor Appl Genet* 85:961-968.
Hillis DM, Moritz C and Mable BK (ed) (1996) *Molecular Systematics (second edition)*. Sinauer Assoc Pub, Sunderland, Massachusetts.
Hipkins VD, Krutovskii KV and Strauss SH (1994) *Forest Genetics*1:179-189.
Hormaza JI, Dollo L and Polito VS (1994) *Theor Appl Genet* 89:9-13.
Kamm A, Doudrick RL, Heslop-Harrison JS and Schmidt T (1996) *Proc Natl Acad Sci USA* 93:2708-2713.
Innan H, Tajima F, Terauchi R and Miyashita NT (1996) *Genetics* 143:1761-1770.
Isabel N, Beaulieu J and Bousquet J (1995) *Proc Natl Acad Sci USA* 92, 6369-6373.
Karhu A, Hurme P, Karjalainen M, Karvonen P, Karkkainen K, Neale D and Savolainen O (1996) *Theor Appl Genet* 93:215-221.
Karvonen P and Savolainen O (1993) *Heredity* 71:614-622.
Karvonen P, Szmidt AE and Savolainen O (1994) *Theor Appl Genet* 89:969-974.
Kaya Z and Neale DB (1995) *Silvae Genetica* 44, 2-3, 110-116.
Kerr RJ, Jarvis SF, Goddard ME (1996) In: Dieters MJ, Matheson AC, Nikles DG, Harwood CE and Walker SM (eds), *Tree Improvement for Sustainable Tropical Forestry, Proc QFRI-IUFRO Conf*, Caloundra, Queensland, Australia, pp 498-505.
Khasa PD, and Dancik BP (1996) *Theor Appl Genet* 92:46-52.
Kinlaw CS, Gerttula SM and Carter MC (1994) *Plant Mol Biol* 26:1213-1216.
Kinlaw CS, Ho T, Gerttula SM, Gladstone E and Harry DE (1996) In: Ahuja MR, Boerjan W and Neale DB (eds), *Somatic Cell Genetics and Molecular Genetics of Trees*, Kluwer Academic Publ, Netherlands, pp 175-182.
Kinloch BB, Parks GK and Fowler CW (1970) *Science* 167:193-195.
Konieczny A and Ausubel FM (1993) *Plant J* 4:403-410.
Konnert M (1995) *Silvae Genetica* 44:346-351.
Kossack DS (1989) PhD dissertation, Univ Calif, Davis. 69 pages.
Kowalski SP, Lan TH, Feldmann KA and Paterson AH (1994) *Genetics* 138:499-510.
Kresovich S, Szewe-McFadden AK, Bliek SM and McFerson JR (1995) *Theor Appl Genet* 91:206-211.
Krupkin AB, Liston A and Strauss SH (1996) *Am J Bot* 83:489-498.
Kubisiak, TL, C D Nelson, W L Nance and M Stine (1995) *Theor Appl Genet* 90:7-8.
Kvarnheden A, Tandre K and Engstrom P (1995) *Plant Mol Biol* 27:391-403.
Lander ES and Schork NJ (1994) *Science* 265:2037-2048.
Lehner,A, M A Campbell, N C Wheeler, T Poykko, J Glossl, J Kreike and D B Neale (1995) *Theor Appl Genet* 91:6-7.
Levin DA and Wilson AC (1976) *Proc Natl Acad Sci USA* 73:2086-2090.
Liou P-C, Gmitter FG Jr and Moore GA (1996) *Theor Appl Genet* 92:425-435.
Lynch M (1996) In: Avise JC and Hamrick JL (eds), *Conservation Genetics, Case Histories from Nature*. Chapman & Hall, New York, pp 471-501.
Lyons LA, Raymond MM, O'Brien SJ (1994) *Animal Biotechnol* 5:103-111.

MacKay JJ, LiuW, WhettenR, Sederoff RR and O'Malley DM (1995) *Mol Gen Genet* 247:537-545.

Mashal RD, Koontz J and Sklar J (1995) *NatureGenetics9*, 177-183.

Michelmore RW, Paran WI and Kesseli RV (1991) *Proc Natl Acad Sci USA* 88:9828-32.

Moore G, Devos KM, Wang Z and Gale MD (1995) *Curr Biol* 5:737-739.

Morgante M, Pfeiffer A, Costacurta A, Olivieri AM, Powell W, Vendramin GG and Rafalski JA (1996) In: Ahuja MR, Boerjan W and Neale DB (eds), *Somatic Cell Genetics and Molecular Genetics of Forest Trees*. Kluwer Academic Publ, The Netherlands, pp 233-238.

Moriyama EN and Powell JR (1996) *Evol* 13(1)261-277.

Morizot DC (1994) *Animal Biotechnol* 5(2):113-122.

Mukai,Y, Y Suyama, Y Tsumura, T Kawahara, H Yoshimaru, T Kondo, N Tomaru, N Kuramoto and M Murai (1995) *Theor Appl Genet* 90:835-840.

Muranty H (1996) *Heredity* 76:156-165.

Namkoong G and Kang H (1990) *Plant Breeding Rev* 8:139-188.

Neale DB, Marshall KA and Harry DE (1991) *Can J For Res* 21:717-720.

Neale DB, Kinlaw CS and Sewell MM (1994) *Forest Genetics* 1:197-206.

Neale DB and Sederoff RR (1996) In: Paterson AH (ed), *Genome Mapping in Plants*. RG Landes Co, pp 311-321.

Nei M and Li W-H (1979) *Proc Natl Acad Sci USA* 76:5269-5273.

Nelson CD, Nance WL and Doudrick RL (1993) *Theor Appl Genet* 87:1-2.

Newcombe G, Bradshaw HD, Chastagner GA and Stettler RF (1996) *Phytopathology* 86:87-94.

Novy RG and Vorsa N (1996) *Theor Appl Genet* :840-849.

O'Brien IEW, Smith DR, Gardner RC and Murray BG (1996) *Plant Science* 115:91-99.

O'Brien SJ (1991) *Current Opinion in Geneticsand Development* 1:105-111.

O'Malley DM, GrattapagliaD, Chaparro JX, Wilcox PL, Amerson HV, Liu B-H, Whetten R, McKeand S, Kuhlman EG, McCordS, CraneB and Sederoff R (1996) In: Gustafson JP and Flavell RB (eds), *Genomes of Plants and Animals:21st StadlerGeneticsSymposium*. Plenum Press, New York, pp 87-102.

O'Malley DM and McKeand SE (1994) *Forest Genetics* 1:207-218.

Orita M, Suziki Y, Sekiya T and Hayashi K (1989) *Genomics* 5:879.

Paran I and Michelmore RW (1993) *Theor Appl Genet* 85:985-993.

Paterson AH, Damon S, Hewitt J, Zamir D, Rabinowwitch H, Lincoln S, Lander E and Tanksley S (1991) *Genetics* 132:823-839.

Paterson AH, Lin Y-R, LiZ, Schertz KF, Doebley JF, Pinson SRM, Liu S-C, Stansel JW and Irvine JE (1995) *Science* 269:1714-1718.

Perry DJ and Furnier GR (1996) *Proc Natl Acad Sci USA* 93:13020-13023.

Phelps RS, Chadwick RB, Conrad MP, Kronick MN and Kamb A (1995) *Biotechniques* 19:984-989.

Plomion C, O'Malley DM and Durel CE (1995a) *Theor Appl Genet* 90:7-8.

Plomion C, Bahrman N, Durel CE and O'Malley DM (1995b) *Heredity* 74:661-668.

Plomion C, Durel CE and O'Malley DM (1996) *Theor Appl Genet* 93:849-858.

Pogson GH, Mesa KA and Boutilier RG (1995) *Genetics* 139:375-385.

Ponoy B, Hong YP, Woods J, Jaquish B and Carlson JE (1994) *Can J For Res* 24:1824-1834.

Powell,W, G C Machray and J Provan (1996) *Trends Plant Sci* 1:215-222.

Powell W, Morgante M, McDevitt R, Vendramin GG and Rafalski JA (1995) *Proc Natl Acad Sci USA* 92:7759-7763.

Qiu YL, Chase MW and Parks CR (1995a) *Am J Bot* 82:1582-1588.

Qiu YL, Parks CR and Chase MW (1995b) *Am J Bot* 82:1589-1598.

Rajora OP and Dancik BP (1995a) *Theor Appl Genet* 90:3-4.

Rajora OP and Dancik BP (1995c) *Theor Appl Genet* 90:3-4.

Rajora OP and Dancik BP (1995b) *Theor Appl Genet* 90:3-4.

Rake AV, Miksche JP, Hall RB and Hansen KM (1980) *Can J Genet Cytol* 22:69-80.

Rand DM (1996) *Conservation Biol* 10:665-671.

Rausher MD (1996) *Trends Genet* 12:212-217.

Rieseberg, LH (1996) *Mol Ecol* 5:99-105.

Ritter E and Salamini F (1996) *Genet Res* 67:55-65.

Ruiz A and Barbadilla A (1995) *Genetics*139: 445-455.

San Miguel P, Tikhonov A, Jin Y-K, Motchoulskaia N, Zahaov D, Melake-Berhan A, Springer PS, Edwards KJ, Lee M, Avramova Z and Bennetzen JL (1996) *Science* 274:765-738.

Schäfer-Pregl R, Salamini and Gebhardt C (1996) *Genet Res* 67:43-54.

Sewell MM, Neale DB (1995) In: *Proc 23rd Southern Forest Tree Improvement Conference* (June 20-22, 1995), Asheville, NC. The National Technical Information Services, US Dept Commerce, Springfield, VA. pp 103-110.

Sewell MM, Parks CR and Chase MW (1996) *Evolution* 50:1147-1154.

Shalom A, Darvasi A, Barendse W, Cheng H and Soller M (1996) *Animal Genet* 27:9-17.

Sherratt EJ, Thomas AW, Gagg JW and Alcolado JC (1996) *Biotechniques* 20:430-432.

Skroch P and Nienhuis J (1995) *Theor Appl Genet* 91:1086-1091.

Smith DN and Devey ME (1994) *Genome* 37:977-983.

Smith JF, and Doyle JJ (1995) *Am J Bot* 82:1163-1172

Soltis DE, Soltis PS, Kuzoff RK and Tucker TL (1992) *Plant Syst Evol* 181:203-216.

Sondur SN, Manshardt RM and Stiles JI (1996) *Theor Appl Genet* 93:547-553.

Strauss SH, Hong Y-P and Hipkins VD (1993) *Theor Appl Genet* 86:605-611.

Steinkellner H, Fluch S, Turetschek, Strieff R, Kremer A, Burg K and Glössl J (1996) In: Ahuja MR, Boerjan W and Neale DB (eds), *Somatic Cell Genetics and Molecular Genetics of Trees*. Kluwer Academic Publ, Netherlands, pp 175-182.

Sutton BCS, Flanagan DJ, Gawley JR, Newton CH, Lester DT and El-Kassaby YA (1991) *Theor Appl Genet* 82:242-248.

Trower MK, Orton SM, Purvis IJ, Sanseau P, Riley J, Christodoulou C, Burt D, See CG, Elgar G, Sherrington R, Rogaev EI, St George Hyslop P, Brenner S and Dykes CW (1996) *Proc Natl Acad Sci USA* 93:1366-1369.

Tsumura Y, Yoshimura K, Tomaru N and Ohba K (1995) *Theor Appl Genet* 91:1222-1236.

Tsumura Y, Ohba K and Strauss SH (1996) *Theor Appl Genet* 92:40-45.

Tulsieram LK, Glaubitz JC, Kiss G and Carlson JE (1992) *Bio/Technology* 10:686-690.

van de Ven W and McNicol RJ (1996) *Theor Appl Genet* 93:613-617.

Vendramin GG, Lelli L, Rossi P and Morgante M (1996) *Mol Ecol* 5:595-598.

Wilcox PL, Amerson HV, Kuhlman EG, Liu BH, O'Malley DM and Sederoff RR (1996) *Proc Natl Acad Sci USA* 93:3859-3864

Villar M, Lefevre F, Bradshaw HD and Ducros ET (1996) *Genetics*143:531-536.

Voo KS, Whetten RW, O'Malley DM and Sederoff RR (1995) *Plant Physiol* 108:85-97.

Vos P, Hogers R, Bleeker M, Reijans M, van de Lee T, Hornes M, Frijters A, Pot J, Peleman J, Kuiper M and Zabeau M (1995) *Nucleic Acids Res* 23:4407-4414.

Watano,Y, M Imazu and T Shimizu (1995) *J Plant Res* 108:493-499.

Watt WB, (1994) *Genetics*136:11-16.

Welsh J and McClelland M (1990) *Nucleic Acids Res* 18:7213-7218.

Williams JGK, Kubelik AR, Livak KJ, Rafalski JA and Tingey SV (1990) *Nucleic Acid Res* 18:6531-6535.

Wu K and Tanksley SD (1993) *Mol Gen Genet* 231:353-359.

Yazdani R, Yeh FC and Rishima J (1995) *Forest Genetics* 2:109-116

Youil R, Kemper B, Cotton RGH (1996) *Genomics* 32:431-435.

Comparative analysis of cereal genomes

Graham Moore[1] and J. Perry Gustafson[2]

Cereals Research Department, John Innes Centre, Colney Lane, Norwich NR4 7UH, United Kingdom[1], and Plant Genetics Research Unit and Plant Science Unit, USDA-ARS, University of Missouri, Columbia, MO, USA[2]

Introduction

There is a large amount of information available about genome evolution in cereals, offering a variety of perspectives. However, no single perspective appears optimal in giving a clear picture of the evolution of a cereal genome. Various studies examining the same question produce different answers because they utilize different techniques, which analyze different aspects of the genome. While some analyze the euchromatic, or coding part of the genome, others look at the ribosomal, chloroplast and mitochondrial parts, and some study the heterochromatic or repeated sequence regions of the genome. Observed differences could most likely arise from different selection pressures and evolutionary rates acting upon the different sectors of the genome. Discrepancies between different analyses do not warrant the acceptance of one genome analysis over another, but rather, offer new insights into the evolutionary processes involved in the particular part of the genome being studied. Because such differences have been observed and it is clear that different parts of a genome evolve at different rates, theories of genome evolution will require an integration of different approaches in order to fully understand the variation in evolutionary rates that have occurred.

More than 60 million years ago, speciation gave rise to the various cereal species. The two which will be discussed in this presentation are the major crops in the world [wheat (*Triticum aestivum* L. em Thell.) and rice (*Oryza sativa* L.)] (Martin et al. 1989, Wolfe et al. 1989). Wheat currently supplies approximately 55% of the carbohydrates consumed world-wide, and rice supplies food for more people than any other crop. In 1993, more than 560 million metric tons of each of wheat and rice were produced in the world. In the last 50 years, these and other cereals have been studied individually, with little exchange of information between breeders or scientists working on each species. However, it is

259

anticipated that in the next 50 years the world population will increase to the point where an additional amount of wheat and rice as has been produced and consumed since the beginning of agriculture, on top of that currently being produced, will be required to feed the population. Clearly it will be vital to use the information generated from the study of all of the cereals more efficiently. Comparative genome analysis involving individual crops like maize *(Zea mays* L.), sorghum *(Sorghum vulgare* L.), rice, barley *(Hordeum vulgare* L.), rye *(Secale cereale* L.), wheat, millet [*Pennisetum typhoides* (Burm. f.) Stapf. & C.E. Hubb], and sugarcane *(Saccharum officinarum* L.), where both the coded and repeated sequences of the genome are analyzed will provide just such a framework.

Since their speciation, cereals such as wheat and rice have evolved with vastly different chromosome numbers and genome sizes. The rice genome is 4.3×10^8 bp, while together, the three genomes of hexaploid bread wheat are about 40 times bigger and encompass a total 1.7×10^{10} bp. It is apparent from genome studies that there are several levels of genome organization. The simplest level is the kilobase and sub-kilobase level, which includes genes and repetitive sequences. The sub-megabase level comprises the repetitive sequences organized into units that have undergone amplification. The multi-megabase level comprises the nonrandom organization of repetitive sequences in satellites or compartments along the chromosome. The sub-chromosomal level includes the linkage blocks containing genes whose linkage is conserved throughout a genome. At the highest level of complexity in genome organization are the chromosomes whose number, gene and recombination distribution vary among the genomes. Comparative genome analysis should provide a large pool of DNA markers for breeding, new strategies for gene isolation, and provide information on the mechanisim by which genomes and traits evolve. These topics will be discussed within the above framework.

Kilobase and sub-kilobase level

Genes

Gene sequence analyses from dicots and monocots reveal that the average GC content of monocot genes is higher (~46%) than that of dicot genes, with the majority of the monocots having a GC content of 60-70%. This difference is because of the use of different synonymous codons. Cereals and mammals share the most commonly used codon for 13 of the 18 amino acids for which synonyms exist, while dicots share only 7 of 18 amino acids. It is therefore not surprising that it is difficult to detect monocot genes by Southern blot analysis using dicot genes as probes. A large number of rice cDNA have been sequenced by the Japanese Rice Genome Program making it possible to compare rice gene protein homologs with those derived from *Arabidopsis thaliana* L., thus allowing for comparative mapping of cereals and dicots. The sequence comparisons reveal that rice has many multigene families and therefore exhibits significant gene redundancy (Sasaki 1995). However, from comparative mapping studies involving various cereals, it was observed that the majority of rice cDNA tested revealed a similar number of homoeologous sequences on Southern blots of wheat, barley, millet, and rye DNA, suggesting strong sequence conservation and a similar gene content. Therefore, whatever the processes which lead to the amplification of the additional DNA in the genomes of wheat and barley as compared to rice, they do not involve amplification of genes/single copy sequences in wheat/barley. However, this does not rule out the occurrence of duplications in the genome of the ancestral grass prior to the speciation of rice and wheat, which may have created the multigene families observed in the rice cDNA sequencing program.

In the case of animal genomes, anchor points have been defined that allow genes to be identified with relative ease (Bird, 1987). These animal genomes contain a CpG content lower than the 4% expected considering they are GC rich (40%). However, CpG islands localized at the 5' end (promoter) of genes have been shown to have the expected CpG content. The CpG dinucleotides in these islands are nonmethylated and provide recognition sites for restriction enzymes such as *Not*I. Of the *Not*I sites, 75% are located within CpG islands and more than half of these are coincident with genes. In contrast to the animal genomes, wheat genomes exhibit no CpG shortage (Swartz et al. 1962, Russell et al. 1971). In plants, 5-methylcytosine is not confined to CpG di-nucleotides, but is also present in over 80% of the trinucleotides CpXpG (Gruenbaum et al. 1981). Despite the lack of CpG suppression, recognition sites for restriction enzymes such as *Not*I provide landmarks for genes. Some wheat genes appear to be located in undermethylated islands, within the regions of repetitive sequences. Since these islands are also rich in highly methylated CpG dinucleotides, they provide sites for restriction enzymes such as *Not*I. For this reason, *Pst*I or *Not*I restriction fragments of less than 0.5 kb that were derived from the wheat genome have been shown to be single-copy sequences. Cloned copies of these fragments are capable of detecting homoeologous sequences in rye, barley, maize and rice, suggesting that these fragments may well contain genes (Moore et al. 1993b).

Repeats

A) Dispersed

A number of repetitive sequences have been defined in the cereal genomes, including several classes of retroelements, such as WIS-2 in wheat, which are also present in rye, barley, and oats *(Avena sativa L.)* (Moore et al. l991a). BARE-1 recently isolated from the barley genome, is a member of the WIS-2 family, having a substantial level of homology at the sequence level (Manninen and Schulman 1993). The hybridization of BARE-1 and WIS-2 to barley DNA yielded two markedly different figures for copy numbers of WIS-2 type elements in this genome. This probably reflects the divergence of these two elements with respect to the most common element in the genome. Elements homologous to WIS-2 have also been isolated from the rice genome (Yano et al. 1996, Ohtsubo et al. 1996). BIS-1 is another retroelement that was isolated from barley (Moore et al. l991b). Analyses indicate that this element could represent more than 6% of the barley genome. When used as probes on Southern blots, BIS-1 and sets of primers detect similar size fragments in a range of Gramineae, from maize, millet, and sorghum to *Lolium.* The evidence suggests that this family of elements may be an important component of these genomes. Sequence comparisons of a "rye-specific" retroelement, 174 (Rogonsky et al. 1992), with BIS-1, not only indicates that 174 has regions of strong sequence homology to BIS-1, but that it has accumulated additional sequences with respect to BIS-1 (Alcaide and Moore, unpublished). It is possible that many cereal repeats apparently specific to particular species when assayed by Southern blot analysis, reveal homology when compared at the sequence level. Such elements likely represent "distant" members of the family being significantly more diverged and possessing additional sequences. It is important to define these additional sequences, which may be genic in origin.

Other types of repetitive sequences have been found to be conserved across the Gramineae. One example would be the "Tourist" elements, stemloop structures, discovered in the maize genome (Bureau and Wessler 1992). Also repetitive sequences in the barley clone Hi-10 have been found in the genomes of rice, sorghum, maize, barley, wheat and *Lolium* (Abbo et al. 1995).

In contrast to the barley and wheat genomes, the small rice genome is undermethylated (Deshpande and Ranjekar 1980). Analysis of rice sequences in short restriction fragments generated by cleavage with *Pst*I or *Not*I is not as clearcut as sequences derived from the wheat and barley genomes. A higher percentage of fragments contain repetitive sequences and far fewer of those will cross-hybridize to DNA derived from other cereals and reveal single-copy hybridization patterns. Using as a model the evolution of methylated vertebrate genomes from undermethylated, invertebrate genomes (Bird 1987), rice would represent the "invertebrate" genome and wheat and barley would be intermediate genomes between the two types. The CpG shortage in the animal genomes is the result of methylated CpG undergoing a transition to TpG. The general lack of CpG suppression in the bulk of the highly methylated, large cereal genomes suggests two possibilities. Either the majority of the additional sequence has arisen relatively recently or there is not the same level or rate of CpG to TpG transitions in cereals. If the majority of the genome has arisen relatively recently, comparative analysis might reveal how this occurred, the nature of the sequences involved, and therefore, provide some explanation for why it occurred.

B) Satellites

Tandem arrays of essentially the same sequence are clustered at many locations throughout a genome. Some of the sequences, such as those present in telomere arrays or in centromeres exceed a kilobase. Unequal crossing-over, replicon slippage, and other mechanisms lead to the variation in copy number observed within arrays. Cleavage with restriction enzymes that do not cut within the arrays, generates polymorphisms. These arrays or variable number of tandem repeats (VNTRs) are beginning to be used as dominant markers to aid molecular mapping. For example, microsatellites are arrays of random di-, tri- or tetra-nucleotides, which most likely result from aberrations in the replication processes and unequal crossing over. Microsatellites are scattered throughout the genome and have been studied in many species (Wu and Tanksley 1993, Pan et al. 1995, Roder et al. 1995, Ma et al. 1996).

There is now considerable information about the portion of the genome consisting of repetitive DNA, which is found in multiple copies ranging in number from as few as two up to several thousand copies, and is found to exist in all eukaryotes. For example, in wheat, it is known that the genomes contain up to 80% repeated sequences (Flavell and Smith 1976). These regions may be much more appropriate for many types of genome analyses than the traditional sources of markers (cDNA and gDNA). It is clear that evolutionary forces acting on repeated sequence DNA can be quite different from those affecting single-copy sequences. In these rapidly evolving regions, mutation and drift would be considerably more important than selection (Begun and Aquadro 1993). Methods utilizing random amplified polymorphic DNA (RAPD), microsatellites, and minisatellites have been used to study repeated sequences in both animal and plant genomes.

Repetitive DNA regions that contain microsatellites and minisatellites are a useful source of markers because they represent a large reservoir of apparently noncoded genetic variation. This amount of variation could not occur in single-copy regions. As has been known for some time most simple, single nucleotide changes that occur in coded regions during evolution often have lethal effects. Even nonlethal changes would be subjected to severe negative selection pressure, greatly reducing the chances of that variation being maintained within a population.

Most of this variation resulting from mutations occurring in repetitive DNA sequences is neutral, and can be retained in a population. In both animals and plants, it has been shown that mutation rates in these regions can be quite high (Jeffreys et al. 1988, Rogstad 1994). Mutation rates at some minisatellite loci can be as high as 5% per gamete (Haymer 1994). Jeffreys et al. (1988) observed that mutation rates between 0.5 and 1.5% are common. It is not known why the mutation rates are so high in these regions, but the fortuitous result is generation of a very large number of polymorphisms. It is for this reason they are often referred to as hypervariable loci (Jeffreys et al. 1985). Using the correct primers flanking these regions coupled with PCR, large numbers of usable polymorphic markers can be generated (Heath 1993, Somers et al. 1996).

In plants, VNTR loci have often shown a high level of mutation when compared to other types of genetic markers (e.g. allozymes, most single-copy RFLP probes). In turnip *(Brassica rapa* L.), three out of 595 VNTR bands (0.5%) transmitted to offspring were non-X (Rogstad 1994a), while mutation frequencies of 0.001 for the human CAC VNTR marker transmission (Nurnberg et al. 1991) and less than 0.0003 for the GATA VNTR marker transmission in a tomato cross *(Lycopersicon esculentum* Mill. × *L. pennellii* Mill.; Arens et al. 1995) have been observed. Although VNTR loci mutation rates can be relatively high when compared to other regions of the genome, they are generally not high enough to make them unusable for many types of analyses. Importantly, VNTR markers have been shown to follow Mendelian inheritance (e.g. Dallas 1988, Rogstad 1994a, Arens et al. 1995, Zhou and Gustafson 1995), making them extremely useful in mapping, population variation studies, and genome analysis.

Variation discrepancies are easily seen when utilizing different molecular approaches to observe genetic variation, such as when comparing isozymes to VNTR markers. Isozyme diversity in several plant species showed a mean heterozygosity (H) for 468 plant taxa of

0.113, and a mean H for long-lived woody perennials of 0.149 (Hamrick and Murawski 1991). One of the highest heterozygosities ever observed using isozymes for any plant species was 0.481 in *Opuntia basilaria* Engelm. & Big. (Hamrick et al. 1979). Jelinski and Cheliak (1992) used isozymes and detected a mean total genetic diversity (H_T) of 0.31 in poplar *(Populus tremuloides)* populations. However, an analysis of 21 population VNTR markers detected an H of 0.760 (Rogstad et al. 1991). In general, heterozygosity estimated from VNTR analyses showed that variation measured across VNTR loci in plants exceeds that generally found with other marker systems. The studies using plants are generally consistent with what has been found for animal populations (Scribner et al. 1994). In studies of cereal species utilizing from one to three different VNTR probes (Zhou and Gustafson 1995, Somers et al. 1996), at least 15 out of 30 bands per individual have been observed to be segregating. Given the high degree of apparent Mendelian variation in banding patterns among individuals within a cereal genome, it should not be difficult to detect 8 or more bands with one VNTR probe. Other marker systems are available which include AFLP in which 50-100 restriction fragments can be displayed on a denaturing polyacrylamide gel (Vos et al. 1995).

Within repetitive-sequence DNA, genetic variation from sequence changes accumulates in the absence of either internal or external selection pressures. Consequently more variation would be found in non-coded DNA than in coded DNA. Differences between humans and rodents in protein coding regions are approximately 30%, while in non-coded regions (including repeated sequences) the differences can vary from 68 to 83% (Palumbi 1989). These non-coded regions (repeated sequences) are an excellent source of polymorphisms. Highly variable loci can be detected in both animals and plants using minisatellites (Jeffreys et al. 1985, Zhou and Gustafson 1995). Stephens et al. (1992) suggested that minisatellites can be used to study cases of endangered species that are thought to be highly limited in their diversity. It should be noted that microsatellites could be used in the same manner as minisatellites. The polymorphism levels would be very useful for developing linkage maps of complex genomes where low levels of polymorphisms using RFLPs have been detected. In human genome analysis, such markers are very important in mapping portions of the genome lacking the more traditional polymorphisms (Weissenbach et al. 1992).

It is clear that the forces of evolution operating within a genome vary tremendously, depending upon whether or not they are acting on coded DNA or DNA present in the repetitive fraction. It appears that in species or genomes where the amount of variation in the coded regions is low, analyses of the variation in the repetitive fraction can produce significantly more resolution for describing genetic relationships.

We now have the potential for marker systems in both the coded region and the non-coded (repetitive) DNA regions of the genome. This will allow for more complete linkage maps and for a better analysis of the different evolutionary rates that are occurring within a genome.

VNTR markers are beginning to be utilized to study many phenomena within the repeated sequence part of the genome, and their applications in plants are increasing. One continuing limitation to their use in studies on plants has been their low availability.

It has been established that VNTRs containing core sequences of from 2 to 15 nucleotides (Ali et al. 1986, Litt and Luty 1989, Vergnaud 1989, Weber and May 1989, Rogstad 1994a, Winberg et al. 1993, Somers et al. 1996) and can be used to reveal considerable variation at microsatellite or minisatellite VNTR loci. It has been shown that VNTR probes can be generated using PCR to create synthesized tandem repeats (STRs). The use of PCR coupled with VNTR core sequences to screen for and isolate VNTR probes (Rogstad 1993, 1994a, Heath et al. 1993, Somers et al. 1996) offers a strong approach for isolating and producing a vaste array of VNTR probes. Screening cereal species with a series of PCR-STR probes to investigate genetic variation is relatively rapid since cloning, clone screening, sequencing, and other steps in alternate methods like the use of RFLPs are not involved. For most self-pollinated cereal genomes which lack significant variation as indicated by RFLP and RAPD analyses, VNTR markers offer a new means of investigating crops with this mode of reproduction by utilizing the variation existing in the repeated sequence part of the genome.

C) Telomeres

The term telomere has often been used loosely. It has been used to describe heterochromatin associated with the terminal regions of chromosomes and regions containing telomere-like repeat sequences. However, several studies in *Arabidopsis* (Richards and Ausubel 1988), tomato (Ganal et al. 1991), and wheat (Cheung et al. 1994) have demonstrated that telomeres consist of repeat units with the sequence $(TTTAGGG)_n$ in *Arabidopsis* and wheat, and $TT[T(A)GGG]_n$ in tomato.

D) Centromeres

Centromeres have been isolated from *Saccharomyces* and *Schizosaccharomyces*. Among higher eukaryotes only a human centromere has been cloned in its entirety (Tyler-Smith et al. 1993). Recently, a repetitive sequence has been identified which is found in the genomes of barley, wheat, maize, and rice. It has been demonstrated by *in situ* hybridization to be located at the centromeres of all wheat, barley, and rye chromosomes as well as being located at the centromeres of rye supernumerary and wheat midget chromosomes (Alcaide et al. 1996).

Sub-megabase level

Examination of cereal genomes at the megabase level also reveals a clear, non-random organization (Moore et al. 1993b, Abbo et al. 1995). Although the most repetitive sequences in wheat and barley are highly methylated, the presence of nonmethylated recognition sites for the restriction enzymes *Mlu*I and *Nru*I at non-random intervals have been observed. Cleavage of barley, rye, and wheat genomes with these enzymes yields discrete fragment size classes (FSCs) ranging from hundreds of kilobases to several megabases in length.

The generation of different fragment sizes indicates that unmethylated recognition sites for *Mlu*I are non-randomly distributed throughout these genomes. In addition, even though the elements BIS-1, WIS-2 and other dispersed repeats have different copy numbers within the various Triticeae genomes, they are distributed similarly between *Mlu*I FSCs. One possible explanation for these observations is that the elements such as BIS-1 and WIS-2 are organized into units that have then undergone amplification.

Multi-megabase level

In situ hybridization of the barley repeat sequence Hi-10 in *a B. sylvaticum* clone is present in many Gramineae genomes (Abbo et al. 1995, Alcaide and Moore, unpublished). Metaphase chromosomes of barley and rye revealed clear compartmentalization of this repeat to specific regions. The existence of such compartments within chromosomes can be explained by the divergence of repetitive sequences and their subsequent amplification, followed by further modification via molecular drive mechanisms. Sequences homologous to this repeat in Hi-10 may have diverged in particular chromosome regions and are no longer detectable by the Hi-10 used for in situ studies. For example, although the retrotransposon (BIS-1) *in situ* hybridization studies detected sequence dispersion along the chromosomes, Southern blot analysis revealed specific variants to be located to particular chromosomes.

Conserved linkage segments

A genetic (RFLP) map of wheat has been constructed (Devos and Gale 1992), which has a number of interesting features. Certain markers when used as probes, detect sequences lying in the same genetic order on all three homoeologous chromosomes, revealing the genes to be essentially colinear within the homoeologous chromosome groups of wheat. The genetic map also shows a clustering of markers derived from the proximal/centromere regions.

The colinearity of markers on homoeologous chromosomes raised the issue of whether markers on barley and rye chromosomes were also colinear with the order in wheat. Although there are several large evolutionary translocations between the various genomes, it is now clear that the linear order of markers on homoeologous chromosomes of barley and rye are the same as those on wheat (Devos et al. 1993a, 1993b).

As earlier stated, rice and wheat started diverging on their separate evolutionary pathways over 60 million years ago. Rice evolved with twelve chromosomes and wheat ended up with three homoeologous sets of seven chromosomes. The average wheat chromosome is approximately 25-fold larger than the average rice chromosome. The important question therefore arises as to the relationship between the gene order within these two species and to the overall genomic constitutions which differ so dramatically. Studies have revealed that nonmethylated, single-copy sequences in the rice genome are still, for the most part, single copy in each wheat genome despite the presence in wheat of considerable additional DNA (Moore et al. 1993b). This suggests that whatever the evolutionary mechanisms involved in the amplification of this additional DNA, it did not generally involve many changes in the single-copy sequences. In other words, the rice and wheat genomes have a similar gene content. Reciprocal mapping involving rice RFLP markers (Kurata et al. 1994b) onto the wheat genetic map and *vice versa* has demonstrated a strong colinearity between the genomes of the two species. Over 500 markers from the rice and wheat genetic maps have so far been analysed in this comparative study (Foote and Moore unpublished).

Comparison of the genetic maps of rice and wheat suggests that rice chromosome maps can be dissected into a series of small linkage groups (like "Lego" blocks). These rice linkage segments could then be used to construct wheat chromosomes. The data comparing markers/genes in the rice and maize genomes (and on duplicated loci in the maize genome) have allowed the genetic maps of maize chromosomes also to be reconstituted from the same set of rice linkage segments (Moore et al. 1995a). By this means, the maize genome was demonstrated to be tetraploid, suggesting that five chromosomes may be derived from one parent and the other five from another parent. A comparison of the colinearity of linkage segments in maize, wheat and rice leads to the construction of an hypothetical "ancestral" chromosome. This ancestral chromosome, when cleaved in different places, could then give rise to all the wheat chromosomes. Cleaved at other positions it would give rise to one set of five maize chromosomes and again at different positions, for the other set of five maize chromosomes. This concept has been extended to include the chromosomes of all the major cereal crops (Moore et al. 1995b). The cleavage of the same ancestral chromosome gives rise to the 56 chromosomes present in these species. Such a comparison makes rice, with its small genome, the model system for cereals.

The implications of this observation are far reaching. All the information generated by studying these species separately in the last 50 years, can be collated and exploited with greater efficiency. By combining all available data on crops grown worldwide, it will be possible to trace the threads of allelic variation during speciation to comprehend plant development within and between species. It will be possible to align the genes controlling stress, disease resistance and yield in these different genomes.

Using markers already mapped on rice chromosomes, it has been shown that the markers are maintained in the same order as on barley and wheat chromosomes (Dunford et al., 1995). Thus, gene-order is conserved at the micro-level. This implies that genes can be identified by map-based cloning of the homoeologue in rice, and then by homology back in the target species. The reconstruction of the rice genome in the form of YACs will reveal the general structure of most cereal genomes. This may provide clues to how and why particular regions of some cereal genomes have undergone extensive expansion by non-random accumulation of repetitive sequences following speciation.

Chromosomes

Although rice linkage segments appear to contain the homoeologous genes present in all-the cereals, the actual DNA content of the segments will vary depending on the extent of amplification of the repeats between the genes. Clearly, the sum of the DNA content in all the different linkage segments accounts for the differences in DNA content of chromosomes. The intense clustering of markers in the proximal/centromere regions on the genetic maps of wheat chromosomes begs the question of what is the relationship between the genetic and physical location of the genes/markers on the chromosome. Studies on the distribution of the sites of recombination along a cereal chromosome using physical markers all show that recombination is predominantly in the distal regions (Linde-Laursen 1982, Dvorak and Chen 1984, Snape et al. 1985, Lawrence and Appels 1986, Gustafson and Dille 1992). As a result of reduced recombination towards the centromere, markers physically located in these regions are clustered on recombination-based maps when compared to those located in the distal chromosome regions. A single rice chromosome contains thousands of genes. The colinearity of markers on several rice chromosomes (for instance, 8, 10 and 11) with the proximal-centromeric regions of various wheat chromosomes suggests that the proximal-centromeric regions of wheat which exhibit reduced recombination, also contain thousands of genes. This raises the important question as to what proportion of genes are being reassorted in wheat and rice breeding programs.

Acknowledgements

JPG would like to thank the Underwood Foundation who funded the collaborative effort required for this manuscript while he was in England.

References

Abbo S, Dunford RP, Foote TN, Reader SM, Flavell RB and Moore G (1995) *Chromosome Res* 3:5-15.
Ahn S and Tanksley SD (1993) *Proc Natl Acad Sci USA* 90:7980-7984.
Alcaide L, Miller T, Schwarzacher T, Reader S and Moore G (1996) Mns. submitted.
Ali S, Muller CR and Epplen JT (1986) *Human Genetics* 74:239-243.
Amos B and Pemberton J (1993) *Fingerprint News* 5:2.
Arens P, Odinot P, van Heusden AW, Lindhout P and Vosman B (1995) *Genome* 38:84-90.
Avise JC (1994) *Molecular Markers, Natural History, and Evolution*. Chapman & Hall, New York.
Batschelet E (1979) *Introduction to Mathematics for Life Sciences* (3rd ed). SpringerVerlag, Berlin.
Bennett MD and Smith JB (1991) *Phil Trans R Soc London* (Biol) 334:309.
Benson G and Waterman MS (1994) *Nucleic Acids Res* 22:4828-4830.
Bird AP (1987) *Trends Genet* 3:342.
Borstnik B, Pumpernik D, Lukman D, Ugarkovic D and Plohl M (1994) *Nucleic Acids Res* 22:3412-3417.
Brown CM, Stockwell PA, Trotman CNA and Tate WP (1990) *Nucleic Acids Res* 18:6339-6346.
Bureau TE and Wessler SR (1992) *Plant Cell* 4:1283-1294.
Cheung WY, Money TA, Abbo S, Devos KM, Gale MD and Moore G (1994) *Mol Gen Genet* 245:349-354.
Dallas JF (1988) *Proc Natl Acad Sci USA* 85:6831-6835.
DePamphilis CW (1988) Hybridization in Woody Plants: Population Genetics and Reproductive Biology in Southeastern Buckeyes. PhD Dissertation. Univ. Georgia, Athens.
Deshpande VG and Ranjekar PK (1980) *Z Physiol Chem* 361:1223.
Devos K and Gale MD ((1992) *Outlook in Agriculture* 22:93-99.
Devos K, Millan T and Gale MD (1993a) *Theor Appl Genet* 85:784-792.
Devos KM, Atkinson MD, Chinoy CN, Francis HA, Harcourt RL, Koebner RMD, Liu CJ, Masojc P, Xie DX and Gale MD (1993b) *Theor Appl Genet* 85:673-680.
Devos KM, Chao S, Li QY, Simonetti C and Gale MD (1994) *Genetics* 138:12871-292.

Dvorak J and Chen K-C (1984) *Genetics* 106:325-333.
Dunford R, Kurata N, Laurie D, Money T, Minobe T and Moore G (1995) *Nucleic Acids Res* 23:2724-2728.
Epplen JT (1988) *J Heredity* 79:409-417.
Feng Y, Zhang F, Lokey LK, Chastain J.L, Lakkis L, Eberhart D and Warren ST (1995) *Science* 268:731-734.
Flavell RB and Smith DB (1976) *Heredity* 37:231.
Ganal WW, Lapitan NLV and Tanksley SD (1991) *Plant Cell* 3:87-94.
Good P (1994) *Permutation Tests*. Springer-Verlag, Berlin.
Grant V (1981) *Plant Speciation* (2nd ed.). Columbia University Press, New York.
Gruenbaum Y, Naveh-Many T, Cedar H and Razin A (1981) *Nature* 292:860.
Gustafson JP and Dille JE (1992) *Proc Natl Acad Sci USA* 89:8646-8650.
Heath DD, Iwama GK and Devlin RH (1993) *Nucleic Acid Res* 21:5782-5785.
Hamrick JL, Linhart YB and Mitton JB (1979) *Ann Rev Ecol Syst* 10:173.
Hamrick JL and Godt MJW (1990) In AHD. Brown, MT Clegg, AL Kahler and BS Weir (eds.), *Plant Population Genetics, Breeding, and Genetic Resources*. Sinauer Associates Inc, Sunderland. pp.
Hamrick JL and Murawski DA (1991) *J Trop Ecol* 7:395.
Jeffreys AJ, Royle NJ, Wilson V and Wong Z (1988) *Nature* 332:278.
Jelinski DE and Cheliak WM (1992) *Amer J Bot* 79:728-736.
Jin L and Chakraborty R (1993) *Mol Biol Evol* 10:1112.
Kaur A, Ha CO, Jong K, Sands VE, Chan HT, Soepadmo E and Ashton PS (1978) *Nature* 271:440.
Kurata N, Moore G, Nagamura Y, Foote T, Yano M, Minobe Y and Gale MD (1994) *Bio/technology* 12:276-278.
Lawrence GJ and Appels R (1986) *Theor Appl Genet* 71:742-749.
Litt M and Luty JA (1989) *Amer J Hum Genet* 44:397-401.
Linde-Laursen I (1982) *Heredity* 49:27.
Ma ZQ, Roder M and Sorrells ME (1996) *Genome* 39: 123-130.
Martin W, Gierl A and Saedler H (1989) *Nature* 339:46.
Manninen I and Schulman AH (1993) *Plant Mol Biol* 22:829.
Mitton JB (1994) *Annu Rev Ecol Syst* 25:45.
Moore G, Lucas H, Batty N and Flavell RB (1991a) *Genomics* 10:461-468.
Moore G, Cheung W, Schwarzacher T and Flavell R (1991b) *Genomics* 10:469-476.
Moore G, Gale MD, Kurata N and Flavell RB (1993a) *Bio/technology* 11:584-589.
Moore G, Abbo S, Cheung W, Foote T, Gale M, Koebner R, Leitch A, Leitch I, Money T, Stancombe P, Yano M and Flavell R (1993b) *Genomics* 15:472-482.
Moore G, Foote T, Helentjaris T, Devos K and Gale MD (1995a) *Trends Genet* 11:81.
Moore G, Devos K, Wang Z and Gale MD (1995b) *Current Biol* 5:737-739.
Nakamura Y, Leppert M, O'Connel P, Wolff R, Holm T, Culver M, Martin C, Fujimoto E, Hoff M, Kumlin E and White R (1987) *Science* 235:1616-1622.
Neuhaus D., Kuhl H, Kohl JG, Dorfel P and Borner T (1993) *Aquatic Bot* 45:357-364.
Nogler GA (1984) In B.M. Johri (ed.), *Embryology of Angiosperms*. Springer-Verlag, Berlin.
Nurnberg P, Barth I, Fuhrmann E, Lenzner C, Lozanova T, Peters C, Poche H and Thiel G (1991) *Electrophoresis* 12:186.
Ohtsubo H, Motohashi R, Nakajima R, Iida Y, Noma K, Kumekawa N and Ohtsubo E (1996) Identification and characterization of mobile genetic elements in the rearranged DNA in rice genomes. Rice Genome Meeting III, Tsukuba, Japan.
Panaud O, Chen XL and McCouch SR (1995) *Genome* 38:1170-1176.
Prodohl PA, Taggart JB and Ferguson A (1994) *Heredity* 73:556-566.
Richards AJ (1986) *Plant Breeding Systems*. George Allen & Unwin, London.
Richards EJ and Ausubel FM (1988) *Cell* 53:127-136.
Ritland K (1990) *J Heredity* 81:235-237.

Roder MS, Plaschke J, Konig SU, Borner A, Sorrells ME and Tanksley SD (1995) *Mol Gen Genet* 246:327.
Rogstad SH (1992) *Taxon* 41:701.
Rogstad SH (1993) *Methods Enzymol* 224:278.
Rogstad SH (1994a) *Theor Appl Genet* 89:824.
Rogstad SH (1994b) *Amer J Bot* 81:14S.
Rogstad SH, Nybom H and Schaal BA (1991) *Plant Syst Evol* 175:115.
Rothuizen J, Wolfswinkel J, Lenstra JA and Frants RR (1994) *Theor Appl Genet* 89:403.
Russell GJ, Follett EAC, Subak-Sharpe JH and Harrison BD (1971) *J Gen Virol* 11:129.
Sasaki T (1995) *Rice genome: Newsletter for rice genome analsis.* Vol. 41. Tsukuba, Japan.
Scribner KT, Arntzen JW and Burke T (1994) *Mol Biol Evol* 11:737.
Shaw DV, Kahler AL and Allard RW. (1981) *Proc Natl Acad Sci USA* 78:1298.
Somers DJ, Zhou Z, Bebeli PJ and Gustafson JP (1996) *Theor Appl Genet,* in press.
Snape JW, Flavell RB, O'Dell M, Hughes WG and Payne PI (1985) *Theor Appl Genet* 69:263.
Stephens JC, Gilbert DA, Yuhki N and O'Brien SJ (1992) *Mol Biol Evol* 9:729.
Swartz MN, Trautner TA and Kornberg A (1962) *J Biol Chem* 237:1961.
Terauchi R and Konuma A (1994) *Genome* 37:794.
Tyler-Smith C, Oakey RJ, Larin Z and Fisher RB (1993) *Nature Genetics* 5:368.
Vergnaud G (1989) *Nucleic Acids Res* 17:7623.
Vos P, Hogers R, Bleeker W, Reijans M, van de Lee T, Hornes M, Frijters A, Pot J, Peleman J, Kuiper M and Zaheau M (1995) *Nucleic Acids Res* 23:4407.
Weber JL and May PE (1989) *Amer J Hum Genet* 44:388.
Wolfe KH, Gouy M, Yang Y-W, Sharp PM. and Li WH (1989) *Prioc Natl Acad Sci USA.* 86:6201.
Weising K, Nybom H, Wolff K and Wieland M (1995) *DNA Fingerprinting in Plants and Fungi.* CRC Press, Boca Raton.
Winberg BC, Zhou Z, Dallas JF, McIntyre CL and Gustafson JP (1993) *Genome* 36:978.
Wu KS and Tanksley SD (1993) *Mol Gen Genet* 241:228.
Yano M, Shomura A, Wang Z, Kurata N, Huang N, Khush G and Sasaki T (1996) *Theor Appl Genet* 91:481.
Zhou Z and Gustafson JP (1995) *Theor Appl Genet* 91:481.

18

Molecular analysis of the rice genome

Nori Kurata and Takuji Sasaki
Rice Genome Research Program, National Institute of Agrobiological Resources, STAFF Institute, Tsukuba, Japan

Introduction

The haploid *Oryza sativa* (rice) genome contains 430 Mb of nuclear DNA (Arumuganathan and Earle 1991) distributed over 12 small chromosomes. These genetic entities have several special features in common with other Graminae species. Rice has the smallest genome size of the cereal crop plants analyzed so far and only around 50% of its DNA is composed of repetitive sequences, a low proportion compared to those of other crops. A large number of cultivated rice lines which possess AA genome and many wild rices carrying other genomes, i.e. BB, CC, DD, EE, FF, GG and so on, would be the most useful sources for genome analysis and applications at both molecular and genetic levels.

In the last decade, a large number of DNA landmarks distributed all along the 12 rice chromosomes have been obtained. Several independent high density molecular maps were constructed by restriction fragment length polymorphism (RFLP) and random amplified polymorphic DNA (RAPD) linkage mapping using both intra-species and inter-species crosses. By the middle of 1996, over 3,000 DNA fragments of genomic, cDNA and RAPD origin had been located on the rice genome scattered at an average interval of 140 kb. These markers play important roles in genome and genetic research, such as map-based cloning of target genes, marker-aided selection in breeding programs, and analyses of genetic and physical structures of the rice genome. The markers used in these studies include amplification fragment length polymorphism (AFLP), RAPD, and RFLP markers, and minisatellite and microsatellite DNA. DNA markers, especially those carrying sequence, genomic localization and expression information, should play vital roles in most areas of rice biotechnology. These markers can also be supplied as landmarks for physical mapping, for dissecting complex genetic traits such as quantitative trait loci (QTL) and for comparative mapping among cereal genomes.

DNA markers used in genome analysis of rice

RFLP, RAPD and AFLP markers

RFLP, RAPD and AFLP markers have been developed recently as powerful tools for marking individual chromosome positions in many organisms. Localization of these DNA markers can be simply determined by genetic mapping of these polymorphic DNA markers. However, the efficiency of genetic mapping largely depends on the genetic variations between the varieties used as cross-pollinating parents. A survey of polymorphisms between parent varieties using these markers, is the first step towards the construction of genetic maps.

Because of their co-dominant nature, RFLP are the most promising markers with which to construct exact maps. Randomly selected cDNA and genomic DNAs are good sources of RFLP markers. A large portion of cDNA clones studied so far in rice showed lower copy number and higher polymorphism content than did genomic DNA fragments (Kurata et al. 1994a). RAPD markers proved to be very useful for rapid and large scale mapping (Reiter et al. 1992, Zhang et al. 1994). After generating RFLP and RAPD markers, sequence tagged sites (STS) and primers derived from single strand conformation polymorphism (SSCP) markers were developed following sequence data obtained from mapped RFLP and RAPD loci (Inoue et al. 1994, Monna et al. 1994, Fukuoka et al. 1994).

The AFLP fingerprinting technique has been developed over several years (Vos et al. 1995) and is now being applied for rapid and efficient construction of plant molecular maps (Becker et al. 1995, Thomas et al. 1995). Suitable enzyme combinations for double digestion, sequence specificities and extensions of degenerate primers vary from organism to organism depending on genome size and ubiquity of DNA sequence. Preferable enzyme combinations and primer sequences for simultaneous mapping of multiple rice fragments have been analyzed (MacKill et al. 1996, Jiahui et al. 1996).

Other kinds of DNA markers derived from minisatellite and microsatellite DNA are also very useful for rapid mapping and phylogenic studies (Dallas 1988, Wu and Tanksley 1993, Yang et al. 1994). However, when using variable numbers tandem repeat (VNTR) sequences and related markers, as well as with most applications of AFLP and RAPD analysis, it is difficult to determine the corresponding allelic DNA fragments among multiple fingerprints of hybridized or amplified polymorphic fragments.

Construction of a high-density molecular map

In rice, the first molecular map which covered all 12 linkage groups was constructed by identifying the loci of about 200 RFLP markers (McCouch et al. 1988). Since then, several high density molecular maps have been constructed using different kinds of F_2 populations (Saito et al.1991, Causse et al. 1994, Kurata et al. 1994a). The most dense map carried about 1,400 DNA markers with sequence tags located at 300 kb intervals, on average (Kurata et al. 1994a). Several other rice maps have been developed and use different kinds of F_2 populations (Jena et al. 1994), backcross populations (Zhao and

Kochert 1993), recombinant inbred (RI; Tsumematsu et al. 1996, Xiao et al. 1996) and doubled haploid (DH; Yamagishi et al. 1996) lines. These maps contained new and previously mapped DNA markers. The total number of DNA markers located by early 1996 on all of these molecular maps exceeded 3,000, dividing the rice genome into DNA fragments of 140 kb average length.

Application of densely distributed DNA markers in genetic studies

The highly populated array of DNA markers located on rice molecular maps appear to be very useful for dissecting complex traits and trait-related multiple loci (Harushima et al. 1996, Yano et al., mns. submitted). DNA markers established on these maps can be applied to the tagging of major and minor trait genes of QTL (Xiao et al. 1995, McCouch and Doerge 1995). At the same time, marker density is very important when searching for closely-linked markers to any target gene one wants to isolate by map-based cloning, or for marker-aided selection. Many examples of gene tagging have been reported using these and newly generated markers (Mackill et al. 1993, Wang et al. 1994, Ishii et al. 1994, Nair et al. 1995). In addition, the molecular linkage map has been integrated with the classical linkage map by locating about 40 trait genes on the molecular map (Ideta et al. 1992, 1993, 1995). An integrated map is shown in Figure 1.

Structural features of the rice genome revealed through genetic mapping with DNA markers

In the process of mapping, especially when using Southern hybridization of cloned DNA fragments for RFLP detection, several noteworthy aspects of rice genome structure were observed. The genetic diversity detected among the many cultivated strains and wild species of rice has advanced phylogenetic studies at the molecular level (Qiau et al. 1995, Wang et al. 1992). Other specific features of the rice genome revealed by mapping studies are discussed below.

Repetitive DNA sequences in the genome

Mapping of microsatellite DNA revealed type and distribution of repetitive DNA sequences in the rice genome. AFLP and RAPD markers showed a high ability to cover those regions in the genome where RFLP markers, which are low copy sequences, can be rarely mapped (Monna et al. 1994, Becker et al. 1995). This stems from the nature of these two marker types which often recognize repetitive DNA sequences. Chromosome mapping with repetitive sequences by *in situ* hybridization showed further information on the distribution of repetitive sequences.

Genetic structure of the genome

Some similarities in the distribution of markers within independent rice RFLP maps were observed. Each of the 12 chromosomes has one or more regions where no markers are mapped along distances of 10 to 20 cM. Highly condensed marker regions with more than 10 markers mapping on single positions are also observed on some chromosomes. The distribution of marker-rare and marker-dense regions correspond well among the 3

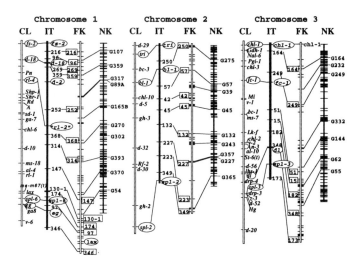

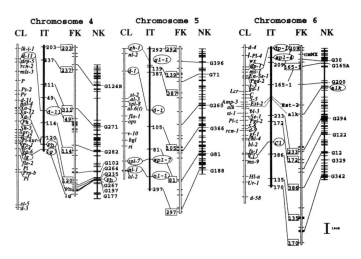

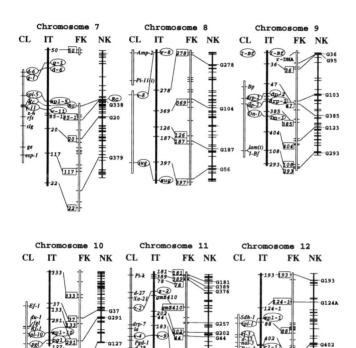

FIGURE 1. The integrated linkage map of rice. The correspondence of the four maps is shown by connecting the same trait genes or DNA markers on the different maps. The left "CL" map is a classical linkage map of rice (Kinoshita 1995), the "FK" map (third from the left) is an RFLP linkage map developed from a cross between FL134 and Kasalath (Saito et al. 1991), and the the"IT" map (second from the left) is an integrated map of the "CL" and "FK" maps constructed by Ideta et al. (1992, 1993, 1995). The right "NK" map is a high-density RFLP linkage map developed from a cross between Nipponbare and Kasalath (Kurata et al. 1994a).

most marker-dense RFLP maps, indicating that the frequencies of genetic recombination in certain chromosomal regions are similar among different crosses. Centromeres were located by secondary and telo-trisomic analysis around the middle regions of 10 chromosomes and on the distal portions of 2 chromosomes (Khush et al. 1995, Singh et al. 1996). The putative centromeric regions are not necessarily dense with DNA markers. In fact, many chromosomes contain regions adjacent to centromeres with sparse markers along genetic distances of many cM.

In the course of RFLP mapping, some genome regions of high structural similarity were evident. The order of many markers arrayed on the distal regions of chromosomes 11 and 12 was conserved along a sequence of more than 10 cM, indicating chromosomal duplication (Nagamura et al. 1995). Certain regions of chromosomes 1 and 5 contained duplications of three gene sequences (Kishimoto et al. 1994).

Physical mapping of the rice genome

Besides the construction of genetic maps, physical mapping are an important resource for molecular biology research, such as positional cloning of genes, sequencing of genomic DNA, and detailed analysis of genome organization. Physical mapping of the rice genome is now being conducted by two groups, working in Japan and China.

Genomic DNA libraries as a resource for physical mapping

Methods for preparation and cloning of high molecular weight genomic DNA of near megabase-size has been accomplished over the past several years (Guidet and Langridge 1992, Hatano et al. 1992, Wing et al. 1993, Zhang et al. 1995). Following the development of these methods, two kinds of rice genomic libraries, one a yeast artificial chromosome (YAC) library and the other a bacterial artificial chromosome (BAC) library, were successfully constructed. The YAC library was made from *japonica* rice which is one of the parents for the mapping population for a high-density molecular map, and now consists of about 7,000 clones with inserts averaging 350 kb in length and covering the rice genome 6 times (Umehara et al. 1995). The BAC library, which contains more than 20,000 clones of 120 kb average length covering 6 genome equivalents, was constructed with an *indica* variety (Tao et al. 1994). Two other BAC libraries constructed with *indica* and *japonica* varieties, parents of 400 RI lines of a permanent mapping population, contain about 14,200 clones and 7,300 clones of 130 kb and 150 kb average length and cover 4.4 and 2.2 genome equivalents, respectively (Zhang et al. 1995). Several other BAC and cosmid libraries derived from specific rice lines or cultivars which contain target trait genes for map-based cloning have also been developed (Kawasaki et al. 1995, Yoshimura et al.1995, Wang et al. 1995).

DNA assembly for physical mapping

Projects to construct overlapping contiguous sequences (contigs) of genomic clones for the whole length of the genome are now underway using YAC and BAC libraries. The strategies used to assemble contigs from YACs are different from those from BACs. A fairly large part of the 12 chromosomes are expected to be covered by arranging YACs of

350 kb length on the corresponding sites of DNA markers, which are distributed at average 300 kb intervals. YAC clones have been "landed" on by more than 1,200 DNA marker sites along all chromosomes. They are estimated to cover about half of the total genome (Wang et al. 1996, Umehara et al. 1996, Kurata et al. 1997). Physical mapping using an *indica* BAC library of about 6 genome equivalents resulted in the construction of a total of 594 BAC clone contigs of 600 to 1500 kb length. These contigs were calculated to cover 430 Mb (98%) of the rice genome but are not yet located on the chromosomes (Tao et al. 1995). Two physical mapping efforts to cover the whole genome with YACs and BACs are now in progress and will soon yield detailed data for analysis of the physical structure of the genome. Several short genomic stretches have also been analyzed by physical mapping with BAC and YAC clones in order to study and target the trait genes for *Xa-1* (Yoshimura et al. 1996), *Pi-b* (Monna et al. 1996), *Pi-b* and *Pi-ta2* (Kawasaki et al. 1995) and *eg* (Van Houten et al., mns. in press).

Physical mapping by in situ *hybridization*

Besides the construction of overlapping clone libraries for each chromosome, the chromosomal distributions of several repetitive and single copy DNA sequences have been analyzed by in situ hybridization. Localization of some repetitive sequences in the distal parts of many chromosomes adjacent to telomeres was first reported by Wu et al. (1991) using *in situ* hybridization (ISH). Fluorescent *in situ* hybridization (FISH) revealed that some transposon-like repetitive sequences specific to the AA, CC and FF genomes were located on the distal ends of certain chromosomes in *O. sativa*, *O. officinalis*, *O. brachyantha*, respectively (Ohmido et al. 1995, Fukui 1996). In addition, a chromosome specific repetitive sequence located near the centromeric region of chromosome 5 was also reported (Wang et al. 1995). Repetitive sequences dispersed all over the genome were only reported in the EE genome species of *O. australiensisi* (Shishido et al. mns. in preparation). Thus the role of repetitive sequences in genome organization and evolution is also being gradually resolved.

As for single copy sequences, a trial to match DNA markers on a genetic map with their chromosome location was done by the ISH method (Gustafson et al. 1993). Single copy sequences carried on cosmid and BAC clones can now be located by FISH on each chromosome position where they occur (Jiang et al. 1995, Ohmido et al. 1995). In addition, a single copy sequence of less than 1 kb length has recently been successfully located on its specific position on a chromosome (Ohmido and Fukui, mns. in preparation).

cDNA analysis as a tool for resolving genome organization

A method of gene isolation and identification

There are several different methods for screening and identifying individual expressed genes. Among these, large scale sequencing and similarity searching are the most promising methods for simultaneous identification of functional genes. Two groups reported large scale cDNA analyses of callus libraries (Uchimiya et al. 1992, Sasaki et al. 1994), presenting many identified cDNA clones through similarity searches of all public protein/gene data of known function. The Rice Genome Research Program in Japan has deposited data for over 11,000 partial sequences of root and shoot cDNA, together with callus cDNA, as an expression sequence tag (EST) data base. The total of more than 17,000 rice cDNA sequences available through public data bases is an important resource for identifying genes and analyzing genome organization for expressed sequences.

Tissue-specific and stage-specific cDNA libraries

Expression specificity of a given gene can be detected by analyzing many cDNA clones in libraries made from specific tissues and specific developmental stages. The abundance of certain cDNA sequences in different libraries suggests when and where the gene is expressed, or in some cases, the expression of some related family genes. Multiple comparisons of cDNA sequences among different cDNA libraries could provide profiles of expression for corresponding genes in tissues and developmental stages (Yamamoto et al. 1996). These profiles of gene expression suggest functional properties of individual genes, even when they have no known function. Gene profiling by large scale cDNA analysis matches well with the strategy of genome analysis in which one extracts several parameters concerning genome structure and then utilize those results for many other purposes.

Genetic and physical mapping of cDNA clones

In another aspect of cDNA analysis, genetic and physical mapping of cDNA clones for which both sequence tags and expression profiles are available can reveal genome organization in a gene expression context. The location and order of a large number of cDNA sequences on chromosomes is expected to show genomic properties that are related to the molecular architecture that controls gene expression, as well as suggesting an evolutionary significance to diversity and uniformity of genes among genomes.

The Rice Genome Research Program has already mapped around 1,500 cDNA clones including about 500 gene clones which have been identified by similarity searches against protein data bases (Harushima et al., mns. in preparation). The distribution of some gene families and many functionally related genes in the genome was also studied. Fifty seven kinds of ribosomal protein genes were mapped; most were widely dispersed over the genome with some exceptions clustering in specific locations (Wu et al.1995). Several members of the histone protein gene family (containing H1, H2A, H2B, H3 and H4) were also analyzed for their full length nucleotide sequences and for the distribution on chromosomes (Song and Antonio, mns. in preparation). Positional relationships among some genes on the map and between some other organisms were pointed out (Kurata et al. 1994a).

Comparative analysis between rice and other cereal genomes

Macro-synteny among cereal crops revealed by genetic mapping

Many molecular maps constructed with RFLP and RAPD markers in other plant species have also been developed in the last decade. In monocotyledoneous cereal crops, high-density molecular maps containing a large number of DNA markers have become available in wheat (Hart et al. 1993), barley (Kleinhofs et al. 1994), maize (Burr et al. 1993), rye (Schlegel and Melz 1993) and other species. Two intensive comparative mapping efforts were done between rice and wheat and between maize and rice. Almost full length synteny between rice chromosomes and wheat chromosome groups was observed with many common DNA markers co-linearly located on both chromosomes (Kurata et al. 1994b). Comparison of the maize RFLP map with the rice map showed that most of the duplicated segments in the maize genome have a high degree of similarity with certain chromosomal segments of rice (Ahn and Tanksley 1993). Many other synteny relationships among homoeologous chromosomes in cereal crop species were detected by genetic mapping (Devos et al. 1993, Van Deynze et al. 1995a, 1995b). The most impressive study of these synteny relationships was a dissection and alignment of chromosomal segments capable of lining up putative common ancestor segments from six grass species (Moore et al. 1995).

Micro-syntenic relationships revealed by physical mapping

As major syntenies among cereal crop genomes are being revealed at the chromosomal level, several intensive analyses for micro-synteny of segments of some hundreds of kilobases in length have been done. A detailed comparison of the restricted regions of the barley and wheat genomes with the syntenic chromosomal segments in rice was planned in order to identify homoeologous genes for map-based cloning. Micro-synteny analysis, conducted to attempt map-based cloning of the barley stem rust resistance gene, *Rpg-1* (Killian et al. 1995), and the wheat pairing homoeologous gene, *Ph-1* (Dunford et al. 1995), revealed substantial co-linearity in the order of many DNA markers between short stretches of the barley and wheat maps and one, two or more YAC contigs from rice.

The results of rice genome research are expected to be very useful for genome and map-based gene cloning studies of other cereal crops, because of the relatively small size of the rice genome, the high resolution and high quality of the high-density molecular map, the accumulation and availability of information about a large number of expressed sequences and related materials, together with the rapid progress of the physical mapping of rice. Thus, the genetic and physical maps of rice as well as the analysis of cDNA supply important information and materials which greatly facilitate genome analysis of other cereal crop plants. This situation makes rice the model plant for research on all other cereal crops and monocot plants.

References

Ahn S, Bollich C, McClung A and Tanksley SD (1994) *Theor Appl Genet* 87:27-32.
Ahn SN and Tanksley SD (1993) *Proc Natl Acad Sci USA* 90:7980-7984.
Arumuganathan K and Earle ED (1991) *Plant Mol Biol Rep* 9:208-218.
Becker J, Vos P, Kuiper M, Salamini F and Heun M (1995) *Mol Gen Genet* 249:65-73.
Burr B, Burr FA and Matz EC (1993) *Genetic Maps* 6:190-203.
Causse MA, Fulton TM, Cho YG, Ahn N, Chunwongse J, Wu K, Xiao J, Yu Z, Ronald PC, Harrington SE, Second G, McCouch SR and Tanksley SD (1994) *Genetics* 138:1251-1274.
Churchill GA, Giovannoni JJ and Tanksley SD (1993) *Proc Natl Acad Sci USA* 90:16-20.
Dallas JF (1988) *Proc Natl Acad Sci USA* 85:6831-6835.

Devos KM, Millan T and Gale MD (1993) *Theor Appl Genet* 85:784-792.

Dunford RP, Kurata N, Lairie DA, Money TA, Minobe Y and Moore G (1995) *Nucleic Acids Res* 23:2724-2728.

Fukui K and Iijima K (1993) *Theor Appl Genet* 81:589-596

Fukui K (1996) *Plant Genome IV*. Abstract W17, p.19.

Fukuoka S, Inoue T, Miyao A, Monna L, Zhong HS, Sasaki T and Minobe Y (1994) *DNA Res* 1:271-277.

Guidet F and Langridge P (1992) *Methods Enzymology* 216:3-12.

Gustasfson PJ and Dille JE (1992) *Proc Natl Acad Sci USA* 89:8646-8650.

Hart GE, Gale MD and McIntosh (1993) *Genetic Maps* 6:204-219.

Harushima Y, Kurata N, Yano M, Nagamura Y, Sasaki T, Minobe Y and Nakaghara M (1996) *Theor Appl Genet* 92:145-150.

Hatano S, Yamaguchi J and Hirano A (1992) *Plant Sci* 83:55-64.

Havukkala I, Ichimura H, Nagamura Y and Sasaki T (1995) *J Biotechnol* 41:139-148.

Ideta O, Yoshimura A, Matsumoto T, Tsunematsu H and Iwata N (1992) *Rice Genet Newslet* 9:128-129.

Ideta O, Yoshimura A, Matsumoto T, Tsunematsu H, Satoh H and Iwata N (1993) *Rice Genet Newslet* 10:87-89.

Ideta O, Yoshimura A and Iwata N (1995) *3rd Intl Rice Genet Symposium*. Abstract, p. 5.

Inoue T, Zhong HS, Miyao A, Ashikawa I, Monna L, Fukuoka S, Miyadera N, Nagamura Y, Kurata N, Sasaki T and minobe Y (1994) *Theor Appl Genet* 89:728-734.

Ishii T, Brar DS, Multani DS and Kush GS (1994) *Genome* 37:217-221.

Jena KK, Khush GS and Kochert G (1994) *Genome* 37:382-389.

Jiang J, Gill B, Wang G, Ronald P and Ward D (1995) *Proc Natl Acad Sci USA* 92:4487-4491.

Kawasaki S, Nakamura S, Ryobka K, Akiyama K, Sasaki N, Asakawa S, Nobuyoshi S and Ando I (1995) *Plant Genome IV*. Abstract P17, p. 31.

Kinoshita T (1995) *Rice Genet Newslet* 11:9-153.

Kishimoto N, Higo H, Abe K, Arai S, Saito A and Higo K (1994) *Theor Appl Genet* 88:722-726.

Kleinhofs A, Killian A, Saghai-Maroof MA and Biyashew RM (1993) *Theor Appl Genet* 86:705-712.

Kurata N, Nagamura Y, Yamamoto K, Harushima Y, Sue N, Wu J, Antonio BA, Shomura A, Shimizu T, Lin SY, Inoue T, Fukuoka A, Shimano T, Kuboki Y, Toyama T, Miyamoto Y, Kirihara T, Hayasaka K, Miyao A, Monna L, Zhong SH, Tamura Y, Wang ZX, Momma T, Umehara Y, Yano M, Sasaki T and Minobe Y (1994a) *Nature Genetics* 8:365-372.

Kurata N, Moore G, Nagamura Y, Foote T, Yano M, Minobe Y and Gale M (1994b) *Bio/Technology* 12: 276-278.

Kurata N, Umehara Y, Tanohue H and Sasaki T (1997) *Plant Mol Biol*, in press.

Khush GS, Singh T, Ishii A, Parco A, Huang N, Brar DS and Multani DS (1995) *3rd Intl Rice Genet Symposium*, Abstract, p. 7.

Killian A, Kudrna DA, Kleinhofs A, Yano M, Kurata N, Steffenson B and Sasaki T (1995) *Nucleic Acids Res* 23:2729-2733.

Kleinhofs A, Killian A, Saghai-Maroof MA et al. (1994)*Theor Appl Genet* 85:705-712.

Lin SY, Nagamura Y, Kurata N, Yano M, Minobe Y and Sasaki T (1994) *Rice Genet Newset* 10:108-109.

McCouch SR and Doerge RW (1995) *Trends Genet* 11:482-487.

McCouch SR, Kochert G, Yu ZH, Wang ZY, Khush GS, Coffman WR and Tanksley SD (1988) *Theor Appl Genet* 76:815-829.

Mackill DJ, David J, Zhang Z, Redona ED and Colowit PM (1996) *Plant Genome IV*. Abstract P72, p. 44.

Mackill DJ, Salam MA, Wang ZY and Tanksley SD (1993) *Theor Appl Genet* 85:536-540.

Monna L, Miyao A, Inoue T, Fukuoka S, Yamazaki M, Zhong HS, Sasaki T and Minobe Y (1994) *DNA Res* 1:139-148.

Monna L, Miyao A, Zhong HS, Yano M, Iwamoto M, Umehara Y, Kurata N, Hayasaka H ans Sasaki T (1996) *DNA Res*, in press.

Moore G, Devos KM, Wang ZX and Gale MD (1995) *Current Biology* 5:737-739.

Nair S, Bentur JS, Parasada Rao U and Mohan M (1995) *Theor Appl Genet* 91:68-73.

Nagamura Y, Inoue T, Antonio BA, Shimano T, Kajiya H, Shomura A, Lin SY, Kuboyi Y, Harushima Y, Kurata N, Minobe Y, Yano M and Sasaki T (1995) *Breeding Science* 45:373-376.

Ohmido N, Ohtsubo H, Ohtsubo E and Fukui K (1995) *3rd Intl Rice Genet Symposium*. Abstract, p. 42.

Qian HR, Zhuang JY, Lin HX, Lu J and Zhung KL (1995) *Theor Appl Genet* 90:878-884.

Ramakrishnan W, Gupta VS and Ranjekar PK (1994) *Proc 17th Meeting Intl Program Rice Biotechnology* 17:18.

Saito A, Yano M, Kishimoto N, Nakagahra M, Yoshimura A, Saito K, Kuhara S, Ukai Y, Kawase M, Nagamine T, Yoshimura S, Ideta O, Ohsawa R, Hayano Y, Iwata N and Sugiura M (1991) *Jpn J Breeding* 41:665-670.

Sasaki T, Song J, Koga-Ban Y, Matsui E, Fang F, Higo H, Nagasaki H, Hori M, Miya M, Murayama-Kayano E, Takiguchi T, Takasuga A, Niki T, Ishimanu K, Ikeda H, Yamamoto Y, Mukai Y, Ohata I, Miyadera N, Havukkala I and Minobe Y (1994) *Plant J* 6:615-624.

Schlegel R and Melz G (1993) *Genetic Maps* 6:235-255.

Singh K, Multani DS and Khush GS (1996) *Genetics* 143:517-529.

Tanksley SD (1993) *Annu Rev Genet* 27:205-233.

Tao QZ, Zhoo H, Qiu L and Hong G (1994) *Cell Res* 4:127-133.

Tao QZ, Qian Y, Zhao S, Yu S, Qiu L, Wu B, Zhu J, Yu D, Liu X and Hong G (1995) *Cell Res* 5:263-271.

Thomas CM, Vos P, Zabeau M, Jones DA, Noroott KA, Chadwick BP and Jones JD (1995) *Plant J* 8: 785-794.

Tsumematsu H, Yoshimura S, Hasegawa H, Nagamura Y, Kurata N, Yano M, Sasaki T and Iwata N (1993) *Breeding Science* 40:279-284.

Uochimiya H, Kidou SI, Shimazaki T, Aotsuka S, Takamatsu S, Nishi R, Hashimoto H, Matsubayashi Y, Kidou N, Umeda M and Kata A (1992) *Plant J* 2:1005-1009.

Umehara Y, Inagaki A, Tanoue H, Yasukouchi Y, Nagamura Y, Saji S, Fujimura Y, Kurata N and Minobe Y(1995) *Mol Breeding* 1:79-89.

Umehara Y, Tanoue H, Kurata N, Ashikawa I, Minobe Y and Sasaki T (1996) *Genome Res*, in press.

Van Deynze AE, Nelson JC, Yglesias ES, Harrongton SE, Braga DP, McCouch SR and Sorrells ME (1995a) *Mol Gen Genet* 248:744-754.

Van Deynze AE, Nelson JC, O'Donoughue LS, Ahn SN, Siripoonwiwat M, Harrington SE, Yglesias ES Braga DP, McCouch SR and Sorrells ME (1995b) *Mol Gen Genet* 249:349-356349-356.

Van Houten W, Kurata N, Umehara Y, Sasaki T and Minobe Y (1996) *Genome* 39: in press.

Vos P, Hogers R, Bleeker M, Reijans M, van de Lee T, Hornes M, Frijters A, Pot J, Peleman J, Kuiper M and Zabeau M (1995) *Nucleic Acids Res* 23:4407-4414.

Wang G, Holsten T, Wang H and Ronald P (1995) *Plant J* 7:525-533.

Wang G, Mackill DJ, Bonman JM, McCouch SR, Champoux MC and Nelson RJ (1994) *Genetics* 136:1421-1434.

Wang ZX, Kurata N, Katayose Y and Minobe Y (1995) *Theor Appl Genet* 90:907-913.

Wang ZX, Umehara Y, Tanoue H, Saji S, Shimokawa T, Yoshino K, Idonuma A, Emoto M, Koike K, Antonio BA, Wu J, Ashikawa I, Minobe Y, Kurata N and Sasaki T (1996) *Plant Genome IV*. Abstract P26, p. 33.

Wang ZY, Second G and Tanksley SD (1992) *Theor Appl Genet* 83:565-581.

Wing RA , Rastogi VK, Zhang HB, Paterson AH and Tanksley SD (1993) *Plant J* 4:893-898.

Wu HK, Chung MC, Wu TY, Ning CN and Wu R (1991) *Chromosoma* 100:330-338.

Wu J, Matsui E, Yamamoto K, Nagamura Y, Kurata N, Sasaki T and Minobe Y (1995) *Genome* 38:1189-1200.

Wu KS and Tanksley SD (1993) *Mol Gen Genet* 241:225-235.

Xiao J, Li J, Yuan L and Tanksley SD (1995) *Genetics* 140:745-754.

Xiao J, Li J, Yuan L and Tanksley SD (1996) *Theor Appl Genet* 92:230-244.

Yamagishi M, Yano M, Fukuta Y, Fukui K, Otani M and Shimada T (1996) *Genes Genet Syst* 71:37-41.

Yamamoto K, Song J, Nagasaki H, Shinozuka Y, Hamamatsu C, Itadani H, Kojima S, Yamanouchi U, Hamada M, Okubo N, Ichimura H, Ohta I, Mukai Y, Nagamura Y and Sasaki T (1996) *Plant Genome IV*. Abstract P7, p. 28.

Yang GP, Maroof MA, Xu CG, Zhang Q and Biyashev RM (1994) *Mol Gen Genet* 245:187-194.

Yoshimura S, Katayose Y, Yamanouchi U, kono I, Toki S, Yano M, Yamamoo K, Kurata N and Sasaki T (1996) *Plant Genome IV*. Abstract P28, p. 33.

Yoshimura S, Umehara Y, Kurata N, Nagamura Y, Sasaki T, Minobe Y and Iwata N (1996) *Theor Appl Genet* 93:117-122.

Zhang HB, Zhao XP, Ping XD, Paterson AH and Wing RA (1995) *Plant J* 7:175-184.

Zhang HB, Choi S, Woo SS, Li Z and Wing RA (1996) *Mol Breeding* 2:11-24.

Zhang Q, Shen BZ, Dai XK, Mei MH, Saghai-Maroof MA and Li ZB (1994) *Proc Natl Acad Sci USA* 91:8675-8679.

Zhao X and Kochert G (1993) *Plant Mol Biol* 21:607-614.

Zhu J, Gale M and Bryan G (1996) *Plant Genome IV*. Abstract P72, p. 46.

Cultivar identification and varietal protection

Stephen Smith
Pioneer Hi-Bred International, Inc., Johnston, IA 50131-1004, USA

Introduction

Plant breeding significantly improves agricultural productivity. Genetic change has increased yield at annual rates of 0.7-2.0% in the United States (Fehr 1984). Quality attributes for pasta, dough strength, malting, amino acid content and levels of unsaturated fats have also been improved by breeding. An increasingly productive agriculture that exchanges chemical inputs with increasing reliance upon genetic resources is essential to prevent famine, malnutrition and environmental degradation. Continued investment in research and product development are essential to achieve these objectives (Plucknett and Smith 1986). Varietal description and identification can help meet these goals.

Varietal identification

Cultivar identification allows farmers and processors to be assured that varieties offered for sale are of the correct genotype and specified provenance. The fundamental importance to good husbandary of recognizing specific distinct varieties is shown by practices of indigenous people of the Amazon basin in Peru who describe varieties of manioc in great detail and require new varieties to be distinct (Boster 1985). Morphological characters have traditionally provided signatures of varietal genotype and purity. However, molecular characters that more quickly and accurately reveal genetic differences without the obscurance of environment, provide significant advantages in discrimination, reliability, timeliness and reduced cost.

Intellectual property protection

Strong intellectual property protection (IPP) is a prerequisite for private investment into plant breeding. Private investment will not occur without effective measures to prevent misappropriation. Nor will investment be maintained if competitors commercialize products that are reverse engineered or otherwise only cosmetically different from the original variety. Use of a proprietary variety to create an improved variety will also limit investment if there is no sharing of revenues to cover research and product development of the initial variety.

There have been some criticisms that IPP could reduce genetic diversity by restricting access to genetic resources for further breeding. However, evidence is to the contrary. Plant Variety Protection (PVP) maintains free public access for breeding to commercially available varieties. Pray and Knudsen (1994) found that passage of the U.S. PVP Act of 1970 "had a positive but statistically insignificant impact upon diversity" in wheat. Sneller (1994) found that "proprietary (soybean) lines had a positive impact upon genetic diversity...with genes from an exotic accession having been introduced solely through a proprietary variety". Pioneer Hi-Bred International, Inc. shut down hard red winter wheat research and product development because revenues were lost due to extensive farmer re-use of seed and "brown-bag" sales of proprietary varieties. In *Asgrow vs Winterboer* (1992), the U.S. Supreme Court ruled subsequently that farmer sale of varieties for re-seeding in excess of the amount sufficient to replant the farm upon which it was originally produced would be prohibited to protect investments into breeding improved varieties. Lack of effective IPP could stagnate agricultural productivity by causing investors to retreat from funding plant breeding.

Plant variety protection

Detailed explanations of the legal basis and requirements for IPP or Plant Breeders' Rights (PBR) can be found elsewhere (UPOV, 1992; Baenziger et al., 1993). A brief outline is as follows. The ability to maintain inbred parents of hybrids as trade secrets, supported the initiation of commercial hybrid maize breeding in the 1920's but is now less relied upon for IPP. Molecular methods make it increasingly difficult to maintain a trade secret. The most widely used means to obtain IPP is Plant Variety Protection (PVP). The Union Protection des Obtentions Végètales (UPOV) prescribes guidelines that member states utilize to set PVP laws. A set of characters with detailed parameters is established for the *de novo* description of new varieties. New varieties must be distinct, uniform and stable (DUS). Distinctness is determined by comparisons to descriptions of previously released varieties. Difference and distinctness often are not synonymous. For example, in the system currently used in France, differences for more than one simply inherited trait are required for distinctness.

Essentially derived varieties

The revised (March, 1991) UPOV treaty has provisions that can bolster innovative plant breeding. Means are provided to manage and restrict commercialization of close copies or plagiarized versions of initial varieties (IV) through the concept of an Essentially Derived Variety (EDV). Molecular profiling makes the finding and creation of minor variants from initial varieties a quick and relatively cheap prospect (Giovonnoni et al. 1991, Michelmore et al. 1991; Waycott and Fort 1994). Capabilities to isolate genes, synthesize oligonucleotides, construct vectors, affect translation and insert exotic germplasm by transformation offer additional means to rapidly extract germplasm from commercial varieties and effect changes in the genome. It would harm plant breeding if new varieties with relatively small genetic changes had the side effect of destroying the potential for originators of initial varieties to recoup investment. Genetics imparting pest resistance, for example, cannot benefit agriculture until they are partnered with germplasm of a productive variety. However, implementation of the EDV concept can do much more than defensively maintain the position of the classical breeder *vis à vis* the well equipped biotechnologist. An EDV system provides a basis for productive collaborations between breeders and suppliers of technologies and of new germplasm thereby promoting continued investment into more innovative research and product development. This would encourage the use of more genetic diversity and help decrease dangers of crop losses and genetic erosion that can be associated with a narrow base of germplasm.

Utility patents

IPP for varieties can be obtained through Utility Patents in the U.S. but not currently, at least, in the European Union. For eligibility *via* Utility Patents, a variety must not only meet the criteria of DUS but also the standards of unobviousness and utility. Seed represents the enabling disclosure and detailed pedigrees and agronomic performance data are provided (Chapman 1995, Noble 1995).

Utility Patents provide stronger incentives for investment and enhanced access to germplasm. The holder of a Utility Patent logically might issue a license allowing freedom to breed but with commercialization subject to the degree of dependency upon a protected variety. Utility Patents increase access since varieties are freely available in the public domain after expiration of protection. Twenty years of patent protection represent less than twice the period of time needed to create a new variety from elite germplasm and about the period of time needed to craft exotic germplasm into new competitive varieties. Using trade secrets, proprietary inbred lines were protected so long as they could be maintained confidentially; usually they only became available to other breeders or seed producers when they were stolen. Trade secrets, so long as they were effective, really did "lock-down" germplasm and failed to allow formalized transference or sharing of intellectual property through licensing.

Variety description

There is a subtle but important difference between cultivar identification and *de novo* description of varieties. IPP requires *de novo* description according to standards established by national or regional bodies acting upon UPOV recommendations. Utility Patent applications include similar morphological descriptions, often with additional isozymic data (Chapman 1995, Noble 1995).

In contrast, cultivar identification is the determination of identity of a previously described variety. Cultivar identification does not require detailed international standardization according to UPOV guidelines. Choice of characters for cultivar identification rests upon demonstrations of repeatable and effective discrimination, often by published procedures.

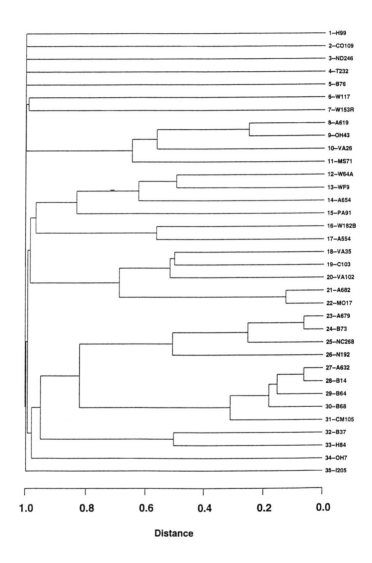

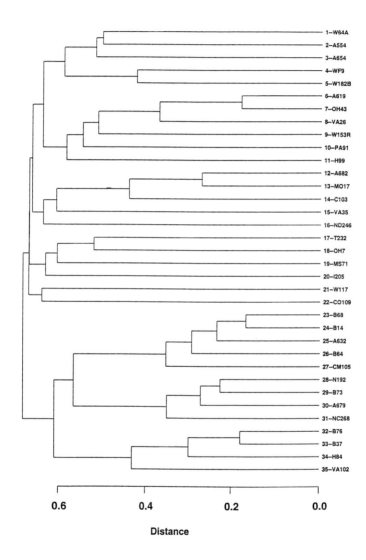

FIGURE 1. A. Associations among inbred lines of maize as revealed by pedigree distance data (1-Malecot's Coefficient of Relatedness). B. Associations among inbred lines of maize as revealed by 100 RFLP probes.

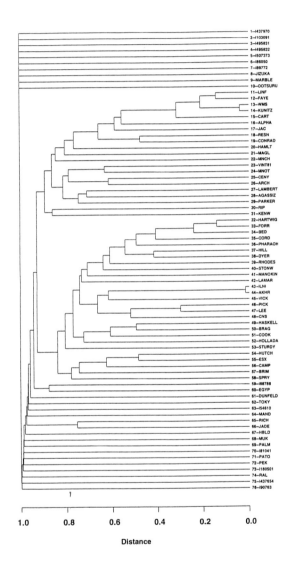

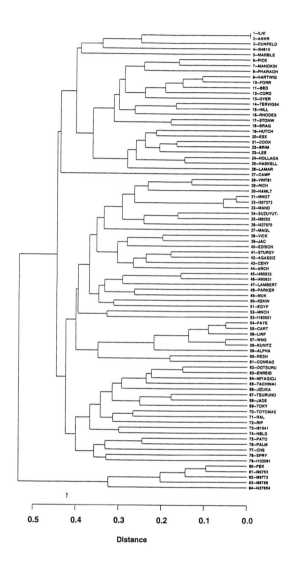

FilGURE 2. A. Associations among soybean varieties as revealed by pedigree distance data (1-Malecot's Coefficient of Relatedness). B. Associations among soybean varieties as revealed by 51 RFLP probes.

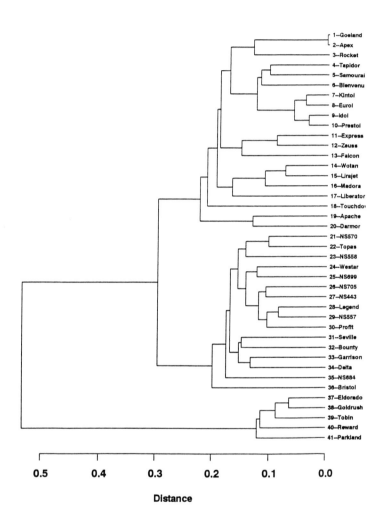

FIGURE 3. Associations among oil seed Brassica spp. revealed by 50 RFLP probes. Entries 1-20 and 38 are winter canola varieties, entries 21-36 are spring canola varieties and entries 37-41 are varieties of B. Compestris.

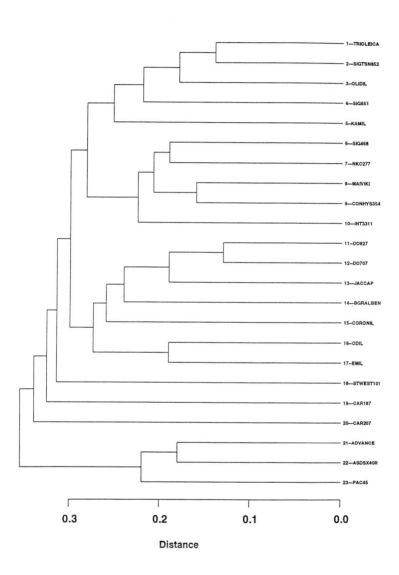

FIGURE 4. Associations among varieties of sunflower revealed by 34 RFLP probes.

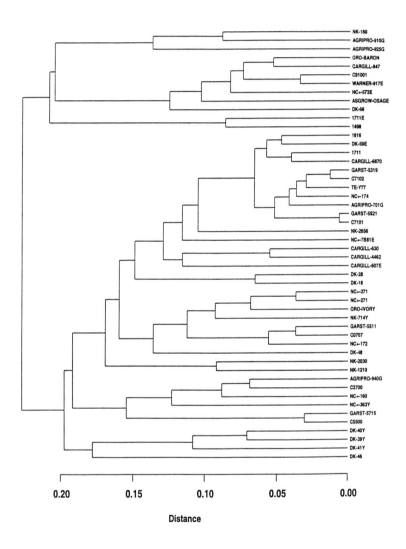

FIGURE 5. Associations among hybrid sorghum varieties revealed by 80 RFLP probes.

BRAND92 Cluster

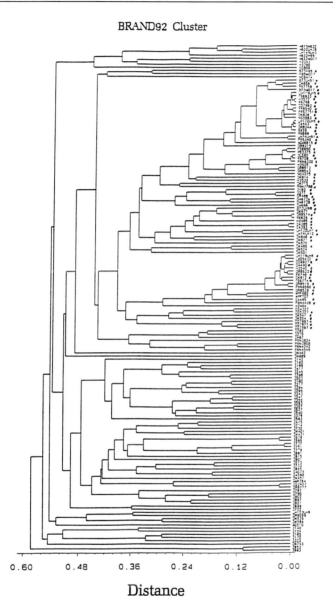

0.60	0.48	0.36	0.24	0.12	0.00

Distance

FIGURE 6. Associations among maize hybrids that are widely used in the U.S. as revealed by 60 RFLP probes. Hybrids are indicted with a dot when a profile ≥80% that of the public line B73 was found within the profile of the hybrid.

Molecular methods

Molecular methods used to identify varieties

Biochemical methods using storage proteins or isozymes have been widely used since the late 1970's (Cooke 1984, 1988, 1995, Smith and Smith 1992). These methods continue in routine testing of parentage, monitoring of genetic purity and as additional descriptors in DUS. However, these traits do not provide sufficient discrimination ability nor do they allow a sufficiently thorough sampling of the genome to provide estimates of genetic distance that can be routinely useful in helping to provide IPP or that can be fully effective in variety identification.

Consequently, technologies that utilize DNA directly and which allow more complete sampling of the genome with greater powers of discrimination, are increasingly in use. Restriction fragment length polymorphisms (RFLP) can provide tens or hundreds of mapped polymorphic loci for many species of cultivated plants (Smith and Smith 1992, Smith 1995). RFLP can be highly discriminative and, provided genomic sampling is sufficiently thorough, they allow associations to be revealed among related varieties that reflect pedigree connections (Figures 1, 2 and 3). Even in crops such as soybean, where the elite germplasm base is relatively narrow, data from just 51 RFLP markers that collectively sample each linkage group differentiated all but two varieties that were very closely related by pedigree (Figure 2). Other examples of discriminatory power are shown for varieties of sunflower and sorghum (Figures 4 and 5).

However, RFLP technology is a relatively slow process; usually requiring three months to profile 50-100 varieties for 60-100 individual probes. Consequently, procedures that use the polymerase chain reaction (PCR) process to amplify DNA are being introduced. Several widely used methods employ arbitrary primers. These include random amplified polymorphic DNA (RAPD; Williams et al. 1990), arbitrarily primed-PCR (AP-PCR; Welsh and McClelland 1990), and DNA amplification fingerprinting (DAF; Caetano-Anollés et al. 1991). One method, amplification fragment length polymorphism (AFLP; Vos et al. 1995) does not require specific sequence data but does utilize high stringency annealing conditions. Simple sequence repeat (SSR) or microsatellite analyses (Akkaya et al., 1992; Senior and Heun, 1993; Plaschke et al., 1995; Rongwen et al., 1995) that use pairs of primers with specifically designed sequences can also be very effective. Data from these methods can help resolve disputes on germplasm identity and ownership. Provided the genome is adequately surveyed, data from different methods provide very equivalent information on genomic conformity for varieties that are related by pedigree (Smith and Helentjaris 1996).

New methods that provide greater discrimination and sample throughput efficiencies are emerging rapidly. These include those capable of detecting single nucleotide differences (Nickerson et al. 1990, Baranyi 1991, Nickerson et al. 1992), genome scanning methods such as genome mismatch scanning (GMS; Nelson et al. 1993), representational difference analysis (RDA; Lisitsyn et al. 1993), restriction landmark genomic scanning (RLGS; Kawase, 1994), and non-gel based systems using oligomer hybridization (Erlich et al. 1995) or automated assays (Gu et al. 1995) that include use of silicon chips or microlaboratories (Chetverin and Kramer 1994; Grossman et al. 1994; Jacobson and Ramsey 1995). It is possible to identify varieties and specific quality attributes in mixtures of grain or flour provided appropriate sequence-based assays are available.

Molecular methods that could be considered for use in variety description

If varietal profiles using molecular data *per se* were to be used for DUS then it would be likely that genetic control would have to be established, even though this is not a prerequisite for morphology. Also, use in DUS would require international standardization in methodologies. These criteria would be challenging to achieve for most arbitrarily primed profiling methods. However, RAPD analysis provided complementary data to show distinctness (Rod Peakall, personal communication). AFLP can perhaps be generated and scored more repeatedly across laboratories than can RAPD, AP-PCR or DAF profiles. However, map positions for amplified segments of the same apparent molecular weight using AFLP technology still might differ according to germplasm background. This feature could limit the use of all arbitrary primer methods in DUS.

Profiling procedures that are more sequence specific, such as SSR analysis, may be more suitable to evaluate distinctness. SSRs are especially useful in crops such as tomato, wheat and soybean (Plaschke et al. 1995; Rongwen et al. 1995) where they reveal extensive diversity among elite varieties; diversity that has been recalcitrant using other methods. SSR data are routinely used to improve genetic purity in high value seed crops such as tomato (Mark van Grinsven, personal communication), and by the U.S. PVP Office to help evaluate distinctness in soybean (John Grace, personal communication). SSRs can provide highly discriminative, standardized and cost-effective varietal profiles without the need for radioactivity. They reveal numerous co-dominantly expressed alleles that can be readily scored in terms of mapped genetic loci.

The use of varietal identification technologies to help maintain IPP

Screening of competitor varieties

Many organizations exercise diligent vigilance in IPP by screening varieties sold by competitor breeders, seed producers, seed conditioners and farmers who market and sell more than incidental volumes of seed for replanting purposes. Varieties are profiled in order to identify any that are identical or closely similar to proprietary products. Morphological comparisons can be misleading because of environmental interactions and the largely unknown genetic basis of inheritance. Molecular data allow genetic diversity among varieties to be compared. These data are prerequisites for the examination of parental inbred lines of F_1 hybrids because the identities of parental inbreds cannot be deduced from morphological examination of the hybrid. Using molecular profiles, it is possible to estimate dependencies of hybrids upon inbred lines (Figure 6). Dependency could be from use of an inbred as a parent *per se* or as the parent of a derivative line.

The use of molecular data to help determine EDV

An EDV must retain the essential characteristics of the variety from which it is derived. Varieties of self-pollinated crops can be observed in the field. However, molecular data allow more reliable, faster and cost effective comparisons. In hybrid crops, molecular data will often be the first line of evidence that an EDV could exist. The measurement of genetic similarity or genetic conformity is one important criterion that contributes toward a determination of EDV status. Varietal profiles can be compared using molecular data from markers that sample the genome in sufficient number and thoroughness to estimate genetic conformity with acceptable precision. The availability of reliable profiling technologies such as RFLP, AFLP and SSR analyses, with published records, provide a sound technological base from which to help determine EDV status.

The American Seed Trade Association (ASTA) has developed an informal set of guidelines for resolving disputes on germplasm identity and ownership that include the use of molecular data. The ASTA is developing a standard set of maize inbred lines from which molecular, performance and pedigree data can be available to help in dispute resolution. The Association Internationale des Selectionneurs (ASSINSEL), of which ASTA is a member, has conducted model studies in tomato and maize that highlight the utility of molecular data and help resolve statistical issues. UPOV, while not mandated to determine EDV status, has fostered a greater awareness of the rapidly increasing capabilities of molecular technologies. National registration authorities and PVP Offices have supported research pertinent to the EDV issue, for example, by showing the effectiveness of SSRs in description of soybean and demonstrating that linkages between molecular marker loci should be factored into measures of genetic conformity (Dillmann et al. 1995). Determination of EDV status will usually require that other data be examined. These include breeding records, morphological comparisons, and, where appropriate, test-crosses of combining ability.

Issues concerning the potential use of molecular data in variety description

The introduction of molecular methods to characterize and define varieties for DUS is an ongoing and controversial process. There is reluctance to supplant traditional morphological characters or to add molecular traits as complementary characters. However, the development of more reliable, cost effective and discriminatory profiling methods represents an inescapable driving force for change. The characters used for DUS are now in question. The definition of distinctness is under scrutiny. The continued effectiveness of the registration process itself has been thrown into debate.

Molecular methods offer distinct advantages in discriminatory power, speed and cost effectiveness. However, there are legitimate concerns that the ease of finding minor genetic differences could erode the amount of genetic difference needed for distinctness. This might promote "breeding" of varieties with minor cosmetic changes rather than significant agronomic performance differences that would then divert resources away from productive breeding. Other negative consequences include an increase of genetic uniformity with greater potential risks of susceptibilities to pests, diseases and unforeseen weather problems and the erosion of genetic resources. Additional concerns relate to extra costs entailed in using molecular methods to evaluate uniformity and stability. Finally, some argue that the UPOV Treaty should be interpreted so that descriptions of varieties should only be based upon characters that are expressed following transcription and translation of DNA. Counter arguments are that sections of the genome, which are not themselves coded into proteins, nevertheless play important roles in regulating gene expression and affecting meiosis. For example, Furman et al. (1995) show that untranscribable portions of the genome markedly affect phenotype and introns, once termed "junk DNA" have been found to have active coding functions (Lambowitz et al. 1993, Tycowski et al. 1994, 1996). Thus, relative ignorance of the structure and function of the plant genome should not determine use of descriptors. A supporting argument is that so long as the focus is on distinctness or novelty, rather than upon agronomic performance, then it is the capability of traits to provide reliable determinations of distinctness and novelty that should determine utility. In this light, transcription is not a relevant consideration. DNA sequence data are expressed on an autoradiogram or *via* the PCR reaction even though they may not represent genes that are expressed *via* the transcription process. A sequence-based molecular polymorphism has as much import as a single gene-based morphological difference. Indeed, molecular polymorphisms have greater reliability since they are unaffected by environmental interaction. Molecular polymorphisms have additional utility for IPP since their genetic control can readily be established. This is not the case for most morphological characters. Molecular polymorphisms are increasingly more relevant for DUS because they can be interpreted genetically.

The potential negative consequences of using additional simply inherited descriptors are real concerns and must be addressed. However, rapid technological developments will make DNA polymorphisms *per se* cost effective for assessments of uniformity and stability. Also, guidelines could be envisioned that would be protective of a devaluation of distinctness (already in use for biochemical and morphological characters in the DUS system used in France). At the very least, the level of distinctness should not fall below the level of difference that is routinely found within single varieties that are bred to currently accepted standards of genetic purity.

Determination of EDV status will mostly occur in discussions between the interested parties, in arbitration, or in law courts. However, the 1994 Australian PVP Act legislates a role for the PVP examiner in the determination of EDV status (Plant Breeders' Rights Act 1994). The EDV concept moves plant breeders towards the type of protection and flexibility of integrating technologies in product development that are provided by patents. More radical change might be envisioned that could involve an elimination of registration and further movement toward a more market driven environment of varietal release. Protection by Utility Patents might become available in more countries, especially if practical implementation of the EDV concept fails to promote further investment in breeding and use of greater genetic diversity. Whatever the changes, abilities to describe and identify specific varieties and components of varieties will be increasingly important prerequisites to stimulate the increase of productivity and genetic diversity in agriculture by helping to provide strong IPP.

Conclusions

During the past twenty-five years, plant breeders have witnessed radical changes in the environs of economics, technology and IPP. World population numbers continue to grow, placing immense pressures upon breeders to help provide a more productive agriculture that can stave off famine, protect fragile land and help maintain clean air and water resources. Varietal identification technologies can "fingerprint" individual varieties for most species with increasing effectiveness and economy. These technologies can be harnessed productively to help improve breeding efficiency. They can help support continued investment into breeding by helping to provide strong IPP. They can allow the monitoring of genetic diversity in agriculture. They can promote further and more effective sourcing of genetic diversity from exotic varieties and wild species. As with all technologies, however, molecular marker profiling represents a two-edged sword. For these technologies can also be used to undermine innovation and investment by making reverse engineering and plagiaristic or cosmetic breeding of counterfeit varieties more feasible. It will be imperative that strong IPP is in place to promote the application of plant breeders skills in an environment that is technologically, intellectually and financially supportive of increased and sustainable agricultural productivity. The effective application of molecular marker technologies to variety description and identification in the support of strong IPP is crucial in order to continue improvements in global agricultural productivity.

References

Akkaya MS, Bhagwat AA and Cregan PB (1992) *Genetics* 132:1131-1139.

Asgrow *vs* Winterboer (1992) Asgrow Seed Company *vs* Denny Winterboer and Becky Winterboer, d/b/a DEEBEE'S, Case 92-1048, United State Court of Appeals for the Federal Court.

Baenziger PS, Kleese R, and Barnes RF (1993) *Intellectual property rights: Protection of plant materials*. Crop Sci. Soc. Amer., Madison, WI.

Baranyi F (1991) *Proc Natl Acad USA* 88:189-193.

Boster JS (1985) *Econ. Bot.* 39:310-325.

Caetano-Anollés G, Bassam BJ and Gresshoff PM (1991) *Bio/Technology* 9:553-557.

Chapman MA (1995) Inbred corn line PHTE4, United States Patent No. 5,453,564, United States Patent Office, Washington, DC.

Chetverin AB and Kramer FR (1994) *Bio/Technology* 12:1093-1099.

Cooke RJ (1984) *Electrophoresis* 5:59-72.

Cooke RJ (1988) *Adv Electrophoresis* 2:171-261.

Cooke RJ (1995) *J Chromatog* 698:281-299.

Dillmann C, Charcosset A, Bar-Hen A, Goffinet B, Smith IS, Battée Y and Guiard J (1995) The estimation of molecular genetic distances in maize for DUS and ED protocols: optimisation of the information and new approaches of kinship, BMT/3/6 UPOV, Geneva.

Erlich HA, Horn G, Saiki RK and Mullis KB (1995) Process for detecting specific nucleotide variation and genetic polymorphisms present in nucleic acids. United Sates Patent No. 5,468,613, United States Patent Office, Washington, DC.

Fehr WR (1984) *Genetic contribution to yield gains of five major crop plants*. Crop Sci. Soc. Amer., Madison, WI.

Furman DP, Rodin SN and Kozhemiakina TA (1995) *Theor Appl Genet* 91:1095-1100.

Giiovonnoni JJ, Wing RA, Ganal MW and Tanksley SD (1991) *Nuceic Acids Res* 19:6553-6558.

Grossman PD, Bloch W, Brinson E, Chang CC, Efferding FA, Fung S, Iovannisci DA, Woo S and Winn-Dean ES (1994) *Nuceic Acids Res* 22:4527-4534.

Gu WK, Weeden NF, Yu A and Wallace DH (1995) *Theor Appl Genet* 91:465-470.

Jacobson SC and Ramsey JM (1995) *Electrophoresis* 16:481-486.

Kawase M (1994) *Theor Appl Genet* 89:861-864.

Lambowitz AM and Belfort MA (1993) *Annu Rev Biochem* 62:587-622.

Lisitsyn N, Lisitsyn N and Wigler M (1993) *Science* 259:946-951.

Michelmore RW, Paron I and Kesseli RV (1991) *Proc Natl Acad Sci USA* 88:9828-9832.

Nelson SF, McCusker JH, Sander MA, Kee Y, Modrich P and Brown PV (1993) *Nature Genetics* 4:11-18.

Nickerson DA, Kaiser R, Lappin S, Stewart J, Hood L and Landegren U (1990) *Proc Natl Acad Sci USA* 87:8923-8927.

Nickerson DA, Whitehorst C, Boysen C, Charmley P, Kaiser R and Hood L (1992) *Genomics* 12:377-387.

Noble A (1995) Inbred Corn Line PHHB4, United States Patent No. 5,444,178, United States Patent Office, Washington, DC.

Plant Breeders' Rights Act (1994) Act No. 110 of 1994, Commonwealth Government Printer, Canberra, Australia.

Plaschke J, Ganal MW and Roder MS (1995) *Theor Appl Genet* 91:1001-1007.

Pluncknett DA and Smith NJH (1986) *Econ Bot* 40:298-309.

Pray CE and Knudsen M (1994) *Contemp Econ Policy* 12:102-112.

Rongwen J, Akkaya MS, Bhagwat AA, Lavi U and Cregan PB (1995) *Theor Appl Genet* 90:43-48.

Senior ML and Heun M (1993) *Genome* 36:884-889.

Smith JSC (1995) In: Wrigley CW (ed), *Identification of Food-Grain Varieties*. Amer. Assoc. Cereal Chem., Saint Paul, MN, pp. 131-150.

Smith S and Helentjaris T (1996) In: Paterson AH (ed) *Genome Mapping in Plants*. Landes Biomedical Press, Austin, TX, in press.

Smith JSC and Smith OS (1992) *Adv Agron* 47:85-140.

Sneller CH (1994) *Crop Sci* 34:1515-1522.

Tycowski KT, Shu M-D and Steitz JA (1994) *Science* 266:1558-1561.

Tycowski KT, Shu M-D and Steitz JA (1996) *Nature* 379:464-466.

UPOV (1992) Publication No. 221(E), International Union for the Protection of New varieties of Plants, Geneva.

Vos P, Hogers R, Bleeker M, Reigans M, van de Lee T, Hornes M, Frijteis A, Pot J, Peleman J, Kuiper M and Zabeau M (1995) *Nucleic Acids Res* 23:4407-4414.

Waycott W and Fort SB (1994) *Genome* 37:577-583.

Welsch J and McClelland M (1990) *Nucleic Acids Res* 18:7213-7218.

Williams JGK, Kubelic AR, Livak KJ, Rafalski JA and Tingey SV (1990) *Nuceic. Acids Res* 18:6531-6535.

The status of DNA fingerprinting: population databases

Ranajit Chakraborty
Human Genetic Center, School of Public Health, The University of Texas Houston Health Science Ceneter, Houston, TX 77225, USA

Introduction

DNA typing methods have been widely used in criminal investigations and trials for over ten years in various countries of the world. DNA extracted from evidence samples can be compared with those collected from the victim and/or suspect(s) by a wide variety of methods, all of which are based on sound scientific principles (Pena et al. 1993). Nevertheless, in adversarial judicial systems, their general acceptance as well as validity of the statistical interpretation of DNA evidence are discussed in courts on a regular basis. As the forensic community in the United States started to supplement the Southern blot restriction fragment length polymorphism (RFLP) method of DNA typing with the more recent PCR-based ones, such as reverse dot blot method using sequence-specific oligonucleotide probes and short tandem repeat (STR) typing, it became necessary to demonstrate that the population genetic principles apply equally well to forensic databases based on PCR-based analyses as well as they did for the RFLP databases. The objective of this contribution is to discuss several broad issues with regard to the comparison of forensic databases based on RFLP and PCR-based typing methods, in order to assess the current status of DNA typing databases on which the statistical basis of DNA evidence in courts can be presented. In particular, based on databases gathered through the efforts of the Technical Working Group of DNA Analysis Methods (TWGDAM) and through other ongoing efforts, I discuss how well the population databases at the polymarker loci (LDLR, GYPA, GC, HBGG, D7S8, and HLA-DQA1), the minisatellite locus D1S80, and the six STR loci (CSF1R, THO1, PLA2A1, F13A1, CYP19, and LPL) satisfy the assumptions of allelic independence within and across loci, and how the genetic variation at these loci within and between populations compare with the same based on RFLP analysis of the VNTR loci (D1S7, D2S44, D4S139, D5S110, D10S28, and D17S79). At the outset I must state that in the title I use the phrase "DNA fingerprinting", but the readers should recall that this term was coined for multilocus

DNA profiles detected by using multilocus hypervariable probes, which simultaneously score genotypes at several loci without recognizing the allelic relationships of the individual DNA fragments (Jeffreys et al. 1985). For the purpose of discussing the current status of population databases, the use of the original phrase is still appropriate because now there are several databases for which each individual has been scored by up to 19 loci (all 6 minisatellite RFLP loci, 5 polymarkers, HLA-DQA1, D1S80, and the 6 STR markers mentioned above), so that even though the genotyping was done by examining each of these loci individually, the combined multilocus profile approximates the DNA fingerprint of an individual quite adequately.

In the paragraphs below, I have chosen to present only brief summary conclusions of the findings of analyses of current databases, because the details of databases and the technical aspects of the methods of analyses are quite broad and varied, and all of these are available in the literature (citations given in their specific contexts below). The purpose is to show that, in general, the forensic databases on the PCR-based DNA markers satisfy the assumptions of allelic independence within and among loci. Occasional departures from these assumptions are either due to chance coincidence of studying multiple samples, or due to the presence of rare alleles within specific populations. Furthermore, I conclude that the inter-population variation at all PCR-based loci constitutes a small fraction of the total variation at all loci. Hence, when populations are defined by their broad racial and ethnic description, the use of Hardy-Weinberg and linkage equilibria assumptions provide a reasonable and adequate estimate of multilocus genotype frequencies. In this process I also give estimates of the locus-specific values of F_{ST} (albeit, coefficient of gene diversity, G_{ST}) that may be used when the effect of population substructure has to be taken into account in providing estimates of the frequency of a multilocus DNA profile, according to the recommendation of the recent NRC (1996) report. To guard against over-emphasizing the rareness of an infrequently found genotype, I also provide a prescription for estimating a minimum allele frequency for any locus-population combination, which yields a conservative estimate of the frequency of a rare multilocus DNA profile. Finally, the question of validity of forensic databases is addressed by the use of standard clustering and dendrogram analysis. These studies demonstrate that the forensic databases are representative of their respective anthropologically affine groups, even though the individuals sampled were not collected by a strictly defined random sampling method, and the ethnicity of the persons were not determined based on detailed pedigree information of the subjects sampled.

Materials and methods

The status of the current methods of DNA typing applied in forensic context will not be discussed. It is enough to mention that these techniques (RFLP, as well as the PCR-based ones) generate a machine-readable computerized database consisting of individuals' DNA profiles at one or more RFLP- and/or PCR-based DNA markers. Generally, the individuals' identity is not recorded in the database; the databases are labeled by the laboratory of origin and they are grouped by major racial and ethnic origin of the subjects sampled. However, in all cases it is determined and verified that the laboratory followed the guidelines of product manufacturers when genotyping each locus (as in the case of polymarker and D1S80 loci), or followed the scientifically accepted protocols of DNA co-amplification for typing the STR loci. Technical details of these protocols are given along with their citations elsewhere [see examples for polymarkers (Budowle et al. 1995, Hayes et al. 1995, Woo and Budowle 1995, Scholl et al. 1996), D1S80 (Budowle et al. 1991a, Deka et al. 1994, Budowle et al. 1995), and short tandem repeat loci (Edwards et al. 1992, Hammond et al. 1994, Deka et al. 1995)]. Each database analyzed satisfied the minimum sample size criteria with regard to the number of individuals typed for each locus (Chakraborty 1992, Evett and Gill 1991).

The objective of compiling such databases was to conduct a comprehensive population genetic analysis of forensic databases, encompassing different technologies of DNA typing methods (e.g. Southern blot RFLP analysis of VNTR loci, reverse dot blot analysis of polymarker loci, and HLA-DQA1, and PCR-based analysis of STR loci and D1S80). The technical details of principles and computational methods used in this analysis are given elsewhere (Yasuda 1968, Weir 1990, Chakraborty et al. 1993a, 1994, Budowle et al. 1996). Apart from using standard statistical tests for judging significance of the estimates of population genetic parameters, these analyses also addressed specific biological issues of laboratory methods used in the preparation of forensic databases (such as coalescence and nondetectability of alleles in Southern blot RFLP analysis, and effects of binning of RFLP alleles). In addition, statistical concerns of large number of alleles and under-representing frequencies of rare alleles are also addressed through empirical methods (Chakraborty 1992, Budowle et al. 1996).

Results and discussion

Database size and sampling of individuals for database preparation

In population genetics a population is a group defined by the anthropological and ancestral identity of individuals. Thus, a common concern against a forensic database is that a group of individuals defined as, say, Alabama Blacks, does not describe a population precisely enough to apply population genetic principles for genotype frequency computations. Since in forensic databases an individual's racial/ethnic identity is recorded based on self-declaration and/or physical description of the subjects, it is argued that the sample may occasionally contain individuals who might truly be of a different ethnicity. Furthermore, concerns were also raised in court proceedings about the insufficient size of databases on the basis of which the rarity of specific DNA profiles in a population is judged. Such criticisms have been addressed through three distinct, but related, procedures.

First, note that the forensic approach of 'convenient sampling' based on subject's own assessment of ethnicity is not unique to forensic databases. From its inception, most human population genetic surveys used the same procedure, and data compiled in classic texts of human genetics (e.g., Mourant et al. 1976, Till et al. 1983, Cavalli-Sforza et al. 1994) on various blood groups and protein variation in human populations are collected by essentially the same sampling method. Current knowledge of gene diversity in human populations and its decomposition into within and between race variation is principally based on such information. Furthermore, a National Health Survey (1980) conducted in

US showed that phenotype frequencies at several blood group and protein markers, estimated from such convenient sampling methods, agree convincingly well with those estimated from a strict stratified random sampling of individuals, when the individuals are classified according to their broad racial background.

Second, through a genetic distance and cluster analysis of allele frequencies, it was shown that the forensic databases gathered through the TWGDAM efforts form distinctive groups represented by broad racial and ethnic classification of the samples Chakraborty et al. 1995). Thus, even if the self-assessment of ethnicity of subjects is sometimes error-prone, they do not grossly misrepresent allele frequencies of populations, broadly defined by race and ethnicity. Furthermore, these analyses also showed that the genetic proximity of the forensic database samples perfectly coincides with the known ethnohistory of human populations. Comparison of samples identified by racial and ethnic classes with those defined by more rigid anthropological criteria indicated that the Caucasian, Afro-American, Hispanic, and American Indian samples in the forensic DNA typing databases are appropriate representative of their respective anthropological affine groups.

Third, critiques have often failed to note that the minimum sample size requirement for forensic databases has been studies at least from two different considerations (Chakraborty 1992, Evett and Gill 1991). It has been argued that for the hypervariable VNTR probes, a sample size of 100 to 150 individuals per population per locus is a sufficient size for any forensic database (Chakraborty 1992). For the polymarker loci where the number of segregating alleles range between two to six, the sample size requirement can be much lower. Nei's estimate (Nei 1978) of 50 to 100 alleles sampled (i.e., 25 to 50 individuals/locus/population) would suffice for such loci. For the STR loci, where the number of alleles is in between those of the polymarkers and VNTRs, a conservative sample size estimate is 100 individuals/locus/population. While these estimates of the minimum sample size requirement are based on theoretical population genetic considerations, the above prescribed sample size agrees with the recommendations given by an empirical procedure suggested by Evett and Gill (1991). Furthermore, statistical power calculations have been done to ensure that databases of above sizes are sufficient to detect any gross discrepancies from assumptions of allelic independence within and between loci used in forensic calculations (Chakraborty et al. 1993a, Chakraborty and Zhong 1994).

Furthermore, these analyses show that the sizes as well as diversity of the current DNA typing databases on VNTR, polymarker, and STR loci are adequate, appropriate, and valid for providing statistical estimates of frequency of any specific DNA profile based on the loci covered in the databases. The randomness of sampled DNA profiles are ensured as long as the individuals included in databases are chosen without any prior knowledge of genotypes, and inclusion of relatives in databases does not systematically bias the allele frequency estimates derived as long as the loci studied are not affected by fertility and viability of individuals. Of course, care should be exercised to identity and exclude relatives from databases, wherever possible, since they have impacts on several other statistical properties of databases (e.g. extent of partial or full multilocus matches are influenced by the presence of duplicates as well as relatives in databases).

Allelic independence within and among loci

Most loci used for forensic DNA profiling are chosen from regions of the human genome which do not code for genes, nor do they have any known function. Thus, the independent transmission of alleles from parents to children without any bias of segregation of alleles is a reasonable assumption for these loci. Furthermore, the VNTR, STR, as well as the polymarker loci used in forensic DNA analyses are located either on different chromosomes, or they are sufficiently far on any single chromosome (e.g., D1S80 and D1S7, or, GYPA and GC), so that the allelic independence across loci should also apply. In spite of these biological considerations, it can be argued that the presence of population substructure within a forensic database may violate the assumptions of allelic independence within loci–Hardy-Weinberg expectation rule and allelic independence across loci–linkage equilibrium rule. Thus, examination of adherence to both of these assumptions was the main thrust of all recent analyses of DNA typing databases.

Analyses of all data, , based on both RFLP and major racial and ethnic categories gathered thus far, indicate no evidence of gross departure from these assumptions (Hayes et al. 1995, Woo and Budowle 1995, Scholl et al. 1996, Budowle et al. 1991a, 1995, Deka et al. 1994, Edwards et al. 1992, Hammond et al. 1994, Chakraborty et al. 1993a, 1994). Occasionally, for some regional samples at a specific locus, sporadic statistically significant departures were observed (Rivas et al. 1995). These are either contributed by the presence of very rare alleles in the databases or are due to chance coincidence alone (an artifact of multiple testing). Such sporadic deviations are not critical for forensic evaluation of judging the rareness of any specific DNA profile.

This may be illustrated through an observation made during the analysis of worldwide variations of allele and genotype frequencies at the HLA-DQA1 locus, as conducted by Rivas et al. (1995). Examination of genotype frequency data from over 170 populations in this study revealed that 12 samples exhibited statistically significant (at 5% level) departures of observed genotype frequencies from their respective Hardy-Weinberg equilibrium (HWE) expectations. The survey designs in these specific populations clearly identified the samples as of mixed origin (e.g. Indians from England, etc.), and hence, departures from HWE in such samples were not unexpected. However, when the observed genotype frequencies of all 21 genotypes at this locus (with alleles designated as 1.1, 1.2, 1.3, 2, 3, and 4) were plotted against their HWE expectations, an interesting observation emerged. This is shown in Figure 1, where the 252 (i.e. 12×21) datapoints are shown for all 21 genotypes in these 12 populations. The graph also shows a straight line indicating equality of observed and expected, along with the 95% confidence limit of observed proportions (computed by the binomial theory). Of the 252 datapoints (deliberately taken from population samples which showed significant violation of HWE expectations), only 3 values lie outside the 95% confidence intervals in the direction where the use of expected genotype frequencies would have significantly underestimated the frequencies. Of course, there are many more (over 24) datapoints where the expected is significantly larger than observed. Thus, this observation alone indicate that a statistically significant failure to conform with HWE expectations is, by itself, not an indication of the inappropriateness of the use of HWE in forensic databases. The direction of bias, as well as for which genotypes they occur, must be examined. Furthermore, there are certain measures in the forensic protocol for computing profile frequency (by rebinning data, or by invoking a threshold for a minimum allele frequency) which guard against reporting very small genotype frequencies, which are generally the cause of departures from HWE.

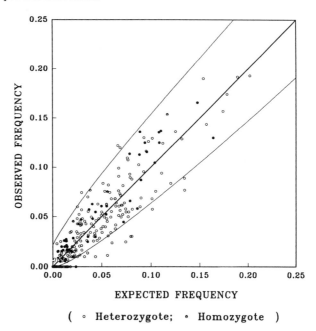

EXPECTED FREQUENCY

(∘ Heterozygote; • Homozygote)

FIGURE 1. Observed versus expected (HWE) frequencies of 21 genotypes at the HLA-DQA1 locus in 12 populations which showed an overall statistically significant deviation from HWE. The straight line represents the equality of observed and expected frequencies, and the two curved lines are 95% confidence limits of observed proportions (based on binomial theory). The open circles represent frequencies of heterozygotes and the closed circles are those of homozygotes.

In the process of testing allelic independence within and between loci, the effect of coalescence (Devlin et al. 1991) and nondetectability of alleles on allele frequency estimates should also be recognized and characterized. This is so, because these factors have substantial impact on the inference regarding independence of alleles, unless appropriate corrective measures are taken. The major conclusions from analyses which deal with this issue are:

(i) Specifically for VNTR loci scored by the RFLP protocol, allelic coalescence and nondetectability of alleles are real and for some loci up to 5% small-size alleles may remain undetected in *Hae*III-digested DNA profiles. Thus, not all single-banded DNA profiles are truly homozygous; any analysis that does not incorporate these facts may yield misleading conclusions regarding lack of allelic independence (such as the ones claimed in: Geisser and Johnson 1992, 1993, Slimowitz and Cohen 1993).

(ii) In the presence of nondetectable alleles, when allele frequencies are estimated by the gene count method (as done in the forensic databases by treating all single-banded phenotypes as true homozygotes), all allele frequencies are overestimated (Chakraborty et al. 1992), as a consequence of which DNA profile frequencies obtained from such allele frequency estimates become conservative (i.e. represented as more common than what they truly are).

(iii) The proportion of apparent heterozygote deficiency is largely explained by the extent of nondetectable alleles, and these are not due to hidden population substructure within forensic databases (Chakraborty et al. 1992, Chakraborty and Jin 1992, Jin and Chakraborty 1995).

Extent of population structure in forensic databases

The overall conclusion of allelic independence within and across loci is not, however, equivalent to the suggestion that population substructure does not exist within forensic databases. The studies relating to the extent of population substructure within forensic databases lead to the following conclusions: (i) the largest component of genetic variation at all loci is due to inter-individual variation within populations and the between-population variation is mostly due to racial classification of populations, making ethnic differences within any major racial group the smallest component of gene diversity (Chakraborty et al. 1995); (ii) for VNTR loci, the later two components of genetic variation account for only 1.4% (0.8% between race, and 0.6% between ethnic groups within a race) of the total variation; (iii) for the polymarkers, of the 6.1% variation ascribed to population difference, 5.4% is due to racial differences; and (iv) for STRs, of the total F_{ST} of 6.4%, 6.0% is explained by inter-racial differences of allele frequencies.

These estimates are consistent with the ones obtained from a re-analysis (Chakraborty 1993) of the U. S. National Health Survey data on eight blood group and protein loci. Also, these qualitatively justify the inference of allelic independence tests; namely, when the forensic databases are categorized by major racial/ethnic groups (such as Caucasians, Africans, Asians, Hispanics, and American Indians), almost all of the existing effects of population substructure within human populations are already taken care of by recognizing that these groups are to be treated separately for profile frequency computations. Within each of these groups, further split of populations does not have any appreciable effect on profile frequencies. These analyses also revealed that the intra-race F_{ST} is not uniform. The tribal differences within the American Indians are larger than the differences between the European populations of Caucasians. Nevertheless, when a range of values of any multilocus DNA profile frequency is given by the above-mentioned major grouping of populations, the effect of inter-tribal differences within the American Indians can be easily inferred.

Also, the extent of inter-population differences of allele frequencies, as measured by F_{ST} or G_{ST} is locus-dependent. Loci with larger levels of polymorphism (e.g. D1S7) tend to exhibit smaller G_{ST} compared with the ones with lesser variability (e.g. any of the polymarker loci). Since, in general, the frequency of a genotype at a locus is inversely related with the level of polymorphism, this last observation implies that the loci more efficient for human identification are lesser affected by the presence of population substructure.

Sample size effect on DNA profile frequency estimates and a threshold for minimum allele frequency

Estimates for any DNA profile frequency are intended to determine its expected frequency in a general population, and thus, the limitation of the sample size of a database should be reflected along with the point estimate provided. The use of multiple databases, defined by broad racial and ethnic classification, does not fully accomplish this. A few misleading suggestions for providing sampling variation of estimates of DNA profile frequencies, resulting from sampling effects, have confused several courts (*see* e.g., NRC, 1992). The suggestion, such as the "ceiling principle", or any of its analogues, has no statistical and/or genetic foundation. To rectify this, a formal procedure of estimating confidence intervals for any specific multilocus DNA profile frequencies has been given by Chakraborty et al. (1993b). This study indicated that for any typical multilocus DNA profile, the width of the confidence interval depends upon the number of loci typed, along with the frequencies of alleles observed in the profile. For example, any one-locus profile will have a smaller range of variation due to the limited size of the database used, while for a five-locus profile the lower and upper bounds of the confidence interval may differ by several-folds. For example, an estimate of 1 in a million (for the frequency of a 5-6 locus profile) may indeed vary between 1 in 300,000 to 1 in 3 million (*see* Chakraborty et al. 1993b for numerical illustrations). Such variations are real, and have no forensic relevance. In spite of such apparent gross fluctuations of frequency estimates, one can safely assert that the specific profile is quite rare in the population to have matched by chance alone with the evidence collected (or, the suspect typed).

Most of the DNA loci typed for forensic purposes consist of alleles that are occasionally very rare in selected populations. Since collecting databases of extremely large size is not cost-effective to precisely estimate their frequencies, a concept of minimum allele frequency is suggested which takes into account the size of the database and the extent of genetic variation at the locus (Budowle et al. 1996). This is more formal than the concept of rebinning for RFLP loci (Budowle et al. 1991b). Determining the threshold of a minimum allele frequency is needed for loci where the concept of binning is not directly relevant (e.g., for polymarkers, or for STRs), yet compensation should be made for sparse sampling of infrequent alleles in population databases. Application of the suggested algorithm to STR and polymarker databases indicates for the D1S80 locus, allele frequencies as low as 0.4% are permissible for some populations, while for the STR loci, the minimum allele frequency can be as low as 2% (Budowle et al. 1996). In general, this suggested method for determining a minimum allele frequency is more conservative than the one dictated only by the sample size of the database, and not by the extent of polymorphism at the locus (Weir 1992).

Future trends and epilogue

In summary, the above observations suggest that like the RFLP databases on VNTR loci, the forensic databases on the polymarkers and STR loci, scored by PCR-based techniques, are of adequate sample size; they are valid and reliable for applying the standard population genetic principles for forensic evaluation; and they generally meet the assumptions of allelic independence within and between loci. The extent of population substructure existing within human populations is small (not more than 7%, between races), and it is even smaller among ethnic groups within major races. These conclusions are in sharp contrast with the premises with which recommendations in the first NRC report (1992) were made, and it is reassuring that the new NRC report (1996) relied heavily on the above-mentioned findings to negate those premises and endorse the use of current RFLP and PCR-based databases for forensic applications. The new report, in its recommendation, also made use of theoretical results (e.g., the method of evaluating confidence intervals, and assessing the extent of population structure effect) and empirical findings (e.g. locus-specific F_{ST} values for VNTR, STR, and polymarker loci). Thus, the summary findings mentioned above have impacted considerably in the re-evaluation of forensic DNA evidence by the National Research Council (NRC 1996).

However, we should be cognizant of future needs for several reasons. First, PCR-based databases do not at present encompass all major populations of the world. Collection of additional databases, particularly for regional populations within as well as outside the U.S., is needed. Second, uniform allelic nomenclature for STR loci should be established for inter-laboratory comparison of data. Allele ladders, as in the case of the D1S80 locus, should be constructed for all STR loci, and be shared by laboratories for uniformity of data comparisons. Third, more polymorphic STR loci should be validated for forensic applications, particularly the ones that specify male DNA. Several Y-chromosome specific STR loci are now known, and they should be implemented in forensics. Introduction of male-specific polymorphic markers will make the analysis of mixed DNA evidence much simpler than the one being practiced now with autosomal markers.

The results described above strengthen the need for future research in these directions. With implemented research, we should soon have a battery of DNA markers that will, with little doubt, be able to exclude any innocent accused individual, or be able to provide evidence "beyond a reasonable doubt" that the suspect is linked to the evidence sample.

Acknowledgements

The findings reported here are from research supported by US Public Health Service Research Grants GM41399, GM45861, and GM58545, and Grants 92-IJ-CX-K024 and 5-7783-TX-IJ from the US National Institute of Justice. The opinions expressed are those of the author, and they do not represent endorsement of the granting agencies.

References

Budowle B, Chakraborty R, Giusti AM, Eisenberg AJ and Allen RC (1991a) *Am J Hum Genet* 48:137-144.

Budowle B, Monson KL, Anoe KS, Baechtel DL, Bergman DL, Buel E, Campbell PA, et al. (1991b) *Crime Lab Digest* 18:9-26.

Budowle B, Lindsey JA, DeCou JA, Koons BW, Giusti AM and Comey CT (1995a) *J Forensic Sci* 40:45-54.

Budowle B, Baechtel FS, Smerick JB, Preseley KW, Giusti AM, Parsons G, Alevy MC and Chakraborty R (1995b) *J Forensic Sci* 40:38-44.

Budowle B, Monson KL and Chakraborty R (1996) *Intl J Legal Med* 108:173-176.

Cavalli-Sforza LL, Menozzi P and Piazza A (1994) *The History and Geography of Human Genes.* Princeton University Press, Princeton.

Chakraborty R (1992) *Hum Biol* 64:141-159.

Chakraborty R (1993) In: Majumder PP (ed.), *Human Population Genetics.* Plenum Press, New York. pp. 189-206.

Chakraborty R and Jin L (1992) *Hum Genet* 88:267-272.

Chakraborty R and Zhong Y (1994) *Hum Hered* 44:1-9.

Chakraborty R, de Andrade M, Daiger SP and Budowle B (1992) *Ann Hum Genet* 56:45-57.

Chakraborty R, Srinivasan MR and de Andrade M (1993a) *Genetics* 133:411-419.

Chakraborty R, Srinivasan MR and Daiger SP (1993b) *Am J Hum Genet* 52:60-70.

Chakraborty R, Zhong Y, Jin L and Budowle B (1994) *Am J Hum Genet* 55:391-401.

Chakraborty R, Jin L, Zhong Y and Deka R (1995) Intra- and inter-population variation of VNTR, short tandem repeat and polymarker loci and their implication in forensic and paternity analysis. Fifth International Symposium on Human Identification, Promega Corp., Madison, Wisconsin, pp. 29-41.

Deka R, DeCroo S, Jin L, Rothhammer F, McGarvey S, Ferrell RE and Chakraborty R (1994) *Hum Genet* 94:252-258.

Deka R, Shriver MD, Yu LM, Ferrell RE and Chakraborty R (1995) *Electrophoresis* 16:1659-1664.

Devlin B, Risch N and Roeder K (1990) *Science* 249:1416-1420.

Edwards S, Hammond H, Jin L, Caskey CT and Chakraborty R (1992) *Genomics* 12:241-253.

Evett IW and Gill P (1991) *Electrophoresis* 12:226-230.

Geisser S and Johnson W (1992) *Am J Hum Genet* 51:1084-1088.

Geisser S and Johnson W (1993) *Am J Hum Genet* 53:1103-1106.

Hammond HA, Jin L, Zhong Y, Caskey CT and Chakraborty R (1994) *Am J Hum Genet* 55:175-189.

Hayes JM, Budowle B and Freund M (1995) *J Forensic Sci* 40:888-892.

Jeffreys AJ, Wilson V and Thein SL (1985) *Nature* 314:67-73.

Jin L and Chakraborty R (1995) *Heredity* 74:274-285.

Mourant SE, Kopec AC and Domaniewska-Sobczak K (1976) *The Distribution of the Human Blood Groups and Other Polymorphisms.* Oxford Univ. Press, London.

National Health Survey (1980) *Selected Genetic Markers of Blood and Secretions.* US Department of Health, Education, and Welfare, Washington DC. Publ. No. (PHS) 80-1664.

Nei M (1978) *Genetics* 89:583-590.

NRC (1992) *DNA Technology in Forensic Science.* National Academy Press, Washington DC.

NRC (1996) *The Evaluation of Forensic DNA Evidence*. National Academy Press, Washington DC.

Pena SDJ, Chakraborty R, Epplen J and Jeffreys AJ (1993) *DNA Fingerprinting: State of the Science*. Birhauser, Basel.

Rivas F, Cerda Flores R, Zhong Y and Chakraborty R (1995) *Am J Hum Genet* 55:A163.

Scholl S, Budowle B, Radecki K and Salvo M (1996) *J Forensic Sci* 41:47-51.

Slimowitz JR and Cohen JE (1993) *Am J Hum Genet* 53:314-323.

Tills D, Kopec AC and Tills R (1983) *The Distribution of the Human Blood Groups and Other Polymorphisms*. Supplement 1. Oxford Univ. Press, London.

Weir BS (1990) *Genetic Data Analysis*. Sinauer, Sunderland, MA.

Weir BS (1992) *Genetics* 130:873-887.

Woo MW and Budowle B (1995) *J Forensic Sci* 40:645- 648.

Yasuda N (1968) *Biometrics* 24:915-936.

Arbitrarily amplified DNA in ecology and evolution

Bernd Schierwater[1,2], Andrea Ender[1], Werner Schroth[1], Hans Holzmann[3], Axel Diez[3], Bruno Streit[1] and Heike Hadrys[1,2]

J. W. Goethe-Universität, Abt. Ökologie und Evolution, D-60054 Frankfurt, Germany[1], Yale University, Department of Biology, New Haven, CT 06511, USA[2], and Klinikum J. W. Goethe-Universität, Dermatologie, D-60590 Frankfurt, Germany[3].

Introduction

The PCR amplification of arbitrary DNA sequences has proven a tremendous success in many fields of evolutionary and ecological research (e.g. Avise 1994). The ease of obtaining arbitrarily amplified DNA markers (AAD; Schierwater 1995) *ad hoc* from any genome of interest is one main reason for the steadily increasing number of studies, which take advantage of AAD technologies. Several techniques for the amplification of AAD markers have been described, which differ mainly by name, technical details, and the number of markers generated (and visualized) per PCR reaction [AP-PCR (Welsh and McClelland 1990), RAPD (Williams et al. 1990), DAF (Caetano-Anollés et al. 1991), AFLP (Vos et al. 1995); for review see Schierwater et al. 1994a]. The dominant feature of these techniques is that they do not require any sequence information of the genome under investigation. Instead, AAD techniques define PCR priming conditions in which a single short arbitrary primer finds a large number of priming sites along any statistically long DNA template, and among the large number of primed PCR products a small fraction will get amplified in an exponential fashion (cf. Figure 9).

Amplification products can be separated according to length by means of gel electrophoresis and made visible as a multi-locus DNA fingerprint of the template DNA. Comparing such fingerprints from different genomes usually reveals monomorphic as well as polymorphic bands with the ratio of polymorphic bands increasing with the degree of dissimilarity between the genomes. Thus the use of AAD markers offers immediate means (i) at the *intra*specific level to population and conservation genetics, to analyze genealogical relationships (e.g. paternity assignment) and to genome mapping, and (ii) at the *inter*specific level to identify diagnostic markers for systematic units and to provide cladistic characters for inferring phylogenetic relationships (e.g. Caetano-Anollés et al. 1993, Hadrys et al. 1993, Huff et al. 1993, Puterka et al. 1993, Rieseberg et al. 1993, Gibbs et al. 1994, Smith et al. 1994, Dawson et al. 1995). In the following we will present examples for case studies to illustrate the principal potential of AAD markers in all of the above areas. Furthermore, we will briefly review a new approach in which AAD markers serve as a rapid means to identify polymorphic microsatellite loci (randomly amplified microsatellites). The type of AAD markers we have been using in the following examples is randomly amplified polymorphic DNA (RAPD), but in principle any other AAD technique could have also been employed.

Randomly amplified microsatellites

Many ecological and evolutionary studies require the use of Mendelian inherited markers to determine genetic variation, both within and between populations (for review see Avise 1994). Ideally, the markers should be highly variable and codominant to allow estimations of allelic and genotype frequencies. One class of genetic markers, "simple sequence repeats" or "microsatellites", remarkably meets these criteria. Microsatellites consist of tandemly repeated sequence motifs with a core sequence of one to six nucleotides. They are abundant, highly polymorphic, and common throughout eukaryotic genomes (Tautz 1989). Microsatellites have been used, for example, to analyze mating systems, population substructures, kinship or social structures in a variety of systems (e.g., whales: Amos et al. 1993; social wasps: Queller et al. 1993; snails: Jarne et al. 1994; chimpanzees: Morin et al. 1994; hairy-nosed wombats: Taylor et al. 1994; bears: Paetkau et al. 1995; swallows: Primmer et al. 1995). The major drawback of applying microsatellites to case studies, however, is the means to identify, clone and sequence loci for any new system under investigation (usually a given locus will work for PCR typing only within closely related species). Microsatellites are normally detected through the construction and screening of genomic libraries, which can be labor- and time-intensive (e.g., Beckmann and Weber 1992, Hughes and Queller 1993, Ashley and Dow 1994).

We have recently introduced a new strategy for identifying microsatellite loci, which is comparatively fast and neither requires prior sequence information nor large amounts of DNA (Ender et al. 1996). This strategy completely circumvents the non-trivial procedure of constructing and screening genomic libraries. Instead, arbitrarily amplified DNA is used to generate genomic DNA fragments for microsatellite screening. RAPD fragments (Williams et al. 1990, Hadrys et al. 1992) harboring microsatellite loci are detected by Southern hybridization to synthetic oligonucleotides. This way one can identify and characterize randomly amplified microsatellites (RAMS; Ender et al. 1996, *see* Chapter 3).

The principal steps of the RAMS protocol for the identification of microsatellite loci are shown in Figure 1. *Step 1*: Total genomic DNA is extracted by means of a standard protocol (e.g., Sambrook et al. 1989). *Step 2*: RAPD fingerprinting with arbitrary primers is performed as described in Schierwater and Ender (1993). *Step 3*: Amplification profiles are blotted onto Nylon membranes using alkaline transfer (Sambrook et al. 1989). *Step 4*: During hybridization analyses digoxigenin-labeled repeat sequences serve as probes to identify amplified products containing microsatellite sequences. *Step 5*: Positive DNA fragments are gel-purified, cloned and sequenced. *Step 6*: Finally, primers can be designed from conserved flanking regions of the repeats for use in PCR typing.

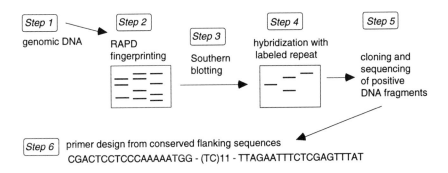

FIGURE 1. Strategy for the identification of randomly amplified microsatellites. Explanations are given in the text. For further details see Ender et al. (1996).

In the following example, we applied the RAMS approach to small waterfleas of the genus *Daphnia* to search for GA and GT repeat sequences. Amplification profiles for several *Daphnia* clones derived from 20 different arbitrary primers for GA screening and 13 primers for GT screening, respectively (Table 1). On average, 13.8 ± 0.68 (mean ± SE) bands per primer and genome were generated, summing up to a total length of 9.3 ± 0.47 kb per primer. With comparatively little effort we amplified a large number of AAD markers and obtained a high yield of positive amplification products (Table 1). The efficiency of the RAMS approach obviously depends on the abundance and type of simple sequence repeats within the study organism. The analyzed microsatellite loci from *Daphnia* show a remarkably high conservation of the flanking sequences between related species. Thus, primers developed for one species can be used in related *Daphnia* taxa (cf. Pépin et al. 1995, Coltman et al. 1996).

One important advantage of the RAMS protocol is that the cloning of positive amplification products is done after the presence of microsatellites has been verified by Southern analysis. Thus, sequencing of different plasmid clones may immediately detect variability within or between genotypes. Another advantage is that the initial step is PCR-based, i.e. it requires only very small amounts of DNA of the study organism. Therefore, RAMS are immediately applicable to small organisms and to endangered species allowing non-destructive sampling methods. The approach provides an effective means to amplify large numbers of microsatellite containing DNA fragments of suitable size for direct cloning and sequencing. Furthermore, the RAMS protocol easily reveals sequences from different individuals and allows to verify the conservation of DNA flanking regions.

As a consequence, suitable primers for PCR typing can be designed from more reliable sequence information; obtaining this information from library cloning would require substantial efforts. The fact that more than 60% of screened amplification profiles contained positive hybridization signals and that in all cases a signal corresponded to a microsatellite locus provides striking evidence for the usefulness and reliability of this approach. By means of RAMS we also found polymorphic microsatellites in the hydrozoan *Eleutheria dichotoma* (Ender and Schierwater, in preparation), the damselfly *Megaloprepus coerulatus*, the dragonfly *Orthetrum coerulescens* (Hadrys and Schroth, mns. in peparation) the wall lizard *Podarcis muralis* and the sea turtle *Caretta caretta*. The number of microsatellite studies using the RAMS protocol will likely increase more rapidly in the future.

TABLE 1. Scores of randomly amplified microsatellites in *Daphnia* genomes.

Probe	Primers (N)	AAD Markers	Positive Markers	Yield	Abundances
(GA)$_{10}$	20	276	19	6.9%	every ≈ 9.8 kb
(GT)$_{10}$	13	180	18	10.0%	every ≈ 6.7 kb

Applications of AAD markers

Conservation genetics of a sea turtle

Genetic information is crucial to conservation biology (e.g. Simberloff 1988; for review see Milligan et al. 1994). Depending on the particular question to be addressed different genetic markers may be used. Maternally inherited mtDNA may be favoured over nuclear DNA (e.g. Dizon et al. 1992, Avise 1994) because of high base substitution rates and its sensitivity to genetic drift (Birkey et al. 1989, Martin and Palumbi 1993). If the contribution of males to the gene flow in populations has to be assessed, nuclear genetic information is required. On the other hand, a discrepancy between estimates of genetic diversity may derive from the use of mtDNA *versus* nuclear DNA sequences, which differ in evolutionary rates (e.g. Pamilo and Nei 1988, Moritz et al. 1992).

Current conservation genetics covers two areas: gene conservation and molecular ecology (species conservation). The first describes the maintenance of biodiversity as a conservation value for itself, and requires long-term studies over several generations (cf. Moritz 1994). Since in most cases conservation efforts demand fast and practical statements on the actual genetic status of a population, molecular ecology approaches are often more important, which aim to describe demographic patterns relevant to wildlife agencies for pragmatic planning.

In the following we present an example for a case study in molecular ecology, relevant to the species conservation of an endangered sea turtle, the loggerhead turtle *Caretta caretta*. Sea turtles are long-living organisms exhibiting a remarkably delayed generation time of up to 20 years (Dodd 1988). They are particularly threatened by the destruction of their nesting habitats (undisturbed sandy beaches). In the Eastern Mediterranean Sea, a massive decline of breeding habitats and rookery sizes over the last years has been observed (Groombridge 1990). In view of the rapid "cultural" developments and due to the long and widely unknown life history of sea turtles, traditional methods like tag recapture records of females fail to resolve questions concerning population substructures in short time. To face the time pressure, the generation of molecular genetic data may provide a solution.

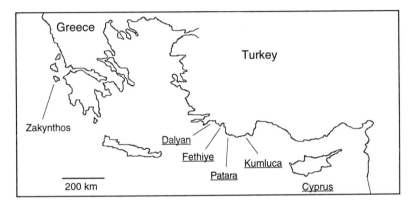

FIGURE 2. Study (nesting) sites of *Caretta caretta* along the coasts of Turkey and Cyprus. Zakynthos indicates the largest rookery in the Mediterranean Sea.

For conservation management, knowledge on the degree of gene flow between neighboured nesting sites would be desirable, i.e. one would like to know whether the loss of individuals from one site may be compensated by migrants from other rookeries (e.g. if females use different nesting sites over their life time). According to the natal homing hypothesis, females return to their natal beaches every 2-3 years for egg deposition (Carr 1972, Meylan et al. 1990, Allard et al. 1994). Maternally inherited mitochondrial DNA markers have been used to verify this hypothesis for rookeries separated by 1000 kilometres and more, e.g. between different ocean basins (Bowen et al. 1992, 1994, Laurent et al. 1993). In order to test the natal homing hypothesis on a higher resolution scale, we aimed to measure gene flow between more closely neighboured nesting sites in the Mediterranean Sea (Schroth et al. 1996). We used RAPD fingerprinting to generate a large set of genetic markers from limited DNA sources in order to estimate the degree of female and male mediated gene flow between neighboured rookeries. We examined tissue samples of dead hatchlings from 5 colonies from Turkey and Cyprus (Figure 2). Out of 30 primers tested, 5 primers (AB4, M4, M9, M12 and P2; Operon Technologies) generated polymorphic amplification profiles, showing genetic variation within and between nesting rookeries, respectively [only reproducible bands were scored in the analyses; Schierwater and Ender (1993)]. In order to increase the resolution, monomorphic DNA fragments were digested with 4-base pair cutters (*Hpa*II, *Rsa*I and *Hae*III). The resulting RAPD-RFLPs were also included in further analyses. All markers were scored in a binary matrix, where 0 codes for the absence and 1 for the presence of a band. The 0/1-matrix was transformed into a symmetrical genetic distance matrix, which was compared to a geographic distance matrix with the same order and number of elements (Mantel 1967). The average genetic distances (dissimilarity coefficients) between the reference colony Dalyan (Western Turkey) and the remaining rookeries increased with increasing geographic distance (Figure 3). The positive correlation between both distance measurements was statistically significant (p = 0.025; matrix comparison test, Mantel 1967, Sokal 1979). This result contradicts the assumption of an existing metapopulation with unlimited interrookery genetic exchange.

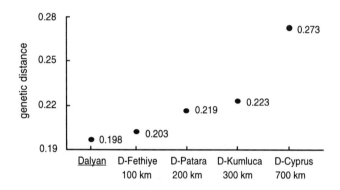

FIGURE 3. Correlation of genetic and geographic distance for five nesting sites. The value for the reference colony Dalyan describes the mean genetic distance between individuals within this colony, all other values represent the mean genetic distance to the reference colony. For further explanations see text. Modified from Schroth et al. (1996).

The data suggest genetic barriers separating closely neighboured rookeries. As migration rates of thousands of kilometres are documented for loggerhead turtles (Meylan et al. 1983, Hughes 1989), geographic barriers cannot explain genetic separation. Precise natal homing behaviour of females remains the most likely explanation for the reduced gene flow between rookeries (Schroth et al. 1996).

Based on nuclear DNA analyses in a survey of global population structure in green turtles (*Chelonia mydas*), Karl et al. (1992) suggested a considerable degree of male-mediated gene flow between rookeries. These findings are in contrast with our data from *Caretta caretta*, which suggest that a potential male-mediated gene flow into non-natal nesting sites is low and cannot compensate the postulated female-mediated genetic isolation. This discrepancy may be due to differences in migration biologies between sea turtle species or to differences in the sensitivity of the DNA markers used (single locus RFLP *versus* multi-locus RAPD). Additional information on maternally inherited mtDNA would be desirable to estimate the male-mediated gene flow from the comparison of mtDNA and ncDNA data.

Kinship analyses in odonates

Reproductive success is the key parameter to behavioral ecology (Clutton-Brock 1988). In many groups, such as insects, the assignment of parent offspring relationships can be a difficult task, especially if species are polygamous and produce large numbers of offspring. Particularly in odonates most species show frequent matings of females with several males before oviposition and furthermore possess cryptic mating tactics (e.g. sperm competition mechanisms) which make kinship determination based on mating observations highly uncertain. This is one reason why behavioral ecologists have eagerly adopted molecular techniques for kinship analyses (cf. Westneat 1993). However, in most cases no genetic information of the study organism is available, and particularly in insects kinship analyses usually rely on small quantities of DNA from large numbers of offspring. Here, AAD markers, which are passed on from the parent to the offspring in a Mendelian fashion (e.g., Bucci and Menozzi 1993), provide an immediate solution. The first demonstration of the use of AAD markers for paternity analyses in animal mating systems derive from some highly promiscuous odonates (Hadrys et al. 1992, 1993). Since then the use of AAD markers to assign paternity, to estimate reproductive success, and to measure relatedness between individuals has become a common tool for behavioural ecologists (for ref. see Scott and Williams 1993, Hadrys and Siva-Jothy 1994, Smith and Williams 1994).

We have been using random amplified polymorphic DNA to assess kinship in odonates, which possess promiscuous mating systems of different complexity, including sperm storage by females and mechanical displacement of alien sperm during copulation by males. Often the investigator will not be able to observe the complete mating history of a female and face the problem to assign paternity to one or more putative fathers from a group of several potential fathers. Even in the absence of any potential father the investigator may seek for evidence for single *versus* multiple paternity in a large offspring sample, and/or look for the presence of sperm from more than one male in the sperm storage organs of a female. If large offspring samples (up to several hundred embryos or larvae are normal for odonates) need to be analysed, the "synthetic offspring" approach of Hadrys et al. (1992, 1993) offers an elegant means for a quick verification of exclusive or almost exclusive paternity for a putative father. The "synthetic offspring" consists of equal amounts of template DNA from the mother and the putative father, which is subjected to AAD fingerprinting and subsequently compared to the fingerprint profile derived from the DNA of a large offspring sample. A mismatch will be indicative of either no paternity or shared paternity for the putative father. If both fingerprints match, this is indicative of exclusive (or at least 90%) paternity success for the putative father. In contrast to a fingerprint from a single offspring, the fingerprint patterns of the real "synthetic offspring" (real and only father plus mother) will be identical to that of a large sample of natural offspring, since in a statistical sample all alleles from the parents will be present. The latter also compensates for possible amplification artifacts. In case of mixed paternity, estimates of the relative paternity ratios for two males may derive from densitometric analyses of diagnostic bands (Hadrys et al. 1992, 1993). The latter, however, depends on the identification of unambigious diagnostic markers for each male and will not detect paternity for a male, if his relative success is below some 10-20%. In several damselflies and dragonflies where females oviposit solitary, often only the mother and her offspring will be available but not the potential fathers. By sampling separate, sequentially oviposited egg clutches and comparing between and within group (clutch) relatedness one may still be able to answer questions about mixed or single paternity. We have also been able to fingerprint the sperm from different locations within the multiple sperm storage organs of females (Figure 4; Hadrys et al., mns. in preparation). Together, these means allow to address questions about the extent of sperm competition in natural populations and the function and priority patterns of multiple sperm storage organs in females.

Since AAD markers are dominant markers the analyses of fingerprint profiles is based on interpreting genetic "phenotypes" rather than allelic "genotypes". Two principally different means may be used to assess paternity from AAD fingerprint profiles; one may use either a small subset of presumably particularly informative diagnostic markers or one may use all generated markers, i.e. without pre-evaluation of whether a marker is informative or not.

(i) Paternity exclusion or inclusion based on diagnostic markers: If a decision has to be made to assign (inclusion) one potential father from a group of potential fathers bands unique to one male (i.e. which are neither present in the mother nor in other potential fathers) may be used in the analysis (cf. Westneat 1993, Jones et al. 1994). The presence of male x diagnostic bands in the offspring is then indicative of parentage for male x, whereas the absence of the band in the offspring is uninformative. To eliminate males from a list of potential fathers (exclusion), diagnostic bands, which are present in the offspring but not in the mother, will be informative. If all potential fathers are known, non-parental males, which lack "offspring diagnostic" markers, can be excluded until the list is reduced to the father candidate, who provides all missing markers found in the offspring but not in the mother. Paternity assignment by exclusion assumes a perfect mendelian inheritance of RAPD markers (cf. Lewis and Snow 1992) and is particularly critical, if it is not possible to collect the entire pool of potential fathers (as this is the case in many field studies on odonates). Levitan and Grosberg (1993) demonstrated in a hydroid system with known parent-offspring samples that based on as little as 5 diagnostic bands non-parents could be correctly excluded for all but one of 30 offspring; 4 diagnostic bands were needed to correctly assign paternity by inclusion. In dragonflies we use diagnostic bands to detect and quantify mixed paternities in large offspring clutches and to detect sperm from different males in the sperm storage organs of females (*see* Figure 4). For both, paternity inclusion and exclusion, probabilities for the statistical presence of diagnostic bands may be estimated from overall null allele frequencies (e.g. Levitan and Grosberg 1993). Besides the latter, however, the use of diagnostic bands is mainly non-statistical and particularly sensitive to any distortion of mendelian segregation patterns and any irregularity in the amplification protocol (e.g. Lewis and Snow 1992, Westneat 1993).

(ii) The use of all markers: Rather than using a subset of diagnostic bands (see above) *all* markers may be compared between each two fingerprint profiles, revealing coefficients, which mirror the similarity (e.g. band sharing) or dissimilarity between profiles (Lynch

1990, Swofford and Olsen 1990). For two different odonate species, Hadrys et al. (1993) have developed a scenario of testable expectations based on pairwise comparisons of band sharing coefficients between known parental kinship (offspring to the mother), known non-first degree kinship (offspring to unrelated females and males) and putative first degree kinship (paternity). This way they could assign paternity to a potential father and also detect mixed paternities. Coefficients of overall similarity in banding patterns have also been used to group parents and offspring by cluster analyses. Several studies (e.g., Apostol et al. 1993, Fondrk et al. 1993) have shown that by using different methods of cluster analyses an accurate clustering of individuals into families can be achieved; offsprings clustered together in correct "siblings clusters", and the parents clustered close to their group of offspring. For dragonflies we have obtained similar results using parsimony analysis (Hadrys and Siva-Jothy 1994, Hadrys and Robertz, mns. in preparation). Several studies suggest that band sharing similarity also reflects the actual genetic similarity in a quantitative manner (e.g. Lynch 1990, Pfennig and Reeve 1993). Thus band sharing scores may also be a means for assigning levels of relatedness beyond

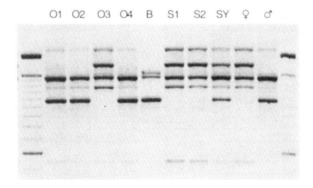

FIGURE 4. RAPD fingerprint profiles (negative image) from family members of the dragonfly *Anax parthenope* (amplified with primer OP-E-11). M = male, O1-O4 = four different offspring (larvae), B = sperm from the bursa of the female, SY = synthetic offspring from the male and the female, S1 = sperm from the spermatheca #1 of the female, S2 = sperm from the spermatheca #2. Note that the synthetic offspring contains all bands present in the offspring, and that the fingerprint profile of the sperm stored in the bursa (B) of the female does not match the offspring.

first degree kinship; for example Reeve et al. (1990, 1992) obtained unambiguous estimates for full and half-sibling relatedness by quantitative comparisons within and among broods. It should be noted, that statistical approaches (the analysis of similarity indices) are little affected by "non-parental" bands, if those are present at low ratios in the data matrix. The latter will increase the random noise in the sample and make statistical testing more conservative, but it will hardly lead to wrong paternity assignments.

Questions about the reliability of the RAPD method with respect to paternity analyses have been raised after a technical note of Riedy et al. (1992) which reports some unusual high levels of "non-parental" RAPD bands. This observation, however, could not be verified in subsequent studies on different organisms, including cnidarians, insects, and plants. Either no or only very low levels of non-parental products were found, which did not significantly affect correct kinship interpretations (e.g. Welsh et al. 1991, Levitan and Grosberg 1993, Milligan and McMurry 1993, Williams et al. 1993, Jones et al. 1994).

Taxonomy of spirochete bacteria

Almost from the beginning of the young history of AAD markers it became obvious that the distinction between monomorphic and polymorphic markers is relative to a given operational taxonomic level, and that these markers might be useful as diagnostic characters for the identification of taxonomic units (e.g. Hadrys et al. 1992). At about the species level often monomorphic bands have been identified, which are shared by all individuals within a sample from a population or species. The latter might be diagnostic for a group at the examined taxonomic level and thus useful to identify this group at the DNA level. The main drawback is that the operational taxonomic level accessible for AAD markers seems to be limited (up to about the genus level). Nonetheless, AAD markers represent a fast, and comparatively easy means to identify systematic units at the DNA level, as this has been shown for various plants, animals and bacteria (for references see Schierwater 1995).

```
              a              b                  b      c
Bor-1 0100101000101100101000010000100000010000001010100010100000000010
Bor-2 1100101000100100100000010000000000010000001010100000100000000000
Tp-1  0000010110101010100010000100001000110010010000001001001000000100
Tp-2  0000100101000101010001000000000000010010000001010000000000000101
Tp-3  0000000100101000101100000000000010111010100000000101000100000000
Tp-4  0000000110000000100100000000000000010000000000100000000000000110
Tsac  0010000101100000100100100000010000000001000000010001000000001010
```

FIGURE 5. Example of a short sequence of scored AAD markers from two *Borrelia* (Bor), four *Treponema pallidum pertenue* (Tp), and one *T. saccharophilum* species (amplified with primers C5 and C2). The complete matrix (9 primers, 251 characters, and 21 taxa, including three *Borrelia* and five *T. pallidum pertenue* species) reveals several specific markers for the *Borrelia* (b), and *Treponema* species (a), as well as for all *T. pallidum pertenue* strains (c) (Schierwater et al., mns. in preparation).

For many groups of bacteria, diagnostic DNA markers may be of particular importance to identify disease causing species by means of PCR-based detection assays (e.g. Myers et al. 1993, Woods et al. 1994). Within the group of spirochete bacteria several species are of clinical relevance to humans, where they cause different diseases, including for example Lyme disease or syphilis. Several spirochetes are hard or not at all to distinguish morphologically or by means of serological assays. Here PCR surveys would be highly desirable to identify the relevant species and subspecies. Welsh et al. (1992) were one of the first who showed that AAD markers are an easy means to distinguish between different bacterial strains. For the causative agent of Lyme disease, *Borrelia burgdorferi,* they distinguished some 20 strains from different countries in North America, Europe and Asia, which could be grouped into distinct phylogenetic clusters. In a similar approach we used RAPD fingerprinting to characterize a variety of strains and species within the genus *Treponema*, which includes the causative agents for syphilis, yaws and pinta.

After screening 20 RAPD primers, 9 primers were identified which distinguished between different strains and species of *Borrelia* and *Treponema*. All of a total of 12 examined *Treponema* and 3 *Borrelia* species could be clearly separated according to their different fingerprint profiles. Also, within one species, different strains of the subspecies *T. pallidum pertenue* from Sumatra, Congo, and Haiti revealed strain specific fingerprint patterns (Figure 5). Within our limited sample size we could immediately identify diagnostic DNA markers at different taxonomic levels, and some of these markers might represent strain, subspecies or species specific markers for the examined taxa in general. To decide on the latter, however, would require to verify the presence of a specific marker in a larger sample within the taxon of interest. Within a universe of examined taxa, a single (or a group of) taxon specific marker(s) could then define a taxon. Alternatively, whole fingerprint profiles may be used to calculate genetic distances (dissimilarity coefficients) and define taxonomic units based on the relative genetic distances (Figure 6). In order to use diagnostic markers for PCR assays of defined

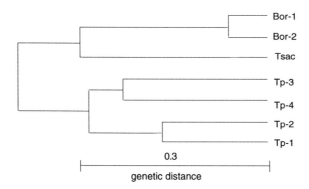

FIGURE 6. UPGMA-phenogram of the same taxa shown in Fig. 6. The analysis is based on the genetic distances calculated from a total of 251 characters. As expected, the *Borrelia* species, the *T. saccharophilum* species, and the *T. pallidum pertenue* strains cluster on separate lineages.

strains or species, one could sequence specific markers to obtain taxon-specific PCR primer sequences (cf. Paran and Michelmore 1993).

Phylogeny of Daphnia

In order to test the usefulness of RAPD markers as cladistic characters in phylogenetic analyses, we examined waterfleas of the genus *Daphnia*. In particular, we concentrated on members of the *Daphnia longispina* complex (*D. galeata*, *D. hyalina*, *D. cucullata*), which have been intensively studied with respect to morphological and life history traits, natural hybridization, allozyme, mtDNA and ncDNA variation (e.g., Wolf and Mort 1986, Hebert et al. 1989, Schwenk 1993, Schierwater et al. 1994, Spaak and Hoekstra 1995). Morphological characters have been used for inferring phylogenetic relationships (Benzie 1986), which, however, bear the problem of genetic variation superimposed by cyclomorphosis (Mort 1989).

For any kind of cladistic character the verification of homology can be a crucial step. In this context RAPD markers bear a potential problem deriving from the comigration of non-homologous DNA fragments. Although the portion of comigrating bands in profiles has rarely been estimated (Williams et al. 1993, Smith et al. 1994), one recent study, which includes 220 RAPD markers from interspecific comparisons, showed that more than 90% were homologous (Rieseberg 1996). If phylogenetic signals occur in a data set, one may detect them by estimating the robustness of the resulting phylogenetic trees. To do so, different methods are available, including estimation of tree-length distributions (e.g. Hillis 1991) and bootstrapping (Felsenstein 1985), which probably is the most commonly used method for measuring support of phylogenetic relationships.

For a total of 25 clones (population samples) from four *Daphnia* species and three hybrid "species" we generated a total of 306 RAPD markers using the protocol of Schierwater and Ender (1993). Four RAPD markers were tested for homology and all of them corresponded to homologous DNA sequences between species. When translating amplification patterns into a 0/1-matrix only unambiguously scorable bands within replicates were included. The original data matrix contained 280 characters (disregarding all markers specific for hybrid clones) generated by 13 primers. Several monomorphic RAPD markers at different taxonomic levels could be observed (cf. Schierwater 1995). To infer phylogenetic relationships we used parsimony procedures with different assumptions and weightings of the binary characters. The data were analyzed with the aid of the computer program PAUP (Swofford 1991). In order to extract noise from the original data set, we modified it in different ways. The original data set was reduced by excluding the uninformative characters [constant characters and autapomorphies (clone-specific characters)]. The remaining data set included only informative characters in a phylogenetic sense of synapomorphies ($N = 237$). After an initial parsimony search, characters which caused homoplasy were recognized by low consistency indices (CI; Farris 1989) and reweighted. After reweighting of the characters, parsimony searches were repeated until the given tree length became stable. In all analyses *Daphnia pulex* served as an outgroup and heuristic searches were run because of the large sample size ($N = 20$). We also changed the character types from unordered to ordered with a weighting of 1:20 between '0' and '1', in order to reflect the asymmetry between the probabilities of gaining and losing a marker. Independent of the procedure used, the overall tree topology was highly reproducible. All clones from one species clustered together and *D. galeata* and *D. hyalina* formed a monophyletic group (Figure 7). An unweighted pair group cluster analysis using arithmetic means (UPGMA; Sneath and Sokal 1973) also revealed the same topology as shown in Figure 7.

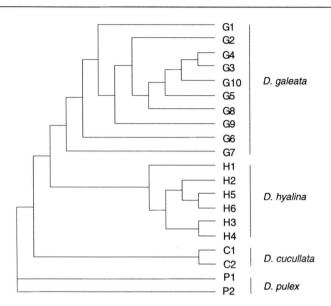

FIGURE 7. Phylogram for 20 clones (from different populations) of four *Daphnia* species derived from parsimony analysis based on 280 RAPD characters. Two *Daphnia pulex* clones serve as an outgroup. Note, that all clones of one species form a single 'species cluster'. The phylogram suggest a sister group relationship for *D. galeata* and *D. hyalina*. The shown topology is not affected by a weighting of the characters. Tree length = 523, CI = 0.54.

Measurements of the consistency index, the percentage of congruent bootstrap replicates and tree-lengths distribution reflect the robustness of the final tree topology. Both, the bootstrap values and the given CIs of each clade improved as the data set was pruned by discarding uninformative characters. In general, growing values of CI were observed as the noisome characters were given a smaller weight. As recommended by several authors (e.g., Fitch 1984, Hillis and Huelsenbeck 1992) the lengths frequency distribution of alternative trees contains information useful for deciding whether the data harbor significant phylogenetic information. An asymmetrical frequency distribution (leftward skew) shows that few solutions exist near the shortest optimal (most-parsimonious) tree. A symmetrical tree-lengths distribution implies many suboptimal solutions indicating random noise inherent in the data. We obtained frequency distributions of tree-lengths (Figure 8) with exhaustive searches by using several different subsets of the entire data set (two randomly chosen samples per species). In all cases leftward skewed distributions were revealed suggesting strong cladistic signals in the data.

Several RAPD studies have examined phylogenetic relationships by means of distance methods, i.e. UPGMA or neighbor joining procedures (e.g., Kazan et al. 1993, Williams and St.Clair 1993, Stammers et al. 1995). Comparatively few studies have used the cladistic approach (Adamkewicz and Harasewych 1994, Baum et al. 1994, Caetano-Anollés et al. 1995, Mackenstedt and Johnson 1995).

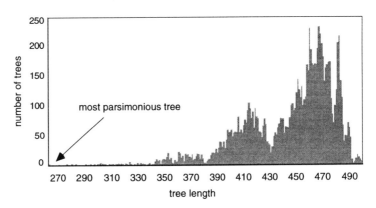

FIGURE 8. Example of a tree-lengths distribution from exhaustive analysis of a random data subset. Explanations are given in the text.

Each given data set harbors a certain level of specific noise (cf. Swofford and Olsen 1990, Miyamoto and Cracraft 1991). For AAD markers all potentially noise causing characters can be simply discarded, since the number of markers, that can be generated, is not limited. In the given example above we have shown that noisy cladistic patterns which are ambiguous or less consistent can be traced and eliminated. In sum, present knowledge suggests that AAD markers are a simple, efficient and reliable means to infer phylogenetic relationships for closely related species.

Remarks

The above examples have illustrated some main potentials of AAD markers to address a broad range of issues in ecological and evolutionary research. It is the arbitrary nature of PCR priming which makes AAD markers the most general tool for fingerprinting genomes applicable immediately to any genome of interest. This ubiquity is bought at the expense of a very low amount of information per marker, which however does not represent a severe limitation since the number of markers generated can be increased to any level of choice (cf. Lynch and Milligan 1994). The researcher has to be more aware of certain aspects concerning the reproducibility and the analysis of AAD fingerprint profiles (e.g. Caetano-Anollés et al. 1993, Meunier and Grimont 1993, Schierwater and Ender 1993, Williams et al. 1993, Smith and Williams 1994). At present it seems that one major drawback of PCR techniques in general, and AAD PCR in particular, is the lack of understanding the kinetics of the PCR reaction, i.e. what is happening in the reaction tube. For example the consequences of competing primer binding sites within an amplifyable DNA region have yet been underestimated. A recent mathematical model has shown that under certain conditions unexpected and not necessarily reproducible markers may get amplified (Figure 9; Schierwater et al. 1996). The model also suggests certain settings for the PCR conditions, which may help to reduce the problem. For the future we expect an increase in empirical and theoretical knowledge about the processes involved in AAD reactions accompanied by a further increase in application studies.

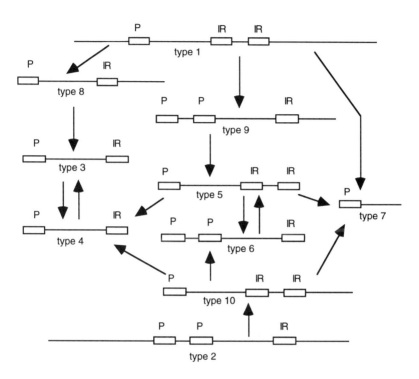

FIGURE 9. The number of different AAD products emerging from a DNA template with nested primer binding sites may be higher than normally expected. The type 1 (top) and type 2 (bottom) molecules represent the two strands of the DNA template. The top strand (type I) contains the primer sequence (P) once and its inverted repeat sequence (IR) twice. During the annealing phase the primer can bind to its inverted repeat sequence (IR) and initiate the polymerisation of the product types shown. For convenience all strands are shown in 5' to 3' orientation. If the primer binds to both binding sites of the DNA template (type 1) simultaneously, one short fragment (type 8) and one residual product (type 7) will be synthesized. Under certain conditions not only the expected type 3(4) product but also the type 5(6) and type 7 products may get amplified in detectable amounts. The shown idealized scenario holds true for all standard DNA polymerases, which lack a 5'-3' exonuclease activity (from Schierwater et al. 1996).

References

Adamkewicz SL and Harasewych MG (1994) *The Nautilus Suppl* 2:51-60.

Allard MW, Miyamoto MM, Bjorndal KA, Bolten AB and Bowen BW (1994) *Copeia* 1:34-41.

Amos B, Schlötterer C and Tautz D (1993) *Science* 260:670-672.

Apostol BL, Black IV WC, Miller BR, Reiter P and Beaty BJ (1993) *Theor Appl Genet* 86:991-1000.

Ashley MV and Dow BD (1994) In: Schierwater B, Streit B, Wagner GP and DeSalle R (eds), *Molecular Ecology and Evolution: Approaches and Applications*. Birkhäuser Verlag, Basel. pp. 186-201.

Avise JC (1994) *Molecular Markers, Natural History and Evolution*, Chapman & Hall, New York, London.

Baum TJ, Gresshoff PM, Lewis SA and Dean RA (1994) *Mol Plant-Microbe Interactions* 7:39-41.

Beckmann JS and Weber JL (1992) *Genomics* 12:627-631.

Benzie JAH (1986) *Hydrobiol* 140:105-124.

Birkey CW, Fuerst P and Maruyama T (1989) *Genetics* 121:613-627.

Bowen BW, Kamezaki N, Limpus CJ, Hughes GR, Meylan AB and Avise JC (1994) *Evolution* 48:1820-1828.

Bowen BW, Meylan AB, Ross P, Limpus CJ, Balazs GH and Avise JC (1992) *Evolution* 46:865-881.

Bucci G and Menozzi P (1993) *Mol Ecol* 2:227-232.

Caetano-Anollés G, Bassam BJ and Gresshoff PM (1991) *Bio/technology* 9:553-557.

Caetano-Anollés G, Bassam BJ and Gresshoff PM (1993) *Mol Gen Genet* 241:57-64.

Caetano-Anollés G, Callahan LM, Williams PM, Weaver KR and Gresshoff PM (1995) *Theor Appl Genet* 91:228-235.

Carr AF and Carr MH (1972) *Ecology* 53:425-429.

Clutton-Brock TH (1988) *Reproductive Success*, University of Chicago Press, Chicago.

Coltman DW, Bowen WD and Wright JM (1996) *Mol Ecol* 5:161-163.

Dawson IK, Simons AJ, Waugh R and Powell W (1995) *Heredity* 74:10-18.

Dizon AE, Lockyer C, Perrin WF, Demaster DP and Sisson J (1992) *Cons Biol* 6:24-36.

Dodd CK (1988) *US Fish Wildl Ser Biol Rep* 88:1-110.

Ender A, Schwenk K, Städler T, Streit B and Schierwater B (1996) *Mol Ecol* 5:437-441.

Farris JS (1989) *Cladistics* 5:417-419.

Felsenstein J (1985) *Syst Zool* 34:152-161.

Fitch WM (1984) In: Duncan T and Stuessy TF (eds), *Cladistics: Perspectives on the Reconstruction of Evolutionary History*. Columbia University Press, New York. pp. 221-252.

Fondrk MK, Page RE Jr. and Hunt GJ (1993) *Naturwissenschaften* 80:226-231.

Gibbs HL, Prior KA and Weatherhead PJ (1994) *Mol Ecol* 3:329-337.

Groombridge B (1990) *Council of Europe Environment Conservation and Management Division, Strasbourg, France*, 48.

Hadrys H, Balick M and Schierwater B (1992) *Mol Ecol* 1:55-63.

Hadrys H, Schierwater B, Dellaporta SL, DeSalle R and Buss LW (1993) *Mol Ecol* 2:79-87.

Hadrys H and Siva-Jothy MT (1994) In: Schierwater B, Streit B, Wagner GP and DeSalle R (eds), *Molecular Ecology and Evolution: Approaches and Applications*. Birkhäuser Verlag, Basel, Boston. pp. 75-90.

Hebert PDN, Schwartz SS and Hrbacek J (1989) *Heredity* 62:207-216.

Hillis DM (1991) In: Miyamoto MM and Cracraft J (eds), *Phylogenetic Analysis of DNA Sequences*. Oxford University Press, New York, Oxford. pp. 278-294.

Hillis DM and Huelsenbeck JP (1992) *J Hered* 83:189-195.

Huff DR, Peakall R and Smouse PE (1993) *Theor Appl Genet* 86:927-934.

Hughes CR and Queller DC (1993) *Mol Ecol* 2:131-137.

Hughes GR (1989) In: Payne AIL and Crawford JM (eds), *Sea Turtles*. Oceans of Life of Southern Africa, Vlaeberg, Cape Town. pp. 230-243.

Jarne P, Viard F, Delay B and Cuny G (1994) *Mol Ecol* 3:527-528.

Jones CS, Okamura B and Noble LR (1994) *Mol Ecol* 3:193-199.

Karl SA, Bowen BW and Avise JC (1992) *Genetics* 131:163-173.

Kazan K, Manners JM and Cameron DF (1993) *Theor Appl Genet* 85:882-888.

Laurent L, Lescure J, Excoffier L, Bowen BW, Domingo M, Hadjichristophorou M, Kornaraki L and Trabuchet G (1993) *CR Acad Sci Paris, Sciences de la vie/Life sciences* 316:1233-1239.

Levitan DR and Grosberg RK (1993) *Mol Ecol* 2:315-328.

Lewis PO and Snow AA (1992) *Mol Ecol* 1:155-160.

Lynch M (1990) *Mol Biol Evol* 7:478-484.

Lynch M and Milligan BG (1994) *Mol Ecol* 3:91-99.

Mackenstedt U and Johnson AM (1995) *Parasitol Res* 81:217-221.

Mantel N (1967) *Cancer Res* 27:209-220.

Martin AP and Palumbi SR (1993) *Proc Natl Acad Sci USA* 90:4087-4091.

Meunier JR and Grimont PAD (1993) *Res Microbiol* 144:373-379.

Meylan AB, Bjorndal KA and Turner BJ (1983) *Biol Cons* 26:79-90.

Meylan AB, Bowen BW and Avise JC (1990) *Science* 248:724-727.

Milligan BG and McMurry CK (1993) *Mol Ecol* 2:275-283.

Milligan BG, Leebens-Mack J and Strand AE (1994) *Mol Ecol* 3:423-435.

Miyamoto MM and Cracraft J (1991) *Phylogenetic Analysis of DNA Sequences*, Oxford University Press, New York, Oxford.

Morin PA, Moore JJ, Chakraborty R, Jin L, Goodall J and Woodruff DS (1994) *Science* 265:1193-1201.

Moritz C (1994) *Mol Ecol* 3:401-411.

Moritz C, Schneider CJ and Wake DB (1992) *Syst Biol* 41:273-291.

Mort MA (1989) *Hydrobiol* 171:159-170.

Myers LE, Silva S, Procunier J and Little PB (1993) *J Clin Microbiol* 31:512-517.

Paetkau D, Calvert W, Stirling I and Strobeck C (1995) *Mol Ecol* 4:347-354.

Pamilo P and Nei M (1988) *Mol Biol Evol* 5:568-583.

Paran I and Michelmore RW (1993) *Theor Appl Genet* 85:985-993.

Pépin L, Amigues Y, Lépingle A, Berthier J, Bensaid A and Vaiman D (1995) *Heredity* 74:53-61.

Pfennig DW and Reeve HK (1993) *Evolution* 47:700-704.

Primmer CR, Moller AP and Ellegren H (1995) *Mol Ecol* 4:493-498.

Puterka GJ, Black WC, Steiner WM and Burton RL (1993) *Heredity* 70:604-618.

Queller DC, Strassmann JE and Hughes CR (1993) *Trends Ecol Evol* 8:285-288.

Reeve HK, Westneat DF, Noon WA, Sherman PW and Aquadro CF (1990) *Proc Natl Acad Sci USA* 87:2496-2500.

Riedy MF, Hamilton WJ and Aquadro CF (1992) *Nucleic Acids Res* 20:918.

Rieseberg LH (1996) *Mol Ecol* 5:99-105.

Rieseberg LH, Choi HC, Chan R and Spore C (1993) *Heredity* 70:285-293.

Sambrook J, Fritsch EF and Maniatis T (1989) *Molecular Cloning*, Cold Spring Harbor Laboratory Press, New York.

Schierwater B (1995) *Electrophoresis* 16:1643-1647.

Schierwater B and Ender A (1993) *Nucleic Acids Res* 21:4647-4648.

Schierwater B, Ender A, Schwenk K, Spaak P and Streit B (1994b) In: Schierwater B, Streit B, Wagner GP and DeSalle R (eds), *Molecular Ecology and Evolution: Approaches and Applications*. Birkhäuser Verlag, Basel, Boston. pp. 495-508.

Schierwater B, Metzler D, Krüger K and Streit B (1996) *J Comp Biol* 3:235-251.

Schierwater B, Streit B, Wagner GP and DeSalle R (1994a) *Molecular Ecology and Evolution: Approaches and Applications*, Birkhäuser Verlag, Basel, Boston.

Schroth W, Streit B and Schierwater B (1996) *Nature* 384:521-522.

Schwenk K (1993) *Mol Biol Evol* 10:1289-1302.

Scott MP and Williams SM (1993) *Proc Natl Acad Sci USA* 90:2242-2245.

Simberloff D (1988) *Annu Rev Ecol Syst* 19:473-511.

Smith JJ, Scott-Craig JS, Leadbetter JR, Bush GL, Roberts DL and Fulbright DW (1994) *Mol Phyl Evol* 3:135-245.

Smith JSC and Williams JGK (1994) In: Schierwater B, Streit B, Wagner GP and DeSalle R (eds), *Molecular Ecology and Evolution: Approaches and Applications*. Birkhäuser Verlag, Basel. pp. 5-15.

Sneath PHA and Sokal RR (1973) *Numerical Taxonomy*, Freeman, San Francisco.

Sokal RR (1979) *Syst Zool* 28:227-232.

Spaak P and Hoekstra JR (1995) *Ecology* 76:553-564.

Stammers M, Harris J, Evans GM, Hayward MD and Forster JW (1995) *Heredity* 74:19-27.

Swofford DL (1991) *PAUP: Phylogenetic Analysis Using Parsimony*, Version 3.0 Illinois History Survey, Champaign.

Swofford DL and Olsen GJ (1990) In: Hillis DM and Moritz C (eds), *Molecular Systematics*, Sinauer Associates, Inc., Sunderland, Massachusetts. pp. 411-501.

Tautz D (1989) *Nuceic Acids Res* 17:6463-6471.

Taylor AC, Sherwin WB and Wayne RK (1994) *Mol Ecol* 3:277-290.

Vos P, Hogers R, Bleeker M, Reijans M, van de Lee T, Hornes M, Frijters A, Pot J, Peleman J, Kuiper M and Zabeau M (1995) *Nucl. Acids Res.* 23:4407-4414.

Welsh J, Honeycutt RJ, McClelland M and Sobral BWS (1991) *Theor Appl Genet* 82:473-476.

Welsh J and McClelland M (1990) *Nucleic Acids Res* 18:7213-7218.

Welsh J, Pretzman C, Postic D, Saint-Girons T, Baranton G and McClelland M (1992) *Int J Syst Bacteriol* 42:370-377.

Westneat DF (1993) *Behavioral Ecol* 4:49-60.

Williams CE and StClair A (1993) *Genome* 36:619-630.

Williams JGK, Hanafey MK, Rafalski JA and Tingey SV (1993) *Methods Enzymol* 218:704-740.

Williams JGK, Kubelik AR, Livak KJ, Rafalski JA and Tingey SV (1990) *Nucleic Acids Res* 18:6531-6535.

Wolf HG and Mort MA (1986) *Oecologia* 68:507-511.

Woods JP, Kersulyte D, Tolan RW, Berg CM and Berg DE (1994) *J Infectious Diseases* 169:1384-1389.

Appendix:
Suppliers of reagents and equipment[1]

Ab Peptides Inc
  2101 S Brentwood Blvd
  St Louis, MO 63144
  (314) 968-4944

Aldrich Chemical Co, Inc
  1001 W St Paul Ave
  Milwaukee, WI 53233
  (414) 273-3850

Amersham Life Science Inc
  2636 S Clearbrook Dr
  Arlington Heights, IL 60005
  (708) 593-6300

Amicon Inc
  72 Cherry hill Dr
  Beverly, MA 01915
  (508) 777-3622

Baxter Scientific Products:
  see VWR Scientific Products

Beckman Instruments Inc
  2500 Harbor Blvd
  Fullerton, CA 92634
  (714) 871-4848

Bellco Glass Inc
  340 Edrudo Rd, PO Box B
  Vineland, NJ 08360
  (609) 691-3247

Bio 101 Inc
  1070 Joshua Way
  Vista, CA 92083
  (619) 598-7299

Bio-Rad Laboratories
  2000 Alfred Nobel Dr
  Hercules, CA 94547, USA
  (510) 741-1000

Boehringer Mannheim Corp
  9115 Hague Rd, PO Box 50414
  Indianapolis, IN 46250
  1 (800) 428-5433

Corning Costar Corp
  One Alewife Center
  Cambridge, MA 02140
  (617) 868-6200

Diversified Biotech
  1208 VFW Parkway
  Boston, MA 02132
  (617) 323-5709

---

[1]Reference to a specific supplier does not constitute endorsement.

Du Pont NEN Research Products
   549 Albany St
   Boston, MA 02118
   (617) 482-9595

Dynatech Laboratories Inc
   14340 Sullyfield Circle
   Chantilly, VA 22021
   (703) 631-7800

E-C Apparattus Corp
   101 Colin Dr
   Holbrook, NY 11741
   (516) 244-2929

Enzo Diagnostics Inc
   60 Executive Blvd
   Farmingdale, NY 11735
   (516) 694-7070

Epicentre Technologies
   1202 Ann St, Madison
   WI 53713
   (608) 258-3080

5 Prime-3 Prime Inc
   5603 Arapahoe Rd
   Boulder, CO 80303
   (303) 440-3705

Fluka Chemical Corp
   980 S Second St
   Ronkonkoma, NY 11779
   (516) 467-0980

FMC Bioproducts
   191 Thomaston St
   Rockland, ME 04841
   (207) 594-3495

Fotodyne Inc
   950 Walnut Ridge Dr
   Hartland, WI 53029
   (414) 369-7000

Fuji Medical Systems Inc
   Bio-imaging group, PO Box 120035
   Stanford, CT 06912, USA
   (203) 353-0300

Genosys Biotechnologies In
   1442 Lake Front Circle, Suite 185
   The Woodlands, TX 77380
   (713) 363-2212

Glen Research Corp
   44901 Falcon Pl, Suite 113
   Sterling, VA 20166
   (703) 437-6191

Hoeffer Scientific Instruments
   654 Minnesota St, PO Box 77387
   San Francisco, CA 94107, USA
   (415) 282-2307

GIBCO BRL:
   see Life Technologies.

Life Technologies, Inc
   8400 Helgerman Ct, PO Box 6009
   Gaitherburg, MD 20884
   (301) 840-8000

E. Merck
   Frankfurterasse 250, Darmstadt D-6100
   Germany
   49-6151-720

Millipore Corp
   80 Ashby Rd
   Bedford, MA 01730
   (617) 275-9200

Molecular Dynamics
   928 E Arques Ave
   Sunnyvale, CA 94086
   (408) 773-1493

Molecular Probes Inc
   4849 Pitchford Ave, PO Box 22010
   Eugene, OR 97402
   (541) 465-8300

Moss Inc
   PO Box 189
   Passadena, MD 21122
   (410) 768-3442

MS Research Inc
   149 Grove St
   Watertown, MA 02172
   (617) 923-8000

National Biosciences Inc
   3650 Annapolis Ln, N Suite 140
   Plymouth, MN 55440-5434
   (612) 550-2012

National Diagnostics Inc
   305 Patton Dr
   SW Atlanta, GA 30336
   (404) 699-2121

New England Biolabs Inc
   32 Tozer Rd
   Beverly, MA 01915-5599
   (508) 927-5054

Operon Technologies Inc
   1000 Atlantic Ave
   Alameda, CA 94501
   (510) 865-8644

Owl Scientific Inc
   10 Commerce Way
   Woburn, MA 01801, USA
   (617) 935-9499

Perkin-Elmer Corp
   761 Main Ave
   Norwalk, CT 06859, USA
   (203) 762-1000

Perkin-Elmer Corp
   Applied Biosystems Division
   850 Lincoln Centre Dr
   Foster City, CA 944404, USA
   (415) 570-6667

Pharmacia Biotech Inc
   801 Centennial Ave, PO Box 1327
   Picataway, NJ 08855
   (908) 457-8000

Promega Corp
   2800 Woods Hollow Rd
   Madison, WI 53711, USA
   (608) 274-4330

Quiagen Inc
   9600 De Soto Ave
   Chatsworth, CA 91311
   (818) 718-9870

Research Genetics Inc
   2310 S Memorial Parkway
   SW Huntsville, AL 35801
   (205) 533-4363

Sigma Bioscience
   3050 Spruce St, PO Box 14508
   St Louis, MO 63178, USA
   1 (800) 325-3010

Sooner Scientific Inc
   Po Box 180
   Garbin, OK 74736, USA
   (405) 286-9408

Stratagene
   11011 N Torrey Pines Rd
   La Jolla, CA 92037, USA
   (619) 535-5400

Sun Bio-Science Inc
   637 Opening Hill Rd
   Madison, CT 06443
   (203) 421-3400

Tomtec Inc
   607 Harborview Rd
   Orange, CT 06477
   (203) 281-6790

Tropix Inc
   47 Wiggins Ave
   Bedford, MA 01730
   (617) 271-0045

United States Biochemical: see USB.

USB Specialty Biochemicals
   26111 Miles Rd
   Cleveland, OH 44128
   (216) 765-5000

VWR Scientific Products
   1310 Goshen Pkwy
   W Chester, PA 19380
   (610) 936-1761

Glossary

The following terms are usually defined in a general sense and avoid details. The objective of this non-comprehensive glossary is to help the general reader dissect some of the language and jargon underlying selected aspects of molecular biology and genetics.

A: Abbreviation for adenine (Ade, 6-aminopurine), purine base characteristic component of nucleic acids.

AAD: *See* arbitrarily amplified DNA.

Abundance: The average number of molecules of a specific mRNA or protein in a given cell at a given time.

Acrylamide: *See* polyacrylamide gel.

Adapter (adaptor, oligonucleotide adaptor): A short synthetic oligonucleotide with a preformed cohesive terminus that can be used to join a blunt end of a DNA duplex with a cohesive end of another. The adapter construct harbors a blunt end with a 5' phosphate group and a non-phosphorilated cohesive end to prevent self-ligation.

Adapter primer: a synthetic oligodeoxynucleotide that functions as a primer for reverse transcription or amplification by for example the polymerase chain reaction, and carries one or several restriction endonuclease sites.

AFLP™: An arbitrarily amplified DNA technique that uses ligation of adaptors to the ends of restricted DNA and amplification with hemi-specific primers. Amplification products are generally visualized by polyacrylamide gel electrophoresis and autoradiography, and fingerprints are highly complex. The AFLP method is also known by selective restriction fragment amplification (SRFA).

Agar: A dried sea-weed extract from red algae such as *Gelidium* and *Gracilaria*, consisting of a mixture of agarose and agaropectin. Agar or agar-agar is used to solidify media in bacteriology and tissue culture.

Agarose: A neutral, linear sulfated galactan polysaccharide used extensively in electrophoresis.

Agarose gel: A polysaccharide matrix for electrophoretic separation of nucleic acid molecules according to size and conformation. These gels are inert, macroporous, and nontoxic. Agarose gels are easy to cast but lack the resolution of polyacrylamide gels and cannot be easily preserved. DNA is usually stained in agarose gels with the dye ethidium bromide.

Allele (allelomorph): One of two or more forms of a gene occupying the same locus on a particular chromosome.

Allele-specific associated primer **(ASAP):** A synthetic oligodeoxynucleotide of 20-30 nucleotides, that is complementary to the 3' or 5' end of an arbitrarily amplified DNA product and is used in a PCR reaction to amplify a specific allele.

AMPGD™: Trivial name for the chemiluminescent β-galactosidase substrate 3-(4-methoxyspiro)-1,2-dioxetane-3,2'-tricyclo (3.3.1.1) decan-(4-yl)phenyl-β-D-galactopyranoside. Enzymatic cleavage of AMPGC at pH 6.5 converts it into a protonated form which, at pH above 10, produces an anion that spontaneously decomposes emitting light.

Amplicon: Nucleic acid region defined by two oligonucleotide primer annealing sites in opposite strands.

Amplification: An increase in copy number of a particular DNA fragment, resulting from replication of a vector into which is has been cloned, replication of specific genes in the absence of general chromosomal replication, or *in vitro* enzymatic activity generally driven by short oligonucleotide segments (primers).

Amplification fragment length polymorphism **(AFLP):** A variant DNA amplification product of different size generally produced by amplification with arbitrary primers (e.g. in AFLP, AP-PCR, DAF, RAPD). The term was originally coined to depict DAF polymorphisms. AFLPs can be used to discriminate between related individuals, localize specific genes or gene regions in complex genomes and establish genetic maps.

Amplimer (*amplification primer*): A synonym for oligonucleotide primer.

AmpliTaq™: *See Taq* DNA polymerase.

AMPPD®: Trivial name for the chemiluminescent substrate of alkaline phosphatase 3-(2'-spiroadamantan)-4-methoxy-1-(3"-phosphoryloxy)-phenyl-1,2-dioxetane. Phosphorilation leads to an unstable intermediate that emits light and can be detected in a radiographic film.

Anchored primer: A specifically designed primer which contains a region with high sequence homology to a specific nucleic acid region, such as a repetitive motif, poly(dA) tail, etc. Primers can be anchored at their 5' and 3' termini.

Aneuploid: Chromosomal number is more or less than the diploid number by sets lesser than the haploid number (e.g. humans with 47 or 45 chromosomes are aneuploids).

Annealing: The spontaneous alignment of two single-stranded nucleic acids to form a duplex molecule. The alignment interaction is based on hydrogen bond formation between complementary base pairs in the individual strands and the formation of a double helix.

AP-PCR: *See* arbitrarily primed PCR.

Arabidopsis thaliana: A small cruciferous weed of the mustard family with a simple genome, well characterized genetics, and advantageous cultural characteristics. The plant has been used as model organism in molecular biological research.

Arbitrarily amplified DNA (AAD): Collective term to describe those nucleic acid amplification-based techniques that use arbitrary primers. Also termed AFLP analysis, arbitrary primer technology (APT), multiple arbitrary amplicon profiling (MAAP), and nucleic acid scanning by amplification.

Arbitrarily primed PCR (AP-PCR): An arbitrarily amplified DNA technique that uses one or more primers of typically between 18 and 32 nucleotides in length and high primer-to-template ratios (1-500). Amplification products are generally visualized by polyacrylamide gel electrophoresis and autoradiography, and fingerprints are moderately complex. See also AAD, AFLP, DAF and RAPD.

Arbitrary sequence oligonucleotide fingerprinting (ASOF): A nontargeted hybridization strategy that uses oligonucleotide arrays to analyze the sequence of arbitrarily amplified DNA (*see* nucleic acid scanning by hybridization) or of a collection of sequence-tagged fragments (*see* sequence tagged site).

Arbitrary signatures from amplification profiles (ASAP): A method that amplifies DNA amplification fingerprints using one or more arbitrary primers, generating a "fingerprint of a fingerprint". This method produces allelic signatures and can be applied to the general amplification of previously amplified nucleic acids, including arbitrarily amplified DNA. *See* allele specific associated primer.

ASAP: *See* arbitrary signatures from amplification profiles *and* allele specific associated primer.

Autoradiography: Method of documentation used to detect radioactively labeled molecules superimposed to a photosensitive emulsion or film. The method can be used on whole organisms, tissue sections, filters or electrophoretic gels.

Bacterial artificial chromosome (BAC): A bacterial cloning vector based on a single copy F-factor of *Escherichia coli* that allows the cloning of fragments of an average size of up to 300 kb.

Bacteriophage: Bacterial virus that either kills or infects latently a bacterium. Often abbreviated as 'phage'. Phage lambda (λ) is the most studied and is frequently used for cloning experiments.

Bacterium: A relatively simple, single cell organism. All bacteria are prokaryotes and therefore lack a defined nucleus and chromosomes characteristic of eukaryotes (such as fungi, plants and animals).

Base: A heterocyclic nitrogen-containing molecule constituent of DNA and RNA. The most common purine and pyrimidine bases are adenine, guanine, cytosine, thymine and uracil.

Base composition: The ratio of the number of adenine and thymine (AT) to guanine and cytosine (GC) bases in a DNA molecule. In double-stranded DNA, A=T and G≡C.

Base mismatch (mismatch, mispairing): The occurrence of incorrectly paired bases in duplex nucleic acid molecules.

Base pair (bp): Any hydrogen-bonded pair of purine-pyrimidine bases in complementary nucleic acid strands. The number of base pairs is function of the length of a nucleic acid molecule and therefore measures its size; e.g. 1 kb = 1 kilo base pair, 1 Mb = 1 mega base pair.

Base stacking: The arrangement of base pairs in parallel planes in the interior or a nucleic acid double helix.

Base substitution: The replacement of one nucleotide by another.

Beta (β) emitter: Any radionuclide that decays and concomitantly emits β-particles, usually negatively charged electrons. The isotopes ^{3}H, ^{14}C, ^{35}S and ^{32}P are beta emitters.

Biotechnology: A scientific discipline that focuses on the development and improvement of biological systems. Biotechnology borrows tools from biochemistry, genetics, microbiology, and engineering to develop useful organisms and products of commercial value.

Blot: A nitrocellulose or Nylon membrane to which a nucleic acid or protein has been transferred or the autoradiograph of a blotting experiment.

Blunt end: A non-cohesive end which is perfectly base-paired terminus of a duplex DNA fragment.

BSA: Abbreviation for bovine serum albumin; also *see* bulked segregant analysis.

Bulked segregant analysis (BSA): Detection strategy used to define molecular regions controlling a certain phenotype. Large populations of organisms are usually pooled (bulked) into two classes representing distinguishing phenotypes. DNA profiling with arbitrary primers (e.g. RAPD, DAF) is then used to generate DNA markers associated with those regions.

C: Abbreviation for cytosine (6-amino-2-hydroxy-pyrimidine), a pyrimidine base chracteristic of DNA and RNA.

Capillary electrophoresis (CE): A technique to separate ionized molecules in silica capillaries by electro-osmotic flow.

CAPS: *See* cleaved amplified polymorphic sequence.

Cathode: Negatively charged electrode in electrophoresis; as compared to anode, the positively charged electrode.

cDNA (complementary DNA): A single or double stranded DNA molecule complementary to an RNA, generally mRNA, generally made by RNA-dependent DNA polymerase (reverse transcriptase).

centiMorgan (cM): Unit of recombination named after the geneticist Morgan. One cM is equivalent to one percent recombination and is a measure of relative distance between two genes or genetic markers in a chromosome.

Centromere: Chromosomal region functioning as the spindle attachment region allowing chromosome and/or chromatid separation during mitosis and meiosis.

Chargaff's (base pairing) rule: The prediction that since in double stranded DNA the number of adenines (A) equals that of thymines (T) and guanines (G) that of cytosines (C). Accordingly A binds to T and G to C by hydrogen bonds, giving the DNA molecule the properties needed for replication and information storage.

Chromatin: Complex form of eukaryotic nuclear material with characteristic staining properties detected within the interphase nucleus (cf. euchromatin, heterochromatin).

Chromosome: Organized structure composed of DNA and proteins, visible through light microscopy during cell division (cf. mitosis and meiosis).

Chromosome (genome) walking: A technique for the sequential isolation of clones present in a genomic library with sequences that overlap and span large chromosomal intervals.

Cleaved amplified polymorphic sequence (CAPS): A technique for the detection of restriction fragment length polymorphisms at a specific genomic locus originally amplified by the polymerase chain reaction. The technique uses restriction endonucleases to cut the amplified DNA and identify sequence polymorphisms.

Clone: Greek word 'klon' meaning 'twig'. A method of vegetative reproduction of an organism. Commonly used in horticulture as cuttings or drafts. Resulting organisms are defined as a clone meaning they were derived from the same original source organism. Commonly clones are presumed to be genetically identical. This may not be the case because of further genetic change after the original duplications. In modern genetic jargon, clone describes an isolated DNA sequence ligated into a bacterial plasmid or virus, so that the sequence can be propagated indefinitely using microbiological means. Such cloned sequences can be used for sequencing, expression studies, or as probes.

Cloning: The use of recombinant DNA technology to propagate the sequence of a DNA fragment in a cloning vector, after its transformation into a suitable host cell.

Cloning vector: Any nucleic acid molecule capable of autonomous replication in a host cell and capable of accepting foreign DNA.

Cluster analysis: The analysis and characterization of subsets of organisms (operational taxonomic units) through search of properties of coherence and isolation.

Cocktail: Laboratory jargon for any mixture of reagents.

Codominance: The phenomenon whereby both alleles at a locus are expressed and influence the phenotype of the heterozygote.

Codon: Arrangement of three nucleotides in mRNA controlling the insertion of an amino acid into a polypeptide.

Cohesive ends: The termini of two DNA duplex molecules capable of annealing to each other.

Colony: A contiguous group of fungal or bacterial hyphae/cells derived from a single ancestor.

Complexity: The length of non-repetitive base sequences in a nucleic acid molecule.

Consensus (conserved) sequence: A defined nucleotide sequence either frequent at a defined position or characteristic for a specific functional region of a gene, occurring generally in the same context in a set of DNA sequences.

Contig: Contiguous segment of DNA used in genome analysis and defined as a group of genomic clones that contain mutually overlapping sequences.

C_0t analysis: The analysis of the kinetics of reassociation of two single stranded DNA molecules as a function of time, where C_0 is the initial single stranded DNA concentration and t the hybridization time.

Cut: A DNase induced break in a double-stranded nucleic acid molecule, as opposed to 'nick', a break in a single stranded molecule.

DAF: *See* DNA amplification fingerprinting.

Denaturation: The loss of native configuration of a macromolecule. In particular, DNA denaturation refers to the conversion of double-stranded to single-stranded molecules by melting, i.e. breaking hydrogen bonds.

Denaturing gel: A gel matrix that contains compounds capable of denaturation.

Denaturing gradient gel electrophoresis (DGGE): The electrophoretic separation of macromolecules in a linearly increasing gradient of denaturant compounds (e.g. urea, formamide, glyoxal). Separation of DNA fragments in these gradients is improved by their differential melting. DNA molecules migrate in the electric field but encounter a gradient concentration that causes partial melting, subsequent branching, and ultimately a decrease in mobility.

DGGE: *See* denaturing gradient gel electrophoresis.

Differential display (DD): An RNA fingerprinting method that uses a 3' poly-T clamped primer to reverse transcribe an mRNA population and a primer of 10 nucleotides in length to fingerprint the resulting cDNA molecules. The technique is also known as DDRT-PCR, and is considered a variant of the PCR capable of detecting differential gene expression, i.e. the expression of a set of potentially active genes in a given cell at a given time and environment. Fingerprint products are usually resolved by sequencing gels and autoradiography.

Digestion: The enzymatic hydrolysis of covalent bonds in a macromolecule. Complete or partial DNA digestion generally refers to the endonucleolytic cleavage of nucleic acids by restriction endonucleases.

Digoxigenin (DIG): The aglycon of the steroid glycoside digoxin extracted from *Digitalis purpurea*, which is derivatized (e.g. Digoxigenin-11-dUTP) and used for the non-radioactive labeling of DNA.

Diploid: The normal somatic chromosome number of an organism (twice haploid).

DNA (deoxyribonucleic acid): A single or double stranded polymer of nucleotides arranged along a deoxyribose and phosphate backbone capable of carrying genetic information in most organisms.

DNA amplification fingerprinting (DAF): An arbitrarily amplified DNA technique that uses one or more primers of typically between 5 and 15 nucleotides in length and high primer-to-template ratios (5-50,000). Primers can harbor secondary structure, such as hairpins or mini-hairpins. Amplification products are generally visualized by polyacrylamide gel electrophoresis and silver staining, and fingerprints are highly complex. See also AAD, AFLP, AP-PCR and RAPD.

DNA array (chip): A tool for the analysis of nucleic acid sequence that uses arrays of immobilized nucleic acid molecules in hybridization experiments, sometimes coupled with amplification. A number of variants have been described whereby the target or the probe molecules (oligonucleotides) are bound to a solid support. *See* oligonucleotide array.

DNA fingerprinting: A method that generates a DNA 'fingerprint' or pattern capable of detecting DNA polymorphism. Typically, genomic DNA is cut with a restriction endonuclease, and the resulting DNA fragments are separated by gel electrophoresis, transferred to a membrane, and hybridized with a labeled 'fingerprint probe' consisting of either a synthetic oligonucleotide or a cloned DNA fragment. In microsatellite fingerprinting, the DNA fragments can be hybridized directly in the gel, obviating membrane transfer.

DNA polymorphism: The variation in length, presence or sequence of a DNA fragment generally produced by restriction or amplification. These polymorphisms are generated by mutation, insertion, deletion or rearrangement of DNA, and are particularly evident in hypervariable genomic regions, such as microsatellite and minisatellite sequences.

DNase (*deoxyribonuclease*): An enzyme that catalyzes the hydrolysis of phosphodiester bonds in single or double-stranded DNA, beginning from either the termini (exonucleases) or from within (endonucleases) the nucleic acid molecules.

DNA sequencing: Any of several methods capable of determining the sequence of bases in a DNA molecule.

DNA synthesis: The chemical or enzymatic linking of deoxyribonucleotide triphosphates to a nascent oligonucleotide polymer.

dNTP: Abbreviation for any 2'-deoxynucleoside-5'-triphosphate.

Dominance: Form of expression of a gene, in which the phenotype of the dominant form is expressed over the recessive form.

Dominant allele: One of a pair of corresponding alleles which manifests phenotypically in the heterozygote condition.

Duplex: A double-stranded form of a molecule.

Electrophoresis: A method used to separate molecules in an electric field according to their net charge, shape and size. Electrophoresis can be done either in free solution or in a matrix such as an agarose or polyacrylamide gel. Smaller, highly charged, or more compact molecules travel faster (i.e. have higher 'electrophoretic mobilities'), and if negatively charged (e.g. DNA) are therefore found further down close to the cathode. Derived from the Greek, *phoro*: I carry.

Element: A consensus, repetitive or transposable sequence.

Elongation: The extension of a polymeric chain by addition of new monomers.

Enzyme: A biological catalyst allowing the completion of biochemical reactions. Most enzymes are proteins, although some RNA can act as enzymes (ribozymes).

Epistasis: Mode of gene interaction where the expression of one gene influences the expression of another at another location in the genome. For example, non-nodulation genes in soybean epistatically suppress supernodulation.

Ethidium bromide: A phenanthridium fluorescent dye [3,8-diamino-6-ethyl-5-phenyl phenantridium bromide] which intercalates between base pairs in double stranded nucleic acids, alters buoyancy, and fluoresces at UV wavelengths of <300 nm.

Euchromatin: The portion of genomic DNA that remains relatively unstained and is transcriptionally active.

Eukaryote: Organism characterized by the presence of a nucleus, organelles such as plastids (e.g. mitochondria, chloroplasts, amyloplasts), and the capacity to undergo division in the form of mitosis. Includes animals, plants, green algae and fungi.

Expressed sequence tag **(EST):** A synthetic oligonucleotide complementary to part of a specific mRNA and usually derived from a cDNA library or RNA fingerprint.

F$_1$ (filia 1): The offspring produced by crossing parental (P) individuals.

F$_2$ (filia 2): The second generation progeny produced by intercrossing or self-fertilization of first generation F$_1$ individuals.

Fingerprinting: The whole repertoire of techniques capable of generating a fingerprint.

Fingerprint tailoring: The design of strategies, constructs or oligonucleotide primers (oligomers) capable of modulating DNA fingerprint complexity, the genomic sites targeted, and the detection of DNA polymorphism in specific applications (e.g. genetic mapping or genotyping).

Fluorescence: The emission of photons by atom or molecule that has been excited by adsorption of a quantum of electromagnetic radiation.

Fluorochrome: A fluorescent chemical substance.

Formamide: An organic molecule that prevents AT base pairing and causes denaturation by reacting with free NH_2 groups of adenine. The chemical is used to decrease melting temperature in hybridization experiments.

G: Abbreviation for guanine (2-amino-6-hydroxy-purine), a purine base characteristic component of nucleic acids.

Gamete: The haploid sex cell, such as a pollen grain, a sperm or an egg.

GC-clamp: A region within a synthetic oligonucleotide with high GC content (>70%) that does not favor melting when establishing a duplex.

Gel electrophoresis: An electrophoretic separation method in a gel matrix usually made of agarose, starch, or polyacrylamide. A great number of gel electrophoretic techniques are available to separate DNA molecules, including crossed field gel electrophoresis, denaturing gradient gel electrophoresis (DGGE), field inversion gel electrophoresis, pulsed-field gel electrophoresis, and temperature sweep gel electrophoresis.

Gene: Physical and functional unit of inheritance. Usually a gene is defined as that region of DNA that controls the synthesis of a polypeptide.

Gene expression: The phenotypic consequence of transcription of one or more specific genes.

Gene probe: A specific single stranded DNA sequence, often labeled with radioactive phosphorous (^{32}P) to help the detection of a gene through the complementary (*see* Chargaff's rules) sequence to which the probe will bind. Hybridization is usually detected by autoradiography.

Genetic bit analysis: A solid-phase method for genotyping single nucleotide polymorphims (SNP) whereby a target is amplified by the polymerase chain reaction using an exonuclease-resistant and a standard primer, one of the PCR product strands digested with an exonuclease and the other one hybridized (captured) by an oligonucleotide complementary to the target sequence up to the variant nucleotide (site), and the bound oligonucleotide labeled and detected by DNA polymerase-mediated extension if its terminal nucleotide is complemetary to the polymorphic site.

Genetic code: The conversion cipher that allows the interpretation of nucleotide triplets (codons) to their matching amino acids.

Genetic engineering (gene technology): The directed genetic manipulation of an organism using recombinant DNA molecules not commonly found in nature.

Genetic fingerprinting: A method that uses DNA technologies to establish genetic relationships between close or distant relatives. *See* DNA fingerprinting.

Genetic marker: A gene that can be detected by a phenotypic effect.

Genetic material: Single or double-stranded nucleic acid that serves as a template of its own replication or for the synthesis of structural or mRNA.

Genome: The entire set of hereditary molecules in an organism.

Genotype: The genetic make-up of an organism which, depending on the environment and other genetic interactions, may be expressed as the phenotype.

Hairpin: A fold-back nucleic acid structure formed by self-annealing that consists of a base-paired stem and a single-stranded loop. The loop can be very short (3 bases; *see* mini-hairpins) or consist of several hundred bases.

Haploid: The gametic chromosome number of an organism, as found in sex cells.

Heterochromatin: Portion of genomic DNA that is highly stained because of high condensation. The heterochromatin is transcriptionally inactive and often contains large amounts of highly repeated DNA (satellite DNA). Heterochromatin comes in two forms. Facultative heterochromatin will alter in staining and activity at different developmental stages (i.e., the inactivated X-chromosome in mammalian females). Constitutive heterochromatin is constantly "turned off". It is often found around the centromere.

Heterosis (hybrid vigor): Interaction of multiple genes that causes an improvement of the heterozygote above the level of both parents.

Holoenzyme: An enzyme composed of two or more subunits and active only if all subunits are assembled.

Homeologous: Related chromosome in a polyploid organism, which shares perhaps a common ancestry, but may contain different genes and alleles. For example, wheat is a hexaploid and carries 3 chromosome sets of 2. Within each set the chromosomes are homologous, between the sets they are homeologous.

Homogenotization: Bacterial genetics procedure used to exchange genetic markers from a plasmid to the recipient linkage group. Also called marker exchange.

Homologous chromosome: Partner chromosome in a diploid organism. Generally the homologous chromosome is similar as it carries the same genes, but is different as many allelic differences may exist.

Homology: The extent of identity between nucleotide or amino acid sequences.

Hybridization: The formation of a duplex molecule from complementary nucleic acid strands in gene technology applications. Hybridization serves to detect sequence homologies between two different nucleic acid molecules by (radioactively or non-radioactively) labeling a single stranded probe and annealing to homologous single-stranded nucleic acid fragments.

Intron: Intervening sequence in genes that is transcribed, but not translated.

Inosine **(I):** The nucleoside of hypoxanthine (6-hydroxipurine) and a normal constituent of nucleic acids. Inosine can be used as a "woble" base in probes or primers and carries degeneracy.

Isozyme (isoenzyme, allozyme): One of multiple forms (usually electrophoretic) of a single enzyme resulting from either allelic variation within one polypeptide or multimeric associations of variants.

Karyotype: The pattern and shape of the complete set of chromosomes belonging to an organism.

Kilobase (kb): One thousand base pairs.

Kit: A set of reagents necessary to perform an experimental protocol.

Labeling: The introduction of radioactive or non-radioactive atoms or chemical groups in biological molecules (e.g. DNA, RNA, proteins) for their identification.

Ladder: A mixture of DNA, RNA or protein molecules that allow the determination of molecular weight within a defined range.

Lambda (λ): Temperate bacteriophage of *Escherichia coli* that contains a double-stranded DNA genome of 49 kb packaged into an icosahedral head.

Ligase: An enzyme used to couple termini of one or more double-stranded DNA molecules together.

Ligation: The formation of a phosphodiester bond between neighboring nucleotides exposing a 5' phosphate and 3' hydroxyl group.

Linkage analysis: Estimation of cross-over frequency or recombination between DNA sequences, as the means to establish the relative position of a locus (e.g. gene, sequence tag) in a chromosome.

Linkage group: A group of genes and/or markers that are contiguous on a linear chromosome map.

Loading buffer: A mixture of reagents that aid in the loading of samples into slots or wells in electrophoretic gels. Loading buffers contain one or more tracking dyes (e.g. bromophenol blue) to monitor the electrophoretic run and compounds that increase the viscosity (density) of the solution (e.g. glycerol, sucrose, Ficoll) and may act as active or passive denaturants (e.g. urea).

Locus: The chromosomal position of a genetic condition as defined by a detectable phenotype or genotype.

MAAP: *See* multiple arbitrary amplicon profiling.

Map: The ordered arrangement of genes or molecular markers of an organism, indicating the position and distance between the markers and loci. Most maps are genetic maps based on the percentage of recombination. Some maps are cytological maps based on the arrangement of chromosomal regions, while others are physical maps based on the amount of DNA between markers and loci.

Map unit: Defined as one centiMorgan.

Marker: A distinguishing feature that can be used to identify a particular part or region of a genome, chromosome or genetic linkage group. *See* genetic marker. Also, any molecule (e.g. DNA, RNA, protein) of known size used as a 'standard' to calibrate the electrophoretic or chromatographic separation of similar molecules. *See* ladder.

***Mega*base (Mb):** One million basepairs.

Meiosis: Cell division that reduces ploidy ($2n \rightarrow n$) in eukaryotes giving rise to gametes of different genetic make-up when compared to the parental cell and to each other.

Melting: The dissociation of the complementary strands of double-stranded nucleic acids, such as homoduplex (dsDNA) or heteroduplex (DNA-RNA) molecules. Melting can be accomplished *in vitro* by heating or alkali treatment or *in vivo* by specific binding proteins involved in replication and translation. *See* denaturation and C_0t.

β-mercaptoethanol: A water soluble thiol compound used to cleave disulfide bridges in proteins and reduce SH groups.

***Messenger RNA* (mRNA):** A single stranded RNA molecule that is the product of the transcription of DNA and the information carrier for translation into proteins.

Methylation: Enzymatic addition of a methyl (CH_3) group from a methyl-donor to DNA that causes inactivation of that region. Usually CpG nucleotide pairs are the target for this addition.

Microsatellite [*short tandem repeat* (STR), *simple sequence repeat* (SSR)]: One of many DNA sequences dispersed throughout fungal, plant and animal genomes, composed of short (2-10 bp) sequences that are repeated in tandem and are usually highly variable and useful for DNA profiling.

***Microsatellite primed polymerase chain reaction* (MP-PCR):** A variant of arbitrarily amplified DNA analysis that uses primers complementary to microsatellite sequences. Several variants of this strategy are available including those that use anchored primers by themselves [*anchored microsatellite primed polymerase chain reaction* (AMP-PCR) or *inter simple sequence repeat polymerase chain reaction* (interSSR-PCR)] or in combination with standard arbitrary primers [random amplified microsatellite polymorphism (RAMP) analysis]. In an AFLP-derived technique (*see* AFLP), primers can be used to amplify adaptor-modified restriction fragment templates [*s*elective *a*mplification of *m*icrosatellite *p*olymorphic *l*oci (SAMPL)].

Minihairpin: An extraordinarily stable helix-turn structure consisting of a loop of 3-4 nucleotides and a stem of only two nucleotide, and naturally occuring in both DNA and RNA molecules. Mini-hairpins have high melting temperatures and unusually rapid mobilities during electrophoresis and polyacrylamide gels, causing artifacts in DNA sequencing.

Minisatellite: One of many DNA sequences that are dispersed throughout human and other eukaryotic genomes, shows substantial length polymorphism, and is composed of a 10-35 bp consensus (core repeat unit) sequence, which is repeated in tandem.

Mismatching (mispairing): The occurrence of incorrectly paired bases in a nucleic acid duplex, generally causing error during annealing of primer to template or probe to target under low stringency conditions.

Molecular marker: A molecular signpost used in eukaryotic gene isolation, usually an RFLP probe or a sequence-tagged site for DNA amplification or hybridization.

***M*ultiple *a*rbitrary *a*mplicon *p*rofiling (MAAP):** A collective term used to describe those nucleic acid amplification-based fingerprinting techniques that use arbitrary primers. Also known as arbitrarily amplified DNA (AAD), arbitrary primer technology (APT), amplification fragment length polymorphism (AFLP) analysis, and nucleic acid scanning by amplification. MAAP include several well-established techniques such as RAPD, AP-PCR and DAF.

Mutant: An organism or gene with an heritable altered phenotype from the wild type.

N: Abbreviation for any one base in a nucleic acid molecule (e.g. N stands for G, C, A or T in DNA).

NASBH: *See* nucleic acid scanning by hybridization.

Near-isogenic line (NIL): A line used frequently to define molecular regions of a genome that cause a phenotype of value in agronomy and plant breeding.

Nick: A break of a phosphodiester bond in only one of two strands of a nucleic acid duplex.

Northern tranfer (blotting, hybridization): RNA detection method whereby RNA molecules separated by size are transferred directly to a nitrocellulose filter or other membranes by electric or capillary force. Single-stranded molecules can be immobilized to the membrane by baking, and are then hybridized to a labeled probe. The procedure is similar to Southern hybridization (DNA), and detects individual RNA species from a complex RNA population.

Nuclease: An enzyme that catalizes the hydrolisis of phosphodiester bonds in nucleic acids.

Nucleic acid: A single or double-stranded polynucleotide containing either deoxyribonucleotides (e.g. DNA) or ribonucleotides (e.g. RNA) linked by 3'-5'-phosphodiester bonds.

***N*ucleic *a*cid *s*canning:** Any nucleic acid strategy capable of providing genetic information from a plurality of sites (loci) in a genome or nucleic acid molecule.

***N*ucleic *a*cid *s*canning *b*y *h*ybridization (NASBH):** A nucleic acid scanning technique that uses arrays of oligonucleotides to analyze by hybridization the sequence of arbitrarily amplified DNA. *See* oligonucleotide arrays and arbitrary sequence oligonucleotide fingerprinting.

Nucleoside: A purine or pyrimidine base covalently linked to ribose or deoxyribose via N-glycosidic bonds.

Nucleotide (nt): A purine or pyrimidine nucleoside that is esterified with one, two or three phosphate groups at the 5' C of ribose or deoxyribose. Nucleotides are component of DNA (deoxyadenylate, deoxyguanylate, deoxycytidylate and thymidylate nucleotides) and RNA (adenylate, uridylate, guanylate, and cytidylate), and their bases paired according to Chargaff's rules.

Nucleotide sequence: Nucleic acid (DNA or RNA) sequence.

Nucleus: Organelle found in cells of all eukaryotes surrounded by a double membrane (envelope) and containing chromatin (i.e. chromosomes associated with proteins).

Oligomer (*oli*gonucleotide pri*mer*): A synthetic oligonucleotide used as a primer for nucleic acid amplification. In broad sense, a polymeric molecule composed of only few monomers.

Oligonucleotide: A short nucleic acid molecule containing up to about 100 deoxynucleotides (oligodeoxynucleotide) or ribonucleotides (oligoribonucleotide), or a mixture of both. These molecules are usually obtained synthetically.

Oligonucleotide array: A collection of immobilized oligonucleotides arranged in two dimensions so that each physical address corresponds to a defined sequence. Oligonucleotides are usually bound to membrane, silicon or glass supports and are used in hybridization or amplification experiments. Oligonucleotides can be arbitrary or target specific sequences. In one specific application, *sequence by hybridization* (SBH), oligonucleotides have overlapping sequences and are used to reconstruct the sequence of the target molecule following computer analysis of their hybridization behavior.

Operational *taxonomic *unit* (OTU): Entity or individual element subjected to phenetic or phylogenetic analysis.

PCR: *See* polymerase chain reaction.

Pellet: Material sedimented by centrifugation.

PFGE: *See* pulse field gel electrophoresis.

Phenetic (pattern, diversity) analysis: Study of patterns of overall resemblance and difference among organisms based on many heritable characteristics. The patterns can reveal groupings with different degrees of variation within and between groups that reflect the relationships between organisms at a given time. *See* phylogenetic analysis.

Phenotype: The observable structural and functional characteristics that result from the interaction of the genotype with the environment.

Phosphatase: An enzyme that removes phosphate residues from substrates (e.g. alkaline phosphatase).

Phylogenetic (pattern, diversity) analysis: Study of the change of phenetic pattern with time. *See* phenetic analysis.

Phylogenetic tree: A relational graphic representation reflecting genealogical or evolutionary relationships in a group of organisms.

Phylogeny: Evolutionary history of lineages.

P$_i$: Symbol for inorganic phosphate.

Plant variety protection (PVP), plant variety rights: Legislation that exists in many countries that protects the commercial use of plant varieties.

Plate: A Petri dish that contains a solid medium for the growth of microorganisms.

Plasmid: A circular covalently-closed DNA molecule commonly found in bacteria and often used as a cloning vector in genetic engineering.

Pleiotropy: Mode of gene expression where the expression of one gene has multiple phenotypes.

poly(A): A homopolymer consisting of adenine nucleotide residues, usually found as a poly(A) tail in most eukaryotic RNA molecules.

Polyacrylamide gel: An insoluble matrix of acrylamide monomers cross-linked with N,N'-methylene-bisacrylamide (or derivatives) by the presence of catalysts (e.g. TEMED) that can be made to harbor varying pore sizes.

Polymerase: An enzyme that catalyzes the assembly of nucleotides into RNA (RNA polymerase) or DNA (DNA polymerase).

Polymerase chain reaction (PCR): An *in vitro* amplification method capable of increasing the mass (number) of a specific DNA region (fragment). The method uses two synthetic oligonucleotide primers (15-30 nucleotides long) that specifically hybridize to opposite DNA strands flanking the region to be amplified. A series of temperature cycles involving DNA denaturation, primer annealing, and extension of the annealed primers by the activity of a thermostable DNA polymerase, specifically and exponentially (at least during first many cycles) amplify the target DNA region many million-fold to produce an amplified fragment with its termini defined by the 5' end of each primer. The method is of extreme value in diagnostics, forensics and general molecular biology (e.g. sequencing, probe preparation, genome mapping). The term polymerase chain reaction has been used in a broad sense to define any amplification technique driven by primers, DNA polymerase and a temperature cycling regimen.

Polymorphism: An alteration in the DNA sequence detected through sequencing or DNA analysis (fingerprinting, profiling). Also defined as the presence of several forms of a genetic characteristic in a population.

Polyploid: More than diploid by multiples of the haploid number (e.g. banana is a triploid and therefore a polyploid).

Positional (map-based) cloning: Experimental approach used to locate and isolate gene sequences for which the gene product (transcript, protein) is not known. Molecular markers tightly linked to the locus of interest are identified and used to link overlapping clones from a genomic library.

Prehybridization: The blockage of nonspecific binding of a probe to a membrane (e.g. nitrocellulose, nylon) during hybridization, by treatment with high concentration of protein and/or detergents.

Primer: A short oligonucleotide complementary to a sequence in a larger nucleic acid molecule which serves as a substrate of a DNA or RNA-dependent DNA polymerase.

Probe: A defined nucleic acid sequence which is labeled and used to detect homologous sequences in DNA or RNA by hybridization.

Prokaryote: Organism characterized by the absence of major organelles (nucleus, plastids). Includes archaebacteria, eubacteria and cyanobacteria.

Pulse field gel electrophoresis (PFGE): A variation on the electrophoresis procedure used to detect and isolate very large DNA molecules (up to 10-20 Mb) including whole chromosomes of organisms such as yeast. Molecules are subjected to alternate electrical fields at different angles that pulsate at specific time intervals.

Quantitative trait loci (QTL): A term given to a genomic region which controls a phenotype by cooperative interaction of several genes (e.g. cyst nematode resistance in soybean).

RAPD: *See* random amplified polymorphic DNA.

Random amplified microsatellite polymorphism (RAMPO) analysis: A technique that uses oligonucleotides complementary to microsatellite sequences as hybridization probes in the analysis of arbitrarily amplified DNA. The labeled oligonucleotides are hybridized to amplification products generated with arbitrary or microsatellite-complementary primers and separated by size. Variants of the technique include *randomly amplified microsatellite* (RAMS) and *random amplified hybridization microsatellite* (RAHM) analysis.

Random amplified polymorphic DNA (RAPD): Amplified DNA polymorphism uncovered by the use of the RAPD technique. *See* amplification fragment length polymorphism.

Random amplified polymorphic DNA (RAPD) analysis: A widely used arbitrarily amplified DNA technique that uses one primer of typically 9 or 10 nucleotides in length, low annealing temperatures and low primer-to-template ratios (<1). Amplification products are generally visualized by agarose gel electrophoresis and ethidium bromide staining, and fingerprints are simple and especially useful for genetic mapping. See also AAD, AFLP, and DAF.

Recessive (allele): Form of expression of a gene in which the phenotype of the recessive form is not expressed over the dominant form.

Recognition site (sequence): Nucleic acid sequence to which a specific protein (e.g. restriction endonuclease) can be bound.

Recombinant inbred line (RIL): The product of an initial cross between two parent lines and the subsequent selfing of the F_2 individuals, generally used for mapping purposes.

Recombination: Natural process of exchanging DNA fragments between different DNA molecules. Occurs in both prokaryotes and eukaryotes, but by slightly different processes. Eukaryotic recombination occurs predominantly during meiosis and gives rise to gametes of non-parental gene combinations.

Repetitive extragenic palindromic polymerase chain reaction (rep-PCR): A variant of the conventional polymerase chain reaction that uses primers complementary to repetitive extragenic palindromic elements to generate fingerprints characteristic of gram-negative bacteria. The strategy has been used with other abundant repetitive motifs for the general identification of bacteria.

Restriction endonuclease: A bacterial enzyme that recognizes specific sites (recognition sequences) in DNA and cuts (restricts) the molecules into (restriction) fragments.

Restriction fragment length polymorphism (RFLP): The variation in the length of a DNA fragment produced by a specific restriction endonuclease and generally detected by Southern hybridization with a probe.

Restriction map: The linear arrangement of restriction sites on a nucleic acid.

Reverse transcriptase (RTase): An RNA dependent DNA polymerase enzyme capable of synthesis of a double-stranded DNA from an RNA template.

RNA (ribonucleic acid): A single-stranded polynucleotide of low molecular weight that contains a ribose backbone and the pyrimidine uracil instead of thymine. Constituent of messenger, transfer and ribosomal RNA and of certain viruses (e.g., tobacco mosaic virus) and viroids. Some RNA molecules may have enzymatic activity (ribozymes).

RNA arbitrarily primed polymerase chain reaction (RAP-PCR): An RNA fingerprinting method that uses a single primer of 10 to 15 nucleotides in length to reverse transcribe an mRNA population and fingerprint the resulting complementary DNA by amplification of arbitrary segments. See differential display.

Ribonuclease (RNase): An enzyme that catalyzes the cleavage of phosphodiester bonds in RNA (e.g. RNase A).

Sequence analysis (sequencing): Determination of sequence of nucleic acids and proteins in bases or amino acids.

Sequence characterized amplified region (SCAR): A molecular marker originating from a DNA fragment defined by amplification with the polymerase chain reaction. The marker is generally generated by the partial sequencing of cloned arbitrarily primed DNA or other DNA amplification products. Also known by sequence-tagged amplified region (STAR).

Sequence tagged site (STS): A small region of DNA on a chromosome detected by hybridization or amplification techniquesand used as a signpost for molecular gene mapping approaches.

Silver staining: A sensitive staining technique for the visualization of nucleic acids, proteins and polysaccharides in agarose or polyacrylamide gels in which ionic silver is preferentially reduced to metallic silver on the surface of the macromolecules.

Simple sequence repeat (SSR): See microsatellite.

Single primer DNA amplification: Any DNA amplification technique that uses a single oligonucleotide primer.

Smear: The inability to distinguish single bands in gel electrophoresis due to digestion, mispriming, etc.

Smiling: The appearance of curved bands of proteins or nucleic acids in gel electrophoresis generally due to heat dissipation effects.

Southern hybridization (blotting, transfer): A gel blotting method whereby DNA fragments are separated by size and transferred to a nitrocellulose, nylon or other membrane or matrix by electrical or capillary action. Single-stranded molecules are immobilized to the membrane and then hybridized to a labeled probe. The procedure has innumerable variants that detect individual DNA species from a complex mixture of DNA fragments.

Stoffel fragment: A truncated *Thermus aquaticus* DNA polymerase lacking the N-terminal 289 amino acids and the intrinsic 5'-3' exonuclease activity, and more resistant to temperature denaturation than the holoenzyme. The Stoffel fragment is the preferred thermostable enzyme for arbitrarily amplified DNA methods.

Stringency: Reaction conditions in nucleic acid hybridization that allow the formation of duplexes from single-stranded molecules. Low stringency allows mismatches to occur while high stringency permit the hybridization of only the perfectly complementary sequences. The term is also applicable to primer-template interactions and describes the influence of reaction conditions on the fidelity of annealing.

Stripping: Laboratory jargon used to describe the removal of labeled probe from the immobilized target.

Supernatant: Liquid left above a pellet after centrifugation.

T: Abbreviation for thymidine (2,6-dihydroxi-5-methyl-pyrimidine,5-methyluracyl), a pyrimidine base characteristic component of DNA.

***Taq* polymerase:** Thermostable DNA polymerase from *Thermus aquaticus*, a thermophylic bacterium found in hot springs and underwater vents, widely used in nucleic acid amplification.

Telomere: Terminal region of chromosomes characterized by repeated DNA sequences.

Template (strand): Any nucleic acid strand that serves as a mold for the synthesis of a complementary nucleic acid.

Template endonuclease cleavage multiple arbitrary amplicon profiling (tecMAAP): A MAAP method that amplifies template nucleic acids that have been restricted with one or more endonucleases.

Thermocycler: An apparatus that allows a programmed and automated change of temperature, and is useful for setting denaturation (94-96°C), annealing (10-65°C) and extension (about 72°C) temperatures during DNA amplification.

Touchdown *polymerase chain reaction* (touchdown PCR): A variant of the polymerase chain reaction whereby mispriming is reduced by setting the annealing temperature very high and then gradually reducing it in subsequent amplification cycles.

Transcription: The process by which DNA is copied into RNA via a DNA or RNA-dependent RNA polymerase. The process is called transcription (cf. translation) because the nucleic acid 'language' stays the same (*see* genetic code).

Translation: The ribosome-based process by which RNA is made into proteins. Called translation because the nucleic acid 4 component language based on triplet codons is changed to a protein 20 component language.

Trisomic: Three chromosomes of one type present within a genome.

U: Abbreviation for uracil (2,4-dihydroxy-pyrimidine) a pyrimidine characteristic of RNA.

Variable number tandem repeats (VNTR): DNA region composed of tandemly repeated short (11-60 bp) sequences, which varies from individual to individual and is thus used in DNA fingerprinting. *See* minisatellites.

Vector: *See* cloning vector.

Wild type: The normal (non-mutant) form of an organism used as a standard.

Yeast *a*rtificial *c*hromosome (YAC): A plasmid cloning vector replicating in yeast, capable of carrying large segments of DNA (up to megabase levels) and used extensively in eukaryotic gene mapping and gene isolation.

Index